The yellow flowers are *Sonchus tenerrimus*, and the lilac flower is *Bituminaria bituminosa*. They are edible and have medicinal purposes.

The word "nourishing" has relation with domesticity, and this image is a detail of a domestic interior — a vase with flowers and a picture on the background wall that, in its blur figuration, furnishes a space for humans to relax at home. Not represented in the photo is "tomorrow," a far-in-the-future utopia. For the near future, our homes will likely not be so different from what they are now. This immediate future is the close reality that we need to nourish. The flowers are not expensive, not limited to the upper crust, nor are they spectacular in beauty. They express the present reality — they are humble flowers that anybody can have, reminding us of the beauty of life as we strive to leave the same legacy to our children. At the center, the seed head, i.e., the blowball, introduces some mystery, enthralling the beholder to observe its complexity, savoring its more intricate beauty. Life is beautiful, even in the thick of it, if we are grateful enough to savor its beauty. All this explains how clean engineering and nature-friendly living can *nourish tomorrow*; the realistic, humble, simple, and vibrant.

Nourishing Tomorrow

Clean Engineering and Nature-friendly Living

Nourishing Tomorrow

Clean Engineering and Nature-friendly Living

editor

David S-K Ting
Jacqueline A Stagner

University of Windsor, Canada

World Scientific

NEW JERSEY · LONDON · SINGAPORE · BEIJING · SHANGHAI · HONG KONG · TAIPEI · CHENNAI · TOKYO

Published by

World Scientific Publishing Co. Pte. Ltd.
5 Toh Tuck Link, Singapore 596224
USA office: 27 Warren Street, Suite 401-402, Hackensack, NJ 07601
UK office: 57 Shelton Street, Covent Garden, London WC2H 9HE

Library of Congress Cataloging-in-Publication Data
Names: Ting, David S-K, editor. | Stagner, Jacqueline A, editor.
Title: Nourishing tomorrow : clean engineering and nature-friendly living /
    editor, David S-K Ting, Jacqueline A. Stagner, University of Windsor, Canada.
Description: New Jersey : World Scientific, [2023] | Includes bibliographical references and index.
Identifiers: LCCN 2022042516 | ISBN 9789811264368 (hardcover) |
    ISBN 9789811264375 (ebook for institutions) | ISBN 9789811264382 (ebook for individuals)
Subjects: LCSH: Sustainable engineering. | Sustainability.
Classification: LCC TA163 .N68 2023 | DDC 666/.14--dc23/eng/20221006
LC record available at https://lccn.loc.gov/2022042516

British Library Cataloguing-in-Publication Data
A catalogue record for this book is available from the British Library.

For any available supplementary material, please visit
https://www.worldscientific.com/worldscibooks/10.1142/13086#t=suppl

Desk Editors: Logeshwaran Arumugam/Amanda Yun

Typeset by Stallion Press
Email: enquiries@stallionpress.com

Mr. Fred Rogers hit the nail on the head when he proclaimed, "It's not the honors and the prizes and the fancy outsides of life that ultimately nourish our souls. It's the knowing that we can be trusted, that we never have to fear the truth, that the bedrock of our very being is firm." This volume is dedicated to those who labor in tree planting, nourishing the soul of tomorrow.

Preface

This endeavor is in part inspired by Harriet Beecher Stowe, who asserted that "[t]here is more done with pens than with swords." The enduring cumulative efforts of numerous champions behind the scenes, along with those on the front stage, have finally come into fruition. Graham T. Reader opens the volume from a global perspective with Chapter 1 "Preparing Today for Nourishing Tomorrow." Nourishing tomorrow is taking care of tomorrow and, to that end, we must make clean water and air, safe sanitation, energy, food, and shelter available to every soul on earth. Reader explains what the United Nations' 2030 Agenda purports and the requirements to overcome current deficiencies in achieving global nourishment needs. The comprehensive chapter ends with a table summarizing the possible achievement of nourishment needs and current action. The next step is for the decision makers around the globe to step up. Talking about the water, energy, and food nexus, Luxon Nhamo et al. enlighten us with Chapter 2 titled "Measuring the Sustainability of Water, Energy, and Food Resources in the Context of the WEF Nexus." They detail whether the water–energy–food (WEF) nexus is appropriate in monitoring progress towards the United Nation's Sustainable Development Goals, particularly Goals 2, 6, and 7 that relate to water, energy, and food security. Among others, they conclude that the WEF nexus is a key pathway to tackling poverty, unemployment, and inequality. To highlight, the WEF nexus approach promotes cross-sectoral cooperation and improving resource-use efficiencies. To nourish tomorrow, there must also be adequate, safe, and nutritious food for everyone around the globe. Alfonso Expósito addresses this in the context of Spain, in Chapter 3 titled "Exploring the Link between Irrigated Agriculture and Food Security: Evidence from the Case of Spain." The growing population living at a higher standard poses a real challenge to both irrigation water and agriculture. Climate change further worsens the irrigation water issue, that is, it renders the reliability of water supply a serious challenge to food security. The fuel for nourishing tomorrow must be renewable. Further progress in renewable energy goes hand-in-hand with advancing energy storage. Mehdi Ebrahimi et al. bring us up to date with Chapter 4 titled "Green Compressed Air Energy Storage Technology." The idea is to harness the associated thermal energy that is wasted in traditional compressed air energy storage. Greening compressed air energy storage can readily gain a few percentages of efficiency, translating into savings of hundreds of tons of carbon dioxide emissions every year for a typical power plant. With the many food challenges on land, we must gaze into the vast volume of water for the nourishing of tomorrow. In Chapter 5 titled "Marinas can Co-function as Biodiversity Sanctuaries," Loke Ming Chou et al. look at revamping marinas to enrich underwater biodiversity. Comprehensive data from three Singapore marinas were presented and analyzed. It is evident from the analysis that these ecological marinas can host biodiversity, serving as sanctuaries for marine life. Proper human intervention can nurture natural life. Purifying water by removing pharmaceutical and antibiotic pollution is the topic of Chapter 6 titled "Biomaterials for Water Purification: Integrating *Chlorella Vulgaris* and *Monoraphidium Conortum* in Architectural Systems for the Biodegradation of Sulfamethoxazole from Wastewater." Yomna K. Abdallah and Alberto T. Estevez applied *Chlorella vulgaris* and *Monoraphidium conortum* separately to test each potency in degrading sulfamethoxazole. *Chlorella vulgaris* came through as a clear winner. With that, they digitally designed, and 3D-printed micro-textured system to immobilize *Chlorella vulgaris* cells inside a pilot-scale photobioreactor. Antibiotics is definitely an important issue, equally so are the chemicals associated with skincare products. Berrak Aksakal et al. create the first steps in tackling the skincare pollution problem by formulating an appropriate evaluation approach in Chapter 7 titled "Assessing Environmental Skincare Products: A Proposed Framework using SWARA, ARAS, and COPRAS methods." They employed the Stepwise Weight Assessment Ratio Analysis (SWARA), Additive Ratio Assessment (ARAS), and Complex Proportional Assessment (COPRAS) methods for deducing the most harmless skincare product. From the deduced framework, based on the multi-criteria decision-making methods, different brands of skincare products can be properly ranked. Studies like this can encourage skincare manufacturers to improve their products. A similar approach can be applied to evaluate household and commercial washing liquids. In Chapter 8 titled "Step-Wise Weight Appraisal Ratio Analysis in the Assessment of Biosurfactant," Figen Balo and Lutfu Sua invoke multi-criteria decision-making methodologies to rank various food cleaning products. The case study involves nine criteria for ranking six washing liquids using the Technique for Order Preference by Similarity to Ideal Solution (TOPSIS) methodology. The message is to always compare and choose the less harmful washing detergents. This volume wraps up fittingly with Chapter 9 titled "Conceptualizations of a Resilient Community for Social Futures" by Eija Meriläinen. The nourishing of tomorrow necessitates a resilient tomorrow and Meriläinen opens the floor for discussion by conceptualizing a resilient community into (i) resilient community of belonging, (ii) resilient community of practice, and (iii) resilient community as an object of governance.

About the Editors

Dr. David S.-K. Ting is the founder of the Turbulence & Energy Laboratory at the University of Windsor. Professor Ting supervises students primarily in the fields of energy and flow turbulence. To date, he has co/supervised over 80 graduate students, co-authored more than 160 journal papers, authored five textbooks and co-edited more than 20 volumes.

Dr. Jacqueline A. Stagner is the Undergraduate Programs Coordinator in the Faculty of Engineering at the University of Windsor. She is also an adjunct graduate faculty member in the Department of Mechanical, Automotive and Materials Engineering. She co-advises students in the sustainability and renewable energy areas, in the Turbulence & Energy Laboratory. She has co-edited 10 volumes.

Acknowledgments

This book would not have materialized if not for the unfailing striving of many champions, authors, and reviewers alike. The editors truly enjoyed the working relationship with the fantastic workforce of World Scientific Publishing, especially Chua Hong Koon and Amanda Yun. Above all, it is grace from above that carried this project from inception to completion.

Contents

Preface   ix

About the Editors   xiii

Acknowledgments   xv

Chapter 1   Preparing Today for Nourishing Tomorrow:
             A Perspective   1
             Graham T. Reader

Chapter 2   Exploring the Link Between Irrigated Agriculture
             and Food Security: Evidence from the Case of Spain   115
             Alfonso Expósito

Chapter 3   Measuring the Sustainability of Water, Energy and
             Food Resources in the Context of the WEF Nexus   131
             Luxon Nhamo, Henerica Tazvinga,
             Sylvester Mpandeli, Stanley Liphadzi,
             Joel Botai, and Tafadzwanashe Mabhaudhi

Chapter 4   Green Compressed Air Energy Storage Technology   159
             Mehdi Ebrahimi, David S.-K. Ting,
             and Rupp Carriveau

Chapter 5   Marinas can Co-function as Biodiversity Sanctuaries   179
            Loke Ming Chou, Chin Soon Lionel Ng, Kok Ben Toh,
            Karenne Tun, Pei Rong Cheo, and Juat Ying Ng

Chapter 6   Biomaterials for Water Purification: Integrating
            Chlorella Vulgaris and *Monoraphidium Conortum*
            in Architectural Systems for the Biodegradation of
            Sulfamethoxazole from Wastewater   213
            Yomna K. Abdallah and Alberto T. Estevez

Chapter 7   Assessing Environmental Skincare Products:
            A Proposed Framework Using SWARA, ARAS,
            and COPRAS Methods   245
            Berrak Aksakal, Zeliha Mahmat, Figen Balo, and
            Lutfu S. Sua

Chapter 8   Step-Wise Weight Appraisal Ratio Analysis in the
            Assessment of Biosurfactant   263
            Figen Balo and Lutfu S. Sua

Chapter 9   Conceptualizations of Resilient Community for
            Social Futures   281
            Eija Meriläinen

Index   317

Chapter 1

Preparing Today for Nourishing Tomorrow: A Perspective

Graham T. Reader

Department of Mechanical, Automotive and Materials Engineering,
University of Windsor, Ontario, Canada

Abstract

Arguably, to nourish or take care of the needs of *all* of humankind —
sustainable and affordable access to clean water, safe sanitation, and
clean air, together with a sufficiency of energy, food, and shelter —
should be universally available. Yet, many humans do not enjoy such
access or availability, even though it has been 70 years since the 1948
United Nations (UN) declaration on human rights proclaimed "that all
human beings are equal, and have inherent rights." However, only food
and shelter were explicitly mentioned in the initial declaration. Others
were recently added to the UN list, but not air and energy. Nevertheless,
basic human needs do not have to be declared as a human right before
national actions are taken. Today's key driver is the UN 2030 Agenda,
a plan to eradicate all global poverty and set the world onto a "sustainable and resilient path," through the achievement of 17 Sustainable
Development Goals (SDGs). Adopted by all UN members, the 2030
Agenda in essence, is a 21st century version of the 1948 proclamation.
The SDGs explicitly detail, or implicitly in the case of clean air, all the
necessary needs for the nourishing of tomorrow. To achieve the plan will

likely require, at least, changes in national cultural values, eliminating inequalities and disparities, developing more appropriate governance strategies, and meaningful technical innovation. In this chapter, these requirements are discussed against a backdrop of presently known deficiencies in global nourishment needs.

Keywords: Air, energy, food, shelter, sanitation, water.

1.1. Introduction

To begin this chapter, which is aimed primarily but not exclusively at an educated public readership, an explanation of the title is given to place the content in perspective. Literal meanings of "today" and "tomorrow" are defined as follows — "today" signifies the period from now to 2030, and "tomorrow" refers to the second half of this century, which is from 2050 to 2100. The rationale of the terms "nourish" or "nourishing" also requires further clarification, as they are frequently associated with the provision of sufficient food or the fostering of a healthier diet, and, more recently, are used to describe particular leadership training and education activities.[1] Both terms have been in use for at least seven centuries, but they have not only been applied to food, and more recently, to business practice. Their original Latin root word, *nūtrīre*, can mean "to feed" but also "to take care of," and there are a number of modern synonyms, such as nurse, foster, cultivate, and so on. So, when using the terms "nourish" and "nourishing," the context of their use must be identified. Some linguistic scholars suggest an element of semantic shift, i.e., when the meaning of a word shifts over time; for example, the word "silly" in its earliest uses referred to things being worthy, but it was then used to describe the weak and vulnerable, and now it is applied to those people or ideas who are considered foolish. In this chapter, "nourish" means to take care of the provisions for energy and food security for all people and to ensure that they have access to clean water and sanitation, clean air, and an affordable, and a safe place to live. This does not simply mean providing more of the basic essentials to account for the forecast in population increases. They are also not universally mandated human rights.

The above-mentioned provisions are embedded, if sometimes indirectly, in the 30 articles of the 1948 United Nations' Universal Declaration of Human Rights (UDHR), especially in Articles 25 and 26.[2] This Declaration, the UN Resolution 217, although not legally binding, was accepted by 80% of the UN members (at the time, 48 out of 58 members). Translated into more than 500 languages, it has yet to gain universal government acceptance or ratification, and yet is often referred to in national documents and quoted in legal decisions.[3] Moreover, the origin of several more recent individual UN human rights resolutions can be traced back to the 1948 declaration, and all current member states have accepted at least one of these resolutions. In terms of global cooperation, this state of affairs may seem encouraging, but, in 2018, the UN High Commissioner for Human Rights, Michelle Bachelet, stated in an open letter that, "[i]n many countries, the fundamental recognition that all human beings are equal, and have inherent rights, is under attack. And the institutions set up by States to achieve common solutions are being undermined".[4] So, the situation is perhaps not as encouraging as may be imagined.

Similarly, although 121 member states have accepted the 2010 UN declaration on water being a human right, there were 41 abstentions, and only a few jurisdictions have changed their constitutions or passed legislation to require the provision of clean water. Clean air has still not been accepted by the UN general assembly as a human right. This does not that no actions are being taken on these issues. Perhaps the most promising outcomes of universal initiatives has been the political acceptance of the 2015 Paris Climate Agreement to limit increases in global surface temperatures to 2°C above pre-industrial levels by the end of this century, and the parallel commitments associated with the Sustainable Development Goals (SDGs) of the UN 2030 Agenda aimed at the eradication of global poverty.[5] The complete achievement of the objectives of these global plans are likely to prove elusive in the stated timelines, regardless of the utterances of some national leaders, the perceivably unceasing cajoling of UN bureaucrats, the lobbying of activists, and other groups with vested interests.

With respect to the 2°C target, the concern with timelines is discernible in the declaration of 2050 net-zero carbon strategies by a growing number of countries.[6] Inevitably, there are also some nuances between setting global targets and their national interpretations. For example, what does "pre-industrial" and "net-zero" mean? Perhaps the more important question is how such nuances could affect the attempts to nourish tomorrow, especially from an engineering perspective. Clearly, the actual strategies adopted nationally will impact the need for the rate of technical innovation. If carbon capture is a key element of such strategies, then there will be a need for a rapid expansion of available existing systems from demonstration units to utility scale, together with an attractive price tag. In terms of fulfilling the 17 SDGs of the UN 2030 Agenda, 169 targets, as tracked by 231 unique indicators, will need to be satisfied in less than a decade — a seemingly impossible task. This does not mean that no progress will be made towards nourishing tomorrow. It simply shows that the timelines appear overambitious for achieving the SDGs and that there could be unsurmountable obstacles in meeting the Paris Agreement temperature limits by the end of the century.

Agreements like the Paris Agreement and the UN 2030 Agenda are indeed necessary to ensure that steps are taken to nourish tomorrow. To some extent, elected governments and the leaders of the richer countries often seek to claim, that they are the best Group of Seven (G7) or Group of Twenty (G20) country on a variety of issues compared to their contemporaries. It seems that 'bragging rights' are important from a political perspective. This was especially obvious during the COVID-19 pandemic with Israel proclaiming that "[w]e will be the first country in the world to emerge from the coronavirus," and, just as prematurely, New Zealand announced it had eliminated the pandemic. Despite such missteps, national political leadership today will still be necessary for global well-being tomorrow, and many have announced, regardless of the UN environmental and sustainable agreements, that they and their counties are committed to a net-zero strategy by 2050. However, are these more instances of rhetorical boastfulness or do these commitments have substance?

In October 2021, just prior to COP26, the annual United Nations Climate Change Conference, the UK government, who hosted the conference, announced and published their detailed and legally enforceable plan for achieving net-zero.[7] No doubt other countries will publish similar plans, although India has recently rejected the net-zero carbon emissions 2050 target. Additionally, some countries have also "taken the pledge" to cut methane emissions by 30% by 2030, but others, such as Australia, have stated that they will not include the pledge in their net-zero pathway.[8]

Notwithstanding the somewhat chaotic milieu surrounding universal acceptances of UN resolutions and what legally constitutes a human right, this should not detract from the global acknowledgment that the basic essentials for a healthy life are air, water, food, shelter, energy, and sanitation. This should not be wholly surprising since humans can only survive a few minutes without air, a few days without water, and maybe a few weeks without food. These are physiological facts and are not dependent on definitions of human rights, anthropogenic climate change, or sustainability. If, however, there are pollutants and contaminants in the breathable air and the drinkable water, then the length of human survival will be severely curtailed. Yet, we have known for centuries that the source of the harmful toxins is predominantly the result of human activities, especially those associated with the lack of adequate sanitation and ventilation. Our ancestors were aware of these issues and used a taste, smell, and sight approach to assess water properties and developed water treatment technologies to combat them, albeit in the absence of scientific evidence. Similarly, although smoke was regarded as more of an annoyance than a health hazard, efforts were made to vent it away from places of dwelling through holes or chimneys, and some fuels used for cooking and heating, such as coal, were banned from as long ago as the 13th century.[9] So, it can be said that the need of clean, or at least cleaner, air and water has long been recognized.

However, it was only in the 19th and 20th centuries that pathogens in water and air were identified, and, subsequently, the shortcomings of those water treatment methods were exposed, together with the recognition of the health hazards posed by airborne particles. These effects have become more pronounced as the global population has rapidly increased resulting in efforts to improve both water and air quality.[10–12] But what about the other identified provisions? All people may not have access to food, and this access cannot be guaranteed for present and future populations. War and conflict, adverse regional weather and climatic occurrences, plant diseases, insect infestations, economic factors, and government level policies have all contributed to famines, for as long as recorded history and, indeed, continue to impact at least 130 million people.[13,14]

Historically, the most dominant cause of famines has been war and conflict. Unfortunately, the last 3.5 millennium have witnessed peace for only about 8% of the time. Since its Declaration of Independence in 1776, the United States of America (US) has only enjoyed 20 years of peace.[15] Moreover, although mitigation of anthropogenic climate change may reduce the frequency and scale of adverse weather incidents, it cannot eliminate natural occurrences such as volcanoes, earthquakes, hurricanes, and oceanic circulations like El Nino. Government policies of afforestation, the encouraged use of plant-based fuels, and wholesale bans on some pesticides and herbicides, although well intended, could reduce the amount of land available for food production and crop yields. Overall, there are many challenges to eliminating global hunger now, what more in the future. A key factor in overcoming these challenges could be reasonably modest amounts of foreign aid to be provided if and when needed (see Sections 1.3.6.4 and 1.4).

Shelter from the weather and predators have always been a priority for survival for humans. Natural features such as caves or rudimentary construction methods provided most of the early hunter–gatherers with some form of shelter, but as agriculture and animal domestication became prevalent, nomadic living was largely replaced by static encampments with permanent buildings. Nonetheless, some groups have continued their itinerant lifestyles, searching for better locations to hunt or to graze their animal herds and taking readily constructed shelters with them, such as the tepees of the Indigenous people inhabiting the North American plains or the yurts of the traveling people living on the Asian steppes. Those without access to shelter were, and still are, unlikely to survive for any length of time. Today, most people live in permanent structures, but there could be as many as 150 million individuals who are homeless and even more, over 1.6 billion, living in inadequate shelter.[16] The situation could be worse because it is proving difficult to accurately quantify these statistics, principally because there are no universally accepted definitions of homelessness and shelter. However, these data give some indication of the shelter problems, with about 2% of the global population needing shelter of any kind and 20% are without permanent and appropriate shelter. Although often associated with extreme poverty, homelessness is a problem also encountered in rich countries, as seen in Figure 1.1.[16] These issues will be further discussed in Section 1.3.5.2.

Having access to energy does not necessarily mean that it is affordable; energy poverty is a very real factor in overall poverty. One of the main indicators of success of SDG 7, namely "to ensure access to affordable, reliable, sustainable and modern energy for all," is the portion of the global population which has access to electricity, i.e., Target Indicator 7.1.1.[17] Although there were noticeable global improvements in access to electricity especially since 2016, in 2019, there were still between 9.9% (World Bank) and 13% (Our World in Data) of the world's population without such access.[18,19] The variation is likely due to slightly different definitions and accounting methodologies. In some instances, although a country may record 100% access to electricity, there could be some indigenous populations and other remote off-grid communities that have either

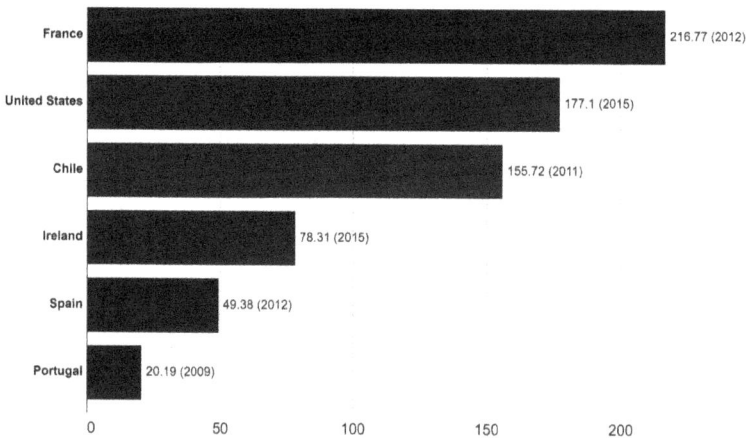

Figure 1.1.　Estimated Homelessness per 100,000 inhabitants.[16]

Note: All of the included countries employ a similar definition of homelessness.
Source: OECD Affordable Housing Database.

no access or reduced access to electricity. Numerically, such instances only slightly reduce the 100% designation by a few tenths of a decimal point, albeit that is no comfort to those affected. The access data also masks the level of affordability in most countries, as seen in Section 1.3.5.

A cursory look at the more recent annual data on the six identified elements needed to nourish tomorrow could result in understandable impressions of hopelessness, but the data for this millennium, prior to the COVID-19 pandemic, indicates that in spite of population increases, substantial global improvements were recorded for most of the elements, with the exception of the number for undernourished people in terms of food, which has been increasing annually since about 2014, after decreasing for the previous decade (Figure 1.2).[20] Is this because of rising populations in poverty-challenged and conflict prevalent regions, or is the problem more widespread? Having sufficient food is different from meeting the UN definition of having access to healthy and nutritious food. If this means enforcing dietary changes, then will that help hungry people and cause angst among others? These are the type of questions that are addressed in Section 1.3.4.

In this chapter, although the needs have been separated individually for the convenience of discussion, they are largely interconnected, i.e., universal access to clean water, safe sanitation, and clean air technologies rely on ensuring that affordable energy is readily available for all. In terms of achieving the UN Sustainable Development Goals 2 to 17, it is likely

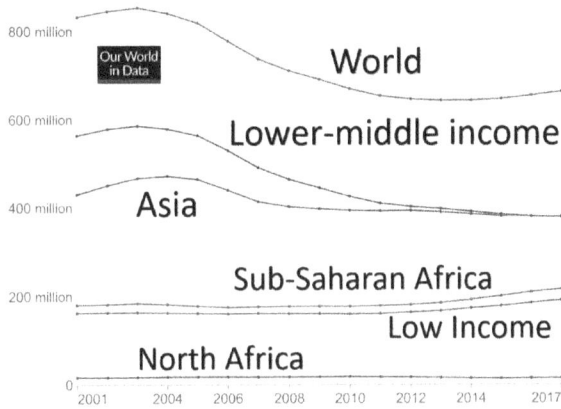

Figure 1.2.   Undernourishment Exemplars 2001–2017.[20]

that only the elimination of poverty, i.e., Goal 1, will enable the rest to become reality, especially those goals associated with the needs of nourishing tomorrow. The presumed timeframe for tomorrow starts in 2050, which coincides with the target date for net-zero emissions. The possible future impacts of the net-zero approaches on the nourishing schedule is discussed later in the chapter. Although, as ever, envisaging the future is fraught with difficulties as recently illustrated by the Prime Minister who announced in June 2020 that New Zealand had eliminated COVID-19, but a few weeks later, new cases were reported, and by November 2021, the number of cases were an order of magnitude higher than at any other time during the pandemic.[21] Maybe, as expressed in the often-quoted expression,[a] "[t]he best way to predict the future is to create it." Nevertheless, it does not need a seer to forecast that the successful achievement of the SDGs of the 2030 Agenda, the climate targets of the 2015 Paris Agreement, and the needs for the nourishing of tomorrow are likely to pave the way to create a better future for all.

1.2. The Nourishing Challenges: Setting the Scene

The present (i.e., today) has been identified as up to 2030 in this chapter. This means that actions must be taken in the next eight years to enable the

---

[a]Attributed to President Abraham Lincoln, the management guru Peter Drucker, and several others.

nourishing challenges to take effect by 2050. The period from 2030 to 2050 will then be crucial for measuring the success of meeting the challenges. However, whether 20 years will be sufficient will all depend on the scale of the tasks and how they are approached. Table 1.1 lists the present challenges for achieving a nourishing tomorrow. Table 1.1 also lists the aims to provide all the elements for all the people and the corresponding obstacles to success. Invariably, the barriers to be overcome will be a complex mix of political, economic, cultural, and technological.

Table 1.1. The present status of the author identified nourishing elements.

Nourishing element	Associated SDG #	UN human rights	Comments/Obstacles
Clean Air	Elements of 3, 7, and 11	No	• Access to modern energy services will contribute to cleaner air but encouraging biomass use could negate any improvements • 7 million estimated premature deaths[22]
Clean Water	6	Yes, UN A/Res/64/292, 2010, A/Res/70/169, 2015	• 2.2 billion have no access to safely managed drinking water[i,23] • Between 5 to 11 million die each year from waterborne diseases[24,25]
Safe Sanitation	6	Yes, UN A/Res/64/292 A/Res/70/169,	• 4.5 billion do not have access to safely managed sanitation services[23]
Adequate Food	2	Yes, UN 1948 charter, Article 25	• 2.37 billion had no access to adequate food • 768 million are undernourished • 928 million are severely food insecure[26]
Adequate Shelter	11	Yes, UN-1948 charter, Article 25	• Estimated 1.6 billion live in inadequate housing and 150 million are homeless[27]
Access to Affordable Energy	7	No	• 940 million do not have access to electricity • 3 billion do not have access to clean cooking fuel • 89% of global primary energy use is not-renewable[28]

Note: [i] Also called potable water.

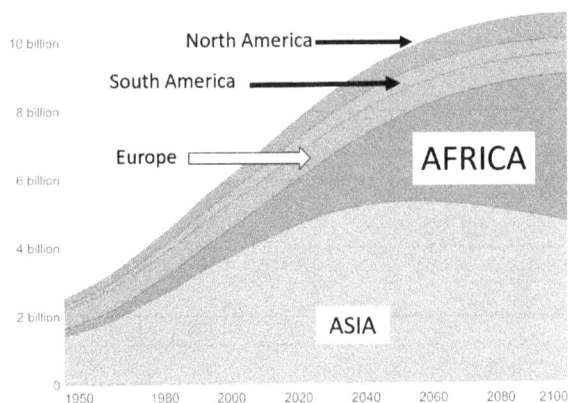

Figure 1.3.   UN projections of regional populations from 1950 to 2100.[31]

The challenges outlined in the final column of Table 1.1 are for the present global population. It is forecasted that by 2050, using the UN's medium population growth scenario, there will two billion more people on earth than today, and by the end of the defined nourishing period, there will be 3.2 billion more.[29,30] Most of this growth is projected to take place in Africa, whose population will be more than three times higher in 2100 at 4.3 billion, compared with 1.3 billion in 2019. This would represent almost 40% of the total global population at the end of this century (Figure 1.3).[31] Arguably then, any needs preparations made in the next two decades will require that they are sufficiently adaptable to cater to expanding regional and global populations. Accurate tracking of the trends in the success of the nourishing needs preparations will be a central factor.

While providing insight into the scale of the specific problem, the numerical values given in the comments column of Table 1.1 cannot be presumed to be wholly definitive, as not all countries regularly report their national situation on individual items, if at all. Nevertheless, several authoritative data collection agencies such as the UN, the World Bank, the US Environment Information Administration (USEIA), the World Health Organization (WHO), and the UK-based Our World in Data (OWD) all go to great lengths to collate worldwide information as accurately as possible. Nonetheless, there are some variations in the precise definitions used by different nations and agencies and even within the various departments of national governments, e.g., homelessness.[32,33] To address some of these data issues, the tracking of particular SDG goals and their associated

Table 1.2.   SDG tracking custodians for nourishing elements.[34]

SDG # Tracking	Nourishing element	Custodian agency
2, 6	WASH[i]	UN Food and Agriculture Organization
3	Clean Air	WHO with United Nations Children's Fund (UNICEF[ii])
7	Energy	Joint Custodians: International Energy Agency (IEA), WHO, World Bank, International Renewable Energy Agency (IRENA), UN Statistics Division (UNSD)[35]
11	Shelter	UNSD (*at time of writing*)

Note: [i] Water, Sanitation, and Hygiene (WASH).

[ii] UNICEF's original name was the United Nations International Children's Emergency Fund.

targets indicators is the responsibility of approved custodian agencies,[34] as indicated in Table 1.2.

The data obtained by these custodian agencies and their subsequent analyses will prove to be essential in determining if the efforts to nourish tomorrow are flourishing. Although the collection processes have been affected by the COVID-19 pandemic, and it will likely have adverse impacts for the next few years, perhaps the main concern at present is that tracking is in its infancy. Tracking is based on the measurements of the indicators pertinent to the SDG goals and targets, but consistent and universally accepted methodologies for acquiring such data will be crucial in helping the decision processes of policymakers, especially those in so-called representative democracies. The establishment of such methodologies is a work-in-progress lead by an international body, the Inter-Agency and Expert Group on Sustainable Development Goal Indicators (IAEG),[b] working under the auspices of the UN.[36] Presently, the criteria for tracking indicators are categorized into three tiers, as described in Table 1.3.[36]

The SDG tracking mechanisms and the intensive efforts to develop clear and universally acceptable methodologies should provide a robust basis on which to measure the progress and success of any provisions taken to nourish tomorrow, at least where the SDGs and nourishing needs are synchronous. However, as no system is perfect, the UN has a safety net for assessing the progress of its human rights resolutions, by appointing Special Rapporteurs, i.e., independent experts to investigate identified aspects and provide reports to the Human Rights Council and the General Assembly. For example, in 2017, the then Special Rapporteur presented a

---

[b] Inter-Agency and Expert Group on Sustainable Development Goal Indicators.

Table 1.3.   Indicator measurement defined tiers.[i,36]

Tier #	Criteria definition
I	Indicator is conceptually clear, has an internationally established methodology and standards are available, and data are regularly produced by countries for at least 50% of countries and of the population in every region where the indicator is relevant
II	Indicator is conceptually clear, has an internationally established methodology and standards are available, but data are not regularly produced by countries
III	No internationally established methodology or standards are yet available for the indicator, but methodology/standards are being (or will be) developed or tested

Note: [i]As of March 29, 2021.

report on the rights of Indigenous people with respect to the impact of climate change and climate finance.[37] More recently, in October 2021, the UN expanded a Special Rapporteur's mandate to include "the promotion and protection of human rights in the context of climate change".[38] Together, all these factors represent a vigorous framework to ensure, or to attempt to ensure, that all sustainable development goals come to fruition and that their targets are scrupulously measured against defined benchmarks. While this appears to be good news for the actions which must be taken to nourish tomorrow, it needs to be acknowledged that UN global aspirations are not legally enforceable and do not ensure compliance. However, nationally legislated standards can be mandatory.

1.3.   The Elements of a Nourishing Tomorrow

Having identified the mix of the constituents required for the nourishing of tomorrow, clarified which are universally acknowledged as human rights and which are not, and discussed how the effects of present actions can be tracked and measured, the focus in this section is on the specific nature of the nourishing needs as summarized in Table 1.1.

1.3.1.   *What is clean air?*

The Earth has only one atmosphere, but its constituents vary by geographical and topological location. These differences are caused by natural phenomena and human-driven activities. For human survival, the constituent of interest is oxygen. In air containing no water vapor, i.e., dry air, the volumetric percentage of oxygen is 20.95%, with the remainder 78.08% being largely nitrogen, with the major greenhouse gases, carbon dioxide and methane contributing 0.00132% and 0.000008% respectively. One may think air would be cleaner and more breathable if it contained more oxygen, but if the concentration is too high, there could be harmful effects. If the concentration of oxygen is below the dry air amount, harmful effects can be experienced if the concentration falls below 19.5%, and humans cannot survive if the concentration is at lower than 6%,. If the oxygen concentration is above 21% in a breathable gaseous mixture, they are known as Oxygen-Enriched Atmospheres (OEA).[39] These situations can be encountered in some marine diving operations and some medical procedures, but they can also produce harmful effects if the OEA is breathed for longer than a few hours.[39]

The US Occupational Safety and Health Administration (OSHA) defines the optimal oxygen concentration levels to be between 19.5% and 23.5%, a very small range. It could be said that humans live in the ideal breathable atmosphere, but as the non-harmful range is so small, any disruptions to these levels are a cause for concern. There is some evidence to suggest that anthropogenic climate change is reducing both the oxygen content of air and the dissolved oceanic oxygen content.[40] The latter can create or exacerbate hypoxia, i.e., dead zones, in large bodies of water such as the Great Lakes and the oceans. The low concentration of oxygen in these zones can cause marine life to either die or move to other areas.[41] This has ramifications for capture fishery activities. Deep-water fishing ships and fleets may be able to move to more fruitful areas, but people living in small coastal communities that rely on fish as the main source of their diet and income will be adversely impacted. Hypoxia can be caused by both natural phenomena, often seasonal, and by human activities. Therefore, for air to be breathable, it should contain an oxygen concentration within the optimal range.

However, having the oxygen concentration within the optimal range does not necessarily mean that the air is clean. Breathable air becomes unclean and unhealthy when pollutants, sometimes referred to as contaminants, enter the atmosphere. Sight and taste were used as the main indicators for air quality until relatively recently. Smoke was largely considered an annoyance until the late-19th century, but as populations grew and coal burning rapidly increased, whether for domestic heating and cooking, and powering railway steam locomotives and boilers in the alkali industry, especially in the United Kingdom (UK), medical and public concerns eventually led to the UK government's passing of the 1926 Public Health (Smoke Abatement) Act.[42] This was perhaps a political gesture because it had little effect, since even its limited measures were rarely enforced.

However, as studies of the causes of disease and their associated health effects became more advanced, particularly with respect to identified airborne and waterborne toxins, many global governments strengthened their legislation, albeit usually only after scientific evidence and public activism could no longer be ignored. The harmful health effects of mixing lead compounds into gasoline which was known since leaded fuel first became commercially available were dismissed for several decades, before being grudgingly accepted by politicians and demurring scientists. In 1970, the US Congress, on the advice the Environmental Protection Agency (USEPA), mandated the phasing out of its use, but it would be more than a quarter of a century before it was partially banned, since the ban only applies to new vehicles and not at all to "aircraft, racing cars, farm equipment, and marine engines".[43] Prior to the Clean Air Act of 1970 in the US, concerns about the health hazards associated with the inhalation of particulate matter from combustion engine exhausts, coal fired power stations, and road dust, resulted both in many national clean air acts being passed into law and intranational air quality guidelines being published, starting in the 1950s and 1960s.[44-46]

As a consequence of these global efforts, the term "clean air" is embedded in many national laws and international guidelines, but there is no explicit definition of what constitutes clean air. Rather, in these official documents, the phrases "air pollution" and "air quality" are used to describe "healthy" air. Thus, the question posed at the start of this section should perhaps be "what is healthy air," in the absence of a universal definition. However, when determining whether air is clean or healthy, the same problem arises, i.e., how do we know? Is air only healthy when there are no constituents in the dry air other than 78% nitrogen, 21% oxygen, and the remainder argon and other trace gases? The answer to the second question is arguably yes, but in reality, such a situation rarely ever exists, at least not on a daily, round-the-clock basis. This lack of occurrence provides an insight into an answer to the first question — if the constituents of the air that is breathed can be measured, and the results do not adhere to the components of dry standard air, then the air may not be healthy, i.e., harmful to human physiology. This can only be verified by identification and compelling medical evidence.

Figure 1.4.   Smoke filled atmospheres (a) Victorian Industrial Period and (b) Canadian Wildfire.[47,48]

Yet, in the absence of such confirmation, visual observation can provide obvious signals of problematic air quality, such as the uncontrolled industrial activities of the Victorian era, as seen in Figure 1.4(a), or large forest wildfires, a recent example occurring in 2016 in Fort McMurray, Alberta, Canada, as seen in Figure 1.4(b).[47,48] However, as previously mentioned, smoke was considered an annoyance rather than a health hazard, a stance that lasted until the second half of the 20th century. The main reason for the change in attitude was that it was finally recognized that smoke contains particulate matter (PM) and inhaling PM causes lung damage, which can be fatal. Now, for more than five decades, PM has been identified as a pollutant along with five to six other major harmful air emissions, although the USEPA had identified a further 188 air pollutants which are known, or suspected, to cause serious health effects.[49] The major pollutants, formally classified as air contaminants by Canada,[50] criteria pollutants by the US,[51] or simply pollutants by the WHO[52] are given in Table 1.4 along with the recognized greenhouse gases (GHG).

The WHO identified the PMs of concern by adding subscripts, in this case $_{2.5}$ and $_{10}$ — these refer to the diameter of the particles measured in microns. Both inhaled $PM_{2.5}$ and PM_{10} can penetrate deep into the lungs where they can cause serious damage, especially the smaller diameter particles, often referred to as fine or ultrafine particles.[53] The relative size of $PM_{2.5}$ is clearly demonstrated in Figure 1.5.[54] In sufficient quantities, PM particles greater than 2.5 microns can also impact visibility, although to the human eyes, the fine particles are usually invisible. An idiosyncrasy of the US system is that PM over 10 microns in diameter comes under individual State jurisdiction, not Federal.[54] However, having identified the pollutants and measured their concentrations, how do the general public

Table 1.4.   Exemplars of major air pollutant categories.[50–52]

US-EPA criteria pollutants	Canadian air contaminants	World health organization pollutants	Greenhouse gases (GHG)
			Water Vapor (H_2O)
Carbon Monoxide (CO)	Carbon Monoxide	Carbon Monoxide	Carbon Dioxide
Particulate Pollution	Particulate Matter (PM)	$PM_{2.5}$ and PM_{10}	
Nitrogen Dioxide (NO_2)	Nitrogen Oxides	Nitrogen Dioxide	Nitrous Oxide (N_2O)
Ozone	Ground Level Ozone	Ozone	Ozone (O_3)
Sulfur Dioxide (SO_2)	Sulphur Oxides	Sulfur Dioxide	
Lead (Pb)	Ammonia		
	Volatile Organic Compounds (VOC)		
Carbon Dioxide	Carbon Dioxide		
			Methane
			Chlorofluorocarbons, Hydrofluorocarbons (CFCs)

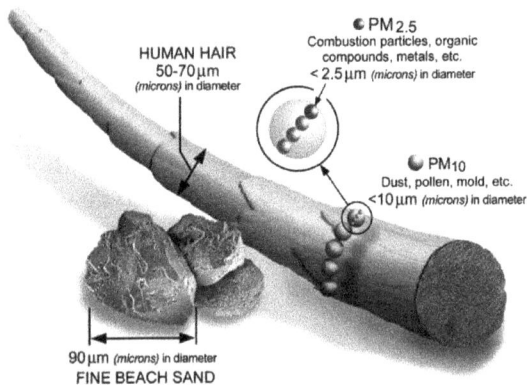

Figure 1.5.   Comparative size of a human hair and a $PM_{2.5}$ particle.[54]

Source: Image courtesy of the U.S. EPA.

know whether the air they are breathing, or breathed, was clean or hazardous?

1.3.1.1. *Air quality assessments and public awareness*

Information about the state of the air is normally provided to the public through localized air quality announcements. In the US, the USEPA recommend the use of an Air Quality Index (AQI) for assessment, which is expressed as a number between 0 and 500.[55] The AQI is divided into six discrete ranges, although sometimes the hazardous category is further sub-divided. Each range is also accompanied by descriptors, e.g., good, unhealthy, hazardous, together with an associated color code from green to maroon, as summarized in Table 1.5.[56] For each category, the criteria pollutant concentrations are also defined. Exemplars of these concentration ranges for $PM_{2.5}$ and Ozone are shown in Table 1.5.

The concentration and AQI ranges can then be used to calculate an index, I_p, for each pollutant actually being measured from a relatively simple expression, i.e.,

$$I_p = (I_{Hi} - I_{Lo})(C_p - BP_{Lo})/(BP_{HI} - BP_{Lo}) + I_{Lo}$$

where: I_p = the index for pollutant p: I_{Hi} = the AQI value corresponding to BP_{HL}

Table 1.5.   USEPA AQI descriptors and exemplar concentration ranges.[55,56]

AQI ranges	Range description	Range color code	Concentration $PM_{2.5}$ $\mu g/m^3$	Concentration ozone ppm
0 to 50	Good	Green	0–12	0–0.054
51 to 100	Moderate	Yellow	12.1–35.4	0.055–0.070
101 to 150	Unhealthy for Sensitive Groups	Orange	35.5–55.4	0.071–0.085
151 to 200	Unhealthy	Red	55.5–150.4	0.086–0.105
201 to 300	Very Unhealthy	Purple	150.5–250.4	0.106–0.200
301 to 500	Hazardous	Maroon	250.5–500.4	0.405–0.604[i]

Note: [i] For Ozone, AQIs between 0 and 300 are from 8-hour measurements, but for 301 to 500, they are from 1-hour measurements. $PM_{2.5}$ is for a 24-hour period.

I_{Lo} = the AQI value corresponding to BP_{Lo}: C_p = the truncated concentration of pollutant p

BP_{HI} = the concentration breakpoint $\leq C_p$: BP_{Lo} the concentration breakpoint $\geq C_p$

using a set of published breakpoints (benchmarks) for the pollutant expressed in parts per million (ppm) for CO and Ozone, parts per billion (ppb) for SO_2 and NO_2, and $\mu g/m^3$ for PM. In the context of the equation, parameters 'truncated' means that data is presented as an integer or as a value to 1–3 places of decimals depending upon the identified pollutant. The breakpoints for the USEPA criteria pollutants, other than lead, are given in Table 1.5 of reference.[56] For example, for the AQI to be announced as "good" for PM over a 24 hour-period, the concentration range would have to be between 0 $\mu g/m^3$ and 12.0 $\mu g/m^3$, whereas for "hazardous," the range would be 425 $\mu g/m^3$ to 604 $\mu g/m^3$. However, what if the calculated index for one pollutant is the highest among the five to six measured pollutants? It is this highest index value that is used to define the overall AQI category and also identify the responsible pollutant. This approach provides anyone in North America with a means of determining their local air quality, but what about globally?

The WHO Air Quality Guidelines (AQGL) published periodically, the latest being in 2021, have recommended levels for pollutants to achieve clean air.[57] They realize that these levels have not yet been attained and so suggest increasingly stringent interim targets as a three-step pathway to obtaining the clean air levels. Their data for all the pollutants identified in Table 1.4 are expressed in $\mu g/m^3$. How do the global AQL recommendations compare to the AQI categories? While not being in precise harmony, they are very similar; for example, the AQGL for $PM_{2.5}$ over a 24-hour period is 15 $\mu g/m^3$ whereas the AQI "good" range is up to 12 $\mu g/m^3$; for CO, the AQGL is 4 mg/m^3; and for the AQI "good" designation, after measurement unit conversion, is 5.13 mg/m^3. Thus, yardsticks exist in determining if the air is clean, or at least, how clean it is. However, there appears to be some confusion in assessing air quality. A location may record a "good" AQI every day of a year but not meet the AQGL annual average target. This is because the annual pollutant concentration targets can be lower than the daily targets. For example, the annual AQGL target for $PM_{2.5}$ is 5 $\mu g/m^3$ whereas a concentration of 12 $\mu g/m^3$ could result in the pronouncement of a daily "good" AQI. However, with several criteria pollutant concentrations contributing to the determination of daily AQI, an averaged compilation of a yearly AQI comparison with annual AQGL could be problematic. Regardless of categorization nuances, if the daily "good" AQIs as well as annual "good" AQGL indices can largely be achieved, this would suggest that the prospects of achieving the clean air nourishing needs of tomorrow are promising. To use AQI type indicators together with AQGLs, locally, regionally, and globally, there will be an obvious need for consistency of measurement and timeliness of reporting.

A possible issue with the current air quality assessment framework is that very recently, some governments such as those of the US and Canada, have declared carbon dioxide (CO_2) as a pollutant. This means that the Federal Canadian[c] government can legitimately impose (i) a fuel charge, popularly known as a carbon tax, over any provincial objections, on "producers, distributors and importers of various types of carbon-based fuel," and (ii) "a pricing mechanism for industrial greenhouse gas ("GHG") emissions by large emissions-intensive industrial facilities."[58,59] Among some commentaries, this is a contentious issue since it could be inferred that it is illegal for humans to breathe, as we exhale CO_2, and vegetation emits CO_2 which it also needs to grow, then it would appear absurd to label CO_2 as a pollutant. However, the communication regarding identifying CO_2 as a pollutant has been unclear and far from exemplary, since the real target of the national declarations is the CO_2 from anthropogenic, not natural, sources such as the burning of fossils fuels. In essence, at least for Canada, all GHGs are considered to be pollutants. As such, do anthropogenic GHG emissions impact air quality?

The answer to the air quality question is yes, in the context of anthropogenic climate change since, according to the USEPA, changes in climate can impact air quality as a result of the potential to increase ground-level ozone (O_3) due to global warming.[60] On the other hand, certain types of particulate matter, e.g., sulfates, can result in atmospheric global cooling.[61,62] There are still many uncertainties with respect to how climate change can affect air quality, or vice versa, and research on these factors continues. Presumably, if it is shown that GHGs do explicitly affect air quality, then changes will need to be made to the pollutant benchmarks used in the current AQI and AQGL assessments. As anthropogenic CO_2 has been identified as the main driver of global warming, at least by surface temperature and concentration correlations, it will highly

---

[c]Canada is a Federation or *Confederation* of 10 provinces and three territories, and constitutionally the central government and provincial governments have different powers.

likely be added to the criteria air pollutants and contaminants lists. This could add an extra layer of complexity to the air quality determinations. However, where would the data and the benchmarks come from?

Before the establishment of the World Meteorological Organization's (WMO) Global Atmospheric Watch Program (GAW), there were only three recognized long-term CO_2 measurement sites — two in the Northern Hemisphere, including the famous Mauna Loa station in Hawaii, and one in the Southern Hemisphere. As such, the opportunities for local measurements of a full suite of GHGs were limited. However, by 2019, there were 30 global stations, 400 regional stations, and around 100 contributing stations reporting to the World Data Centre for Greenhouse Gases (WDCGG), which is operated by the Japanese Meteorological Agency (JMA).[63] Two GHGs, nitrogen oxides and ozone, are already part of 24-hour AQI determinations, and humidity, which gives a measure of the amount of airborne water vapor, is continuously monitored at weather stations. Under certain circumstances, the presence of water vapor can make the air temperature feel higher. Canadian meteorologists use a humidity index referred to as the "humidex," a mathematically defined combination of heat and humidity to determine what the weather will feel like to the average person and use it to express an air temperature equivalent.[64] At high humidex values, i.e., $\geq 45°C$, temperatures experienced can be dangerous to human health, resulting in debilitating heatstroke. Different countries and weather agencies use similar, but not identical, indexes. Despite their mathematical formulation, the use of such indexes is still somewhat subjective, since, for example, in the Canadian case, it has not been defined what constitutes an average person.

In general, the concentrations of three of the six GHGs are already measured at many locations, two of which are included in air quality determinations, water vapor being the exception. The remaining three GHG concentrations are also measured with precision instrumentation and, in the case of CFCs, largely by orbiting satellites.[65] Indeed, at specific locations, CO_2 concentrations are measured on an hourly basis so they could be readily included in AQI calculations, provided that benchmark levels are defined and the data from far more locations than at present are collected and collated. Knowing the pollutant causes of unhealthy air, how to measure them, and how to control their concentrations, if they cannot be wholly eliminated, could provide the framework for meeting the "clean air" part of nourishing needs. However, identifying the causes of air pollution and measuring pollutant concentrations, while drawing attention to the challenges, do not give specific insight into how these requirements can be met politically, scientifically, or technologically. In Section 1.3.2, methods of controlling pollutant levels are examined next.

1.3.2. *Cleaning the air*

In addition to the criteria (or common) air pollutants, the USEPA, as previously mentioned, lists a further 188 hazardous air pollutants. Pollutants can be added or removed from this list. Methods of combating, i.e., eliminating or controlling, these particular pollutants are given by the USEPA.[66] These hazardous pollutants can be very localized, and they are not universally encountered. The criteria air pollutants are the most commonly encountered globally and, thus, the discussion in this section will explore how air can be cleaned of these pollutants. The two most prevalent types of air pollution are particulate matter, or PM, (often called soot) and ground-level ozone, which is closely associated with certain types of smog (i.e., smoke and fog).[d] Subsequently, the main focus will be on these two pollutants, with particular emphasis on PM, since about one-third of the global population uses unclean cooking fuels, which are a major source of PM, and the population of the regions where such fuels are prevalent is forecasted to increase significantly during the remainder of this century.[67,68] A brief overview of how the other criteria air pollutants can be reduced is given in Section 1.3.2.3.

1.3.2.1. *Particulate matter*

Of all the pollutants identified in the first three columns of Table 1.4, $PM_{2.5}$ is the one considered to be the most harmful, although recently, particles with even smaller diameters of less than 0.1 microns ($PM_{0.1}$), termed "ultrafine particles," are attracting increasing attention among the scientific and medical communities.[69,70] According to Green Facts,[71] ultrafine particles make up approximately 90% of all PM, so it is perhaps surprising that $PM_{0.1}$ has not yet been included in air quality standards, especially since their inhalation can be accompanied by serious health effects, including higher risks of heart attacks and strokes, and subsequent

---

[d]A term often attributed to the Anglo-French physician, H. A. Des Voeux, although there are other claims to its invention.

reductions in life expectancy. The problem is the global lack of sufficient data and a dearth of advanced and somewhat costly measurement instrumentation, although research and political efforts are underway to address these issues. The invisibility of such nano-particles can be appreciated by looking at the relative actual size of the fine $PM_{2.5}$ as illustrated in Figure 1.5, and recognizing that ultrafine particles are at least 25 times smaller.

$PM_{2.5}$ is not only one of the most harmful air pollutants, it is also the most prevalent, with more than 90% of the global population living in areas which exceed the WHO's AQGL. Moreover, there are wide discrepancies between nations and regions, as illustrated in the exemplars shown in Figure 1.6 illustrating the share of populations exceeding the AQGL for $PM_{2.5}$.[72] Nevertheless, as shown over the last decade, global $PM_{2.5}$ levels have in general been reducing especially in the US. However, several countries, such as India, Japan, China, and many in Sub-Saharan Africa (SSA) with large populations, still struggle to make any progress in attaining the AQGL. In India, its neighboring countries Pakistan and Bangladesh, and in SSA countries, the main challenge is that large swathes of the population still use wood and solid fuels such as coal and animal dung for cooking, usually on open hearths, both inside and outside their houses, as shown in Figure 1.7.[73]

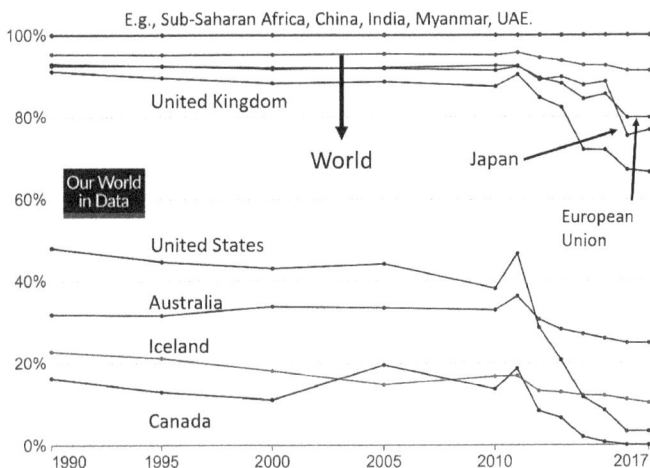

Figure 1.6.   Exemplars of PM concentrations above WHO AQGL by share of population.[72]

Source: Brauer et al. (2017) via World Bank.

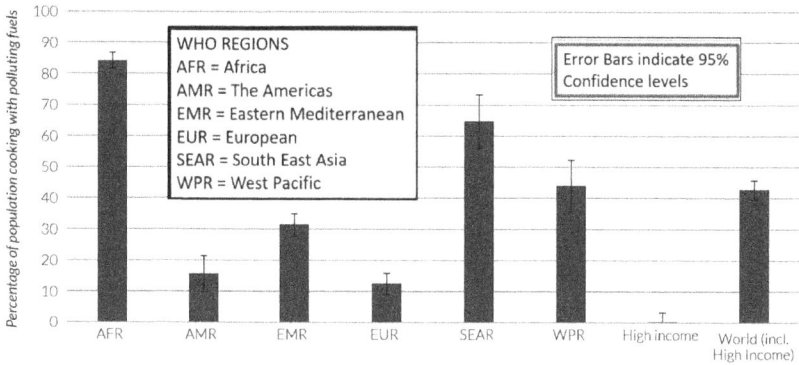

Figure 1.7. Percentage of population cooking with unclean fuels in 2016.[73]

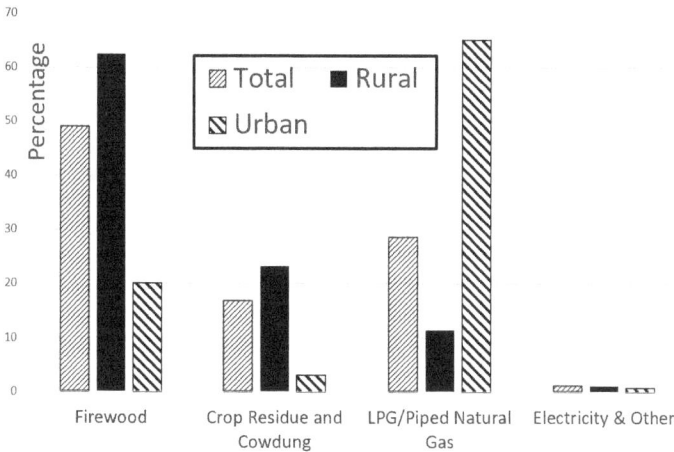

Figure 1.8. Cooking fuel usage in India 2011 updated to 2016.[76]

Ironically, the burning of biomass is strongly encouraged as a fossil fuel replacement, especially by European political leaders, as it is legally recognized as being carbon-neutral based on its eventual sequestration of CO_2,[74] which is achieved by replacing harvested plants with new plantings. However, such a strategy only marginally reduces PM emissions.[75] India is further challenged because of its large rural population, with many in remote, off-grid communities. India's efforts to address the wood burning dilemma and the obstacles they face are illustrated in Figure 1.8.[76] However, it should be noted that the data used in the figure are from a decade-old census and, at the time, the intention was to replace wood with petroleum gases (e.g., propane and butane) and natural gas (CH_4). This intention may have to be revisited since, at the recent UN Climate Change Conference (COP26) lead by the US and the European Union, more than 100 nations pledged to cut methane emissions by 30% by the end of 2030.[77] This commitment followed reports by the Intergovernmental Panel on Climate Change (IPCC), the UN, and various US organizations, such as the National Resources Defense Council, identifying CH_4 as an unclean or "dirty" fuel.[78–80] The rationale being that "[m]ethane is 84 times more powerful in trapping heat than carbon dioxide over a 20-year period and has caused about 30% of global heating to date."[81]

India, along with several other countries, did not commit to this pledge, but stated the intention to achieve net-zero carbon by 2070, two to three decades after most other countries, except China whose commitment for net-zero is by 2060. India and China have growing populations, with India forecasted to overtake China by 2027, and for both countries to reach peak population before 2050 and then experience steep declines by 2100.[30] Globally, as of December 2021, one in three people were living in either India or China. Perhaps, with such large populations, it is not too surprising that both these countries will need more time than others to achieve the net-zero carbon targets, or perhaps they may have other competing and compelling policy priorities they have decided to address.

As shown in Figure 1.6, on the opposite end of the PM compliance are countries such as Canada and the US — both have measurably reduced their PM emissions over the last decade. Is it then easier to reduce PM concentrations in countries like the US? Is domestic wood burning not as prevalent as in India? What are the sources of PM in the US? In Figure 1.9, it can be seen that residential wood burning in the US is still a significant contributor to PM emissions. However, unlike statements often encountered in media commentaries, the main sources were not on-road vehicles or fossil fuel combustion but road dust and wildfires.[82] The latter were largely caused by lightning strikes and other natural phenomenon, such as volcanic eruptions, although human-ignited fires whether intentionally or otherwise were also responsible. These fires can be exacerbated by inadequate land and forestry management. While better stewardship of forests can help reduce the impact of wildfires, they will still occur naturally.

The adverse health effects of road dust PM received more attention in recent years because of the concentration of heavy metals, i.e., chromium,

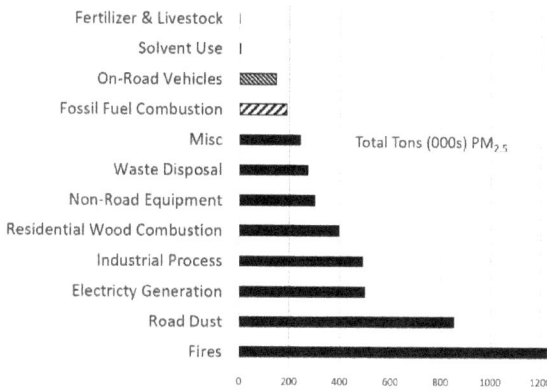

Figure 1.9.   Sources of PM in the US.[82]

lead, copper, zinc, and nickel. There are various sources of these metals, ranging from e-waste recycling to vehicular tire (tyre[e]) and brake wear, and street paint markings.[83] Whether vehicles are powered by diesel or gasoline internal combustion engines, or secondary electrical batteries, tire and brake wear are inevitable as is road surface deterioration. Diesel engine exhausts, once considered a major source of PM, are now significantly cleaner with the legislated treatment of tailpipe emissions using catalytic convertors, diesel particulate filters (DPFs), and selected catalytic reduction (SCR).[84] While mainly using precious metals such as platinum and rhodium, catalytic convertors can also contain the heavy metals nickel and copper, albeit the European Union banned the use of nickel and only Japan outlawed copper.[85] As catalytic convertors and DPFs age and become less efficient, trace amounts of precious and heavy metals are deposited as road dust and then enter the atmosphere. Another unintended consequence of the use of catalytic convertors is the production of ammonia, a PM pre-cursor, a circumstance which has been largely ignored until recently. In some countries, fears have been expressed that the concentrations of such emissions in urban environments may have been markedly underestimated.[86,87]

Moreover, although the transition of nearly all terrestrial transport to electric vehicles can be foreseen, probably by the latter part of this century, such a complete transition for marine freight and aviation transportation is difficult to contemplate without a revolutionary discovery in

---

[e]More common outside of the United States and Canada.

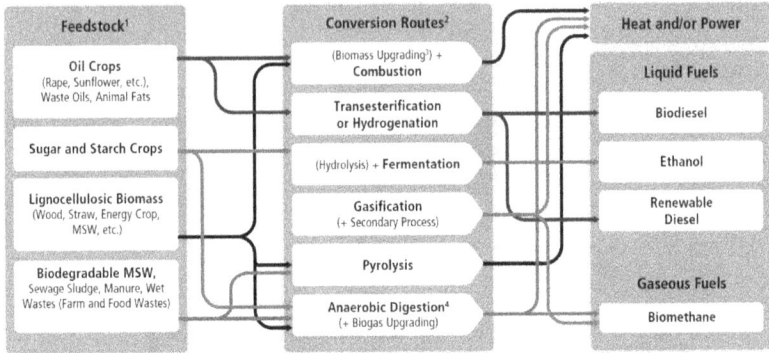

Figure 1.10.   Sources and production of biofuels.[88]

Notes: 1. Parts of each feedstock, for example, crop residues, could also be used in other routes. 2. Each route also gives co-products. 3. Biomass upgrading includes any one of the densification processes (pelletization, pyrolysis, etc.). 4. Anaerobic digestion processes release methane and CO_2 and removal of CO_2 provides essentially methane, the main component of natural gas; the upgraded gas is called biomethane.

battery technology. Nevertheless, hydrogen fuel cells coupled with wind-assisted shipping could make marine vehicles cleaner. Biofuels (Figure 1.10), like biomass, are considered to be carbon-neutral, albeit controversially, and could be used with all modes of transportation (terrestrial, marine, and air). Although some experimental and modeling studies have shown that PM emissions would not be eliminated, they could be reduced.[88,89] Unfortunately, such reductions would be accompanied by increasing NO_x emissions and greater fuel consumption.[90] Furthermore, over at least the past three decades, frequent concerns have been expressed about the crop land-use changes, from food to more profitable feedstock, for fuel production, though some studies suggest that certain types of land not used for crops could be used as a fuel feedstock, thereby not impacting food production.[91]

With regard to aviation, although the future use of biofuels is a distinct possibility, regardless of fuel efficiency decrease thereby reversing the recent improvements, battery-powered planes could possibly replace existing fossil-fuelled aviation engines,[92] and such vehicles already exist for short-haul flights and limited passenger capacity, e.g., two to four people. Moreover, the Danish government announced that by 2030, their intention is to eliminate fossil-fuelled aviation for all domestic flights.[93] It seems unlikely that long-haul flights will adopt electric propulsion because of the lack of energy storage capacity even with state-of-the art batteries. However, PM emissions from aviation does not appear to be a major issue with respect to air quality. According to the US Federal Aviation Administration (FAA), less than 1% of aircraft engine emissions are pollutants such as oxides of nitrogen and sulfur, carbon monoxide, PM, and other trace compounds.[94] Only about 10% of the engine emissions, including water vapor and CO_2, are emitted at 3,000 feet (142 meters) or below. Yet, people living or working close to an airport, and arriving and departing passengers when outside the terminal, will usually sense by smell and taste that the inhaled air is not of particularly good quality. Obviously, the aircraft will be below 142 meters to land and to take-off and their engines will emit PM, but the bulk of the air pollutants around airports will come from ground operations including on-road vehicles collecting and dropping-off passengers. Nevertheless, in the US metropolitan areas where commercial airports are located, the contributions of aviation activities to their area's air pollutant PM levels are quite small, invariably lower by a 0.5%.[94]

The US has already made great strides in reducing PM concentrations from human activity from fossil-fuelled land vehicles and by shifting from coal-fired to natural gas electrical power generation. Unfortunately, the dirty epithet now being applied to natural gas by the same US agencies who strongly recommended its use less than decade ago could present a challenge to nourishing tomorrow as, arguably, a likely outcome could be the curbing of any planned new and refurbished gas-fired utility scale electrical power stations. If the replacement of coal becomes wood rather than natural gas, then despite the political advocacy for biomass as illustrated in Figure 1.9, this strategy will have deleterious impacts on air quality. Indeed, there are weighty arguments for eliminating all biomass burning. What about other coal and natural gas replacement options? Technologically and operationally-proven nuclear fuel remains unpopular as does large-scale hydropower, which to some extent is partially due to the somewhat overstated but not entirely unjustified opposition of researchers, especially in academic institutions.

Consequently, it appears that only solar and wind power, and perhaps geothermal sources, together with somewhat limited oceanic power sources such as tidal, wave, and thermal energy conversion, offer acceptable energy use alternatives, at least in terms of improving air quality by reducing anthropogenic harmful PM concentrations to the WHO's AQGL benchmarks. All these alternatives will also require an exponential increase in energy storage capacities to counteract natural daily and seasonal intermittency as well as unhelpful weather patterns and unexpected volcanic eruptions. Globally, this will mean the replacement of at least 84% of the current energy sources, i.e., fossil fuels.[95] To achieve this and the necessary infrastructure by 2030 or even 2050 seems highly unlikely, although some have argued that it is possible.[96]

Whatever steps are taken to ensure a clean air for the nourishing of tomorrow, as with climate change, regions and countries do not have to be contiguous for the sources of air pollution, especially PM, to be transported from one area to another, as exemplified by the effects of volcano eruptions, sandstorms, and agricultural activities via global atmospheric circulations and seasonal weather patterns.[82] The use of natural and synthetic fertilizers and the generation of livestock waste, i.e., urine and solid manure, also contributes to PM pre-cursors such as ammonia. This is a problem, especially in Canada, and it is likely that as the major sources of PM are reduced, the ammonia precursor issues will become more prominent,[97] as seen in Section 3.2.3. Nevertheless, despite the many obstacles to reducing PM which need to be overcome, some countries, as shown in Figure 1.6, will meet the target AQGL. However, others will not, and that situation will have global as well as regionalized impacts because of airborne pollutant transfers.[98]

A region of particular concern, now and in the future, is the conglomeration of 48 African countries known as Sub-Saharan Africa, as shown in Figure 1.11 (shaded in white), which contains many almost-destitute countries, political instabilities, continued armed conflicts, and growing

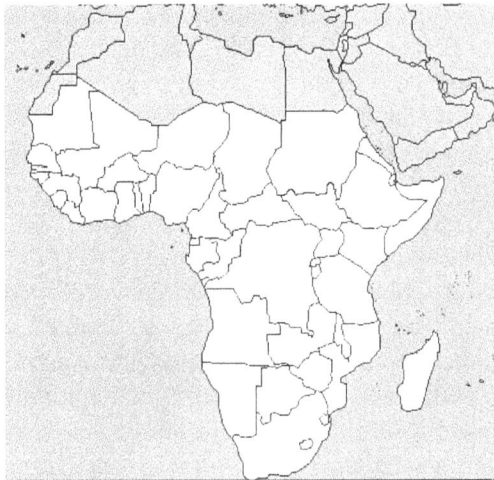

Figure 1.11.   Sub-Saharan Africa.[99]

populations.[99,100] By 2100, the SSA countries are forecasted to have a population of 3.78 billion, accounting for as many as one in three of the global population and more than twice of India and China combined. Of the SSA countries, Nigeria will have a population approaching 800 million and become the second most populous country in the world. The median age in the SSA is much lower than in other countries, not only now, but at the start of the nourishing era and it will remain so up to 2100. There is an abundance of somewhat worrying statistics about SSA, but what have they to do with clean air, and particularly PM, and the nourishing needs to achieve the AQGL? The answer is multifaceted, but the main concerns are: (i) systemic per capita poverty, (ii) energy poverty, and (iii) increasing global share of unclean cooking fuels.[101,102]

By 2030, the SSA is slated to overtake Southern and Central Asia in terms of the proportion of their population using polluting cooking fuels. The latest UN forecast is for the SSA population to rise by almost 350% over the period 2020 to 2100 and, if the same proportion of the population (44%) continue to use open-stove wood burning for cooking and to a lesser extent domestic heating and lighting, then absolute PM concentrations will significantly increase, because a further one billion people or more, in addition to the current 0.5 billion people, will be using unclean fuels. The situation could be exacerbated since the fuels currently being labeled as "clean," as shown in Figure 1.12, include liquid petroleum gas and natural gas, as well as alcohol, leaving only solar and electricity, presumably green electricity, as acceptable options. This could also be problematic because while more than 90% of the global population has access to electricity, less than 47% of SSA enjoys such provisions.[18]

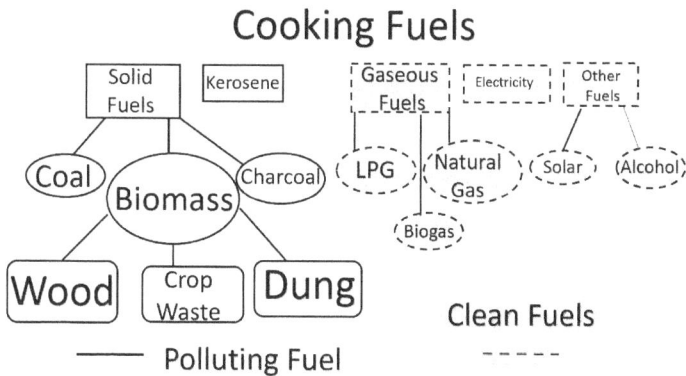

Figure 1.12. Cooking fuels clean and polluting.

The problems facing the people of SSA are many and varied, not just with air quality, but is beyond the limits of this chapter. A recent US congressional report provides a succinct overview of the issues.[102]

1.3.2.2. *Ground level ozone*

There are two basic categories of ozone — good and bad. Good ozone is found in the stratospheric layer of the earth's atmosphere, which starts at an altitude about seven kilometers above the Earth's surface and extends to about 50 kilometers.[103] Within this layer, natural ozone is formed, which acts a barrier to solar ultraviolet radiation at particular light wavelengths. Bad ozone, also known as ground-level ozone, although identified as a criteria air pollutant, and is therefore bad, is a secondary pollutant being produced when ultraviolet sunlight causes a complex series of chemical reactions involving so-called ozone-precursors, such as NO_x and VOCs. Ozone then is not emitted directly from anthropogenic sources. These pre-cursor interactions give rise to photochemical smog,[f] as illustrated in Figure 1.13.[104] In addition to ozone formation, other harmful compounds such as peroxyacyl nitrates (PANs) are also produced, creating a toxic mixture of health hazards, especially lung damage, if the air is

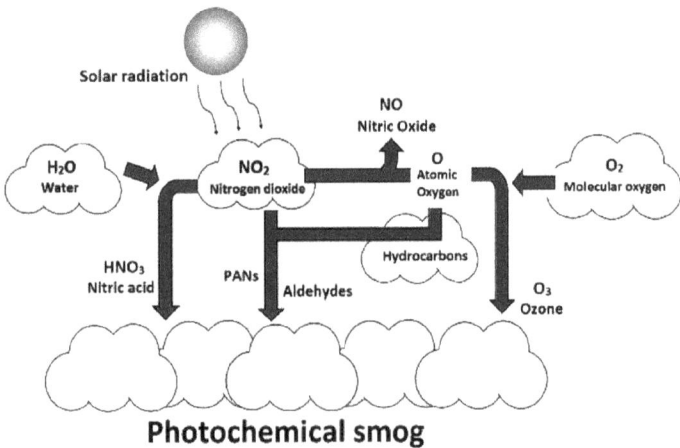

Photochemical smog

Figure 1.13.   Formation of photochemical smog.[104]

---

[f]Sometimes referred to as Los Angeles smog after the reporting of a weather event in the 1940s.

inhaled. This situation can be further aggravated if a temperature inversion overlies the smog, holding it *in-situ* at the location it has formed. Ironically, although the descriptor "smog" is frequently used with "photochemical," this is somewhat of a paradox since such smog type does not involve smoke or fog. This probably explains why the phrase ground-level ozone is becoming more favored.

Smog, a combination of smoke and fog caused by the emissions from burning fossil fuels, contains sulfur, particularly coal, and coincides with high levels of humidity. Historically, this type of smog was referred to as classical, sulfurous, or London smog, following the horrendous Great London Smog of 1952 when a temperature inversion and lack of wind caused clouds of smog over the city areas to remain stationary for five days, resulting in between 4,000 and maybe as many as 12,000 deaths.[105,106] Now, usually called industrial smog, it is the involvement of sulfur dioxide (SO_2,) not the ozone, that leads to the formation of acid rain that forms a coating on the airborne particulates if not carried away by the wind, which results in damaging health effects when inhaled. Ground-level ozone is not involved with industrial smog which has been substantially reduced, with the help of legislation governing the amount of sulfur in fossil fuels or in some instances, severe restrictions on the use of such fuels in large urban conurbations, especially for industrial activities, transportation, and heating, apart from natural gas. In terms of nourishing needs, further restrictions on fossil fuel usage are an obvious way of ensuring that industrial smog is further reduced, if not wholly eliminated.

Restricting or eliminating the combustion of fossil fuels and biomass will also impact the ground-level ozone because pre-cursor production will be reduced. However, this strategy will not wholly eliminate ground-level ozone, although in some situations and locations, the ozone concentrations may be reduced to acceptable levels in terms of defined air quality. Why should this be? Natural sources of ozone pre-cursors exist regardless of whether or not carbonaceous fuels are being combusted. These include NO_x production from the approximately 1.5 billion annual lightning strikes globally, volcanic eruptions, and biological decay, providing an estimated total of between 20 to 90 million tons (18 to 82 tonnes) of emissions annually compared with about 24 tons (22 tonnes) from anthropogenic sources.[107] Similarly, annual emissions from VOCs natural sources, especially vegetation, are considerably more than from human activities.

From the data given earlier, it could be implied that anthropogenic sources of ozone precursors play only a minor role in the ground-level ozone and photochemical smog air quality problems. However, this is not the case, as the situation is largely defined by geography, wind strength and direction, and seasonal variations in sunlight. Photochemical smog is more prevalent in urban areas, particularly, largely populated cities during the summer seasons than it is in either rural or winter seasons. Nevertheless, there have been some concerns expressed, especially by the US Congress and its appropriate committees and echoed by the USEPA, that background ozone concentrations may mask the effects of anthropogenic activities on ground-level ozone concentrations.[108] As yet, there appears to be no method of precisely quantifying the separate amounts of pre-cursors and ground-level ozone emanating from natural and human activities. This can result in measures being taken to reduce the precursors, which do not reduce the ozone in a particular area to give "good" air quality, but the same measures may prove to be an expensive and unnecessary approach to the attainment of a "good" ozone AQI. So, what measures can be taken to reduce human-generated pre-cursors and prevent the production of ozone?

To appreciate the actions that could be taken, it is first necessary to identify the key anthropogenic sources of VOCs and NO_x. As these chemical compounds become pre-cursors in the presence of sunlight, intuitively, it would appear that only external sources need be considered. However, regardless of the formation of ground-level ozone, VOCs in particular play a significant role in indoor air pollution and, according to the USEPA, some types of VOC concentrations can be between twice to 10 times higher indoors than outdoors.[109] These particular VOCs are associated with many residential and commercial building sources, from air fresheners and construction materials to cleaners and disinfectants and even computer printers and dry-cleaned clothing, as well as the storage of a variety of liquids and solids such as fuels, aerosols, and solvents associated with the use and removal of some paints.[110] Many types of commercial products carry warnings that they should only be used outdoors or in well-ventilated spaces such as garages, with the crucial factor being the level of ventilation. However, whether the instructions are followed by residents or occupiers of commercial spaces is almost entirely at their discretion, though household economics can adversely impact the choices, i.e., are the steps needed to reduce indoor air pollution affordable for all? According to a recent publication, it was found that socioeconomic status plays a significant role in indoor air quality, but more investigative research is required on the topic.[111]

The efforts to improve AQI or AQGL for outdoor air pollution are more promising, being the subject of layers of locally, regionally, and nationally enforceable legislation aimed at reducing the formation of anthropogenic ozone pre-cursors. The USEPA identify several sources contributing to the creation of outdoor VOCs and NO_x, including emissions from industrial chemical manufacturers to gasoline-powered lawn mowers. As a general rule, the relative scale of these sources will vary, depending on the geographical area where the AQI measurements are taken and the quality of the instrumentation used. While there are no universally applicable examples of the emissions, a published USEPA summary of the sources of VOCs and NO_x in a specific region, New England,[g] provides an example of identifiable sources.[112,113] For VOCs, the sources are divided into six main categories as shown in Figure 1.14.[112] As seen in Figure 1.4, the dominant sources are solvents and mobile sources, both on-road and non-road, with significant contributions from wood burning. Overall, as with industrial smog, the uses of the combustion of carbonaceous fuels from power plants to all forms of transportation lead to more than 70% of the VOC emissions. Interestingly, although VOCs concentrations are measured in the US, they are not included as a separate criteria pollutant in the USEPA list, but are a distinct category in the Canadian list, as seen in Table 1.4.

Clearly, to reduce VOC emissions to meet the desired nourishing needs of tomorrow will necessitate the increased reduction, if not

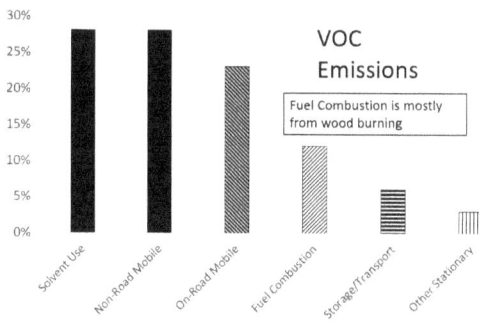

Figure 1.14. Sources of VOCs — New England example.[112]

---

[g]New England is usually considered to be formed from six US states — Connecticut, Maine, Massachusetts, New Hampshire, Rhode Island, and Vermont.

complete elimination, of fossil fuels and wood burning. While it seems unrealistic to achieve this situation in the next decade, it should be possible to eventually address the VOC issue with the development of renewable energy sources such as solar, wind, and, to a lesser extent, tidal power, notwithstanding the current societal aversion to nuclear power. These provisions would also impact the emissions of the other ozone precursor, NO_x, as seen in Figure 1.15, taken from the same USEPA summary data for the same region, since the use of carbonaceous fuels accounts for at least 60% of such emissions. The complete electrification of transportation, with the electrical power coming from renewable or nuclear sources, would have a significant impact on ground-level ozone formation and its pre-cursors. It remains to be seen whether this will happen by 2030.

It seems an implausible scenario; for example, in the US in 2018, there were just one million registered on-road electric vehicles compared with the total vehicle registration of just over 270 million.[114] However, global electric vehicle sales have grown by about 30% year-over-year, so it is forecasted they will reach 28 to 30 million by 2030 and 57% of all passenger vehicle sales will be electric by 2040.[115] Yet, even so, in 2040, two-thirds of all on-road cars will not be electric, and the number of registered vehicles presently may double in the next two decades from the present 1.45 billion. No amount of legislation, including the banning of

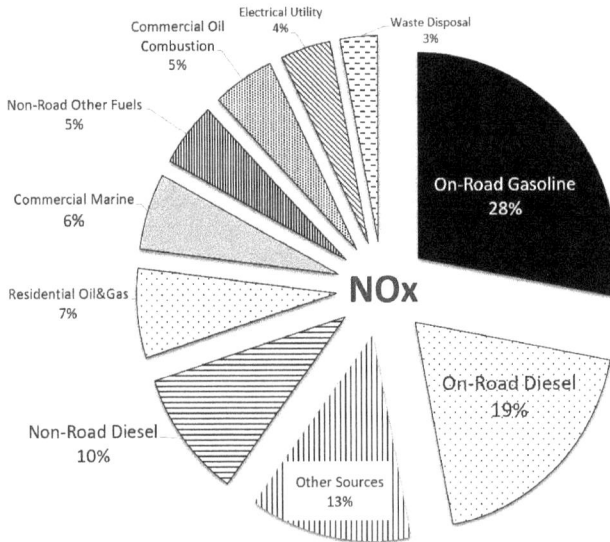

Figure 1.15. Sources of NOx — New England example.[112]

new non-electric vehicles, will ensure complete electrification, especially since the average lifetime of Internal Combustion Engine (ICE) powered vehicles is vehicles close to 13 years. Nevertheless, as electric vehicles become and remain more popular, there should be a noticeable improvement in air quality in the later part of the defined nourishing period, but the effects could be very regionalized. For example, the electrification of on-road vehicles may have little impact on air quality in the majority of African countries, as the latest reliable data indicates that the overall rate of vehicles per thousand people in Africa is only 35, compared to 590 in Western Europe and 816 in the US.[116]

Somewhat ironically, the pre-cursors of photochemical smog are also purposefully generated because they are beneficial to many societies in certain applications. Manufactured ozone is used as a disinfectant in community and personal water treatment systems, although the latter is not considered to be particularly effective. The ozone is manufactured using electrical power to disassociate pure oxygen or the oxygen constituent of air. Nitrogen oxides are used in the manufacture of chemicals, especially fertilizers, which are needed to increase food production for a globally growing population.

1.3.2.3. *Other criteria air pollutants.*

In addition to PM and ozone,[h] the US, Canada, and the WHO also all identify nitrogen dioxide or oxides, sulfur dioxide or oxides, and carbon monoxide as air pollutants. Canada, but not the US nor the WHO, also identifies ammonia as an air pollutant, whereas the US alone identifies lead as a criteria air pollutant. Since the early 1920s, tetraethyl lead (TEL) has been used as an additive to petroleum fuels, such as gasoline, to improve the performance of Internal Combustion (IC) engines, and it is still used in some aviation fuels today. Its inventors had been made aware of the health effects of leaded-gasoline toxic exhausts, but the science community were somewhat divided on the validity of these effects, with the result of TEL-gasoline being the most common automobile fuel from the 1920s onwards, especially in the US. Gradually, the scientific community gathered a large dataset of the adverse health effects and, coupled with public pressure, the Clean Air Act was passed in the 1970s, which restricted the use of leaded-gasoline for most on-road vehicles in

---

[h] In Canada, ozone is listed as ground-level ozone.

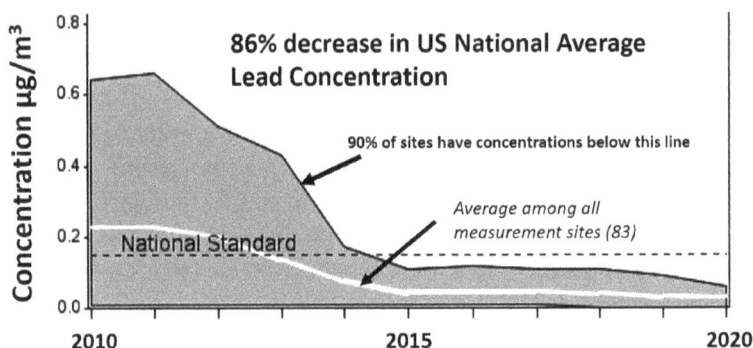

Figure 1.16.   Reduction in US lead concentrations.[119]

the US and in several other countries.[117] However, it would only be in July 2021 that global sales of TEL-gasoline cars were wholly phased out.[117,118] As shown in Figure 1.16, though the steps taken to improve lead air quality in the US have ensured that concentrations now meet, or are well below the USEPA national standard for the last five to six years, lead air pollution remains a global problem, and any level of airborne lead is considered hazardous.[119,120]

Nitrogen oxides, while classified as a separate category, are the precursors of photochemical smog and they are involved in the formation of PM. Such oxides can also react with atmospheric water vapor to produce nitric or nitrous acid, a damaging acid that can contribute to the formation of acid rain, which is often assumed to only involve sulphur oxides. The USEPA and WHO both identify NO_2, nitrogen dioxide, as an air pollutant, but NO_x, or nitrogen oxides, is not used as the pollutant measure as preferred by Canada as NO_2 is used as an indicator for the larger group of nitrogen oxides. The main source of anthropogenic nitrogen oxides is the burning and combustion of fuels in situations which use air as the oxidant. Carbonaceous fuelled on-road and off-road vehicles and power plants are major contributors to NO_x emissions. However, in diesel-fuelled vehicular and power generation systems, NO_x only starts to form when the combustion temperature reaches a certain level. If the temperature is kept below this level, then the emissions will be significantly reduced, if not completely eliminated. This strategy has led to many efforts to achieve efficient low-temperature combustion in IC engines.[121]

The three main challenges to achieving this aim have been the (i) NO_x-PM gap, (ii) maintaining catalytic convertor operational efficiency, and (iii) meeting fuel consumption targets, such as manifest in the US National Highway Traffic Safety Administration's Corporate Average Fuel Economy (CAFE) standards. The NO_x-PM gap is particularly associated with the operation of diesel engines. As combustion temperatures are lowered to reduce NO_x emissions, there is an increase in the potential production of PM. To counteract this situation, exhaust gas recirculation (EGR) is used, whereby part of the engine exhaust is recycled into the engine intake, reducing the amount of airborne O_2 fed to the engine, and subsequently lowering the in-cylinder temperatures thus reducing NOx generation. EGR is now a common provision for most IC Engines.

To address PM production, aftertreatment systems in engine tailpipes downstream from engine exhausts have been developed, such as catalytic convertors and particulate filters. In the initial developments, the lowering of the exhaust temperatures prevented the efficient operation of the catalytic convertors which needed a threshold temperature to activate the catalysts, but acceptable solutions were soon found. To further reduce PM concentrations from the tailpipes, a particulate filter is also used. Although research and development of EGR and aftertreatment systems is continuing, especially in academic institutions, the general strategies have been proven to be very successful in lowering both NO_x and PM emissions, as well as meeting increasingly challenging fuel economy standards.[122] Zero-Emission Vehicles (ZEV) are an essential element in addressing both the mitigation of anthropogenic climate change and improving air quality, and, since modern ICE emissions are approaching ZEV equivalents, perhaps more focus should be directed to other air pollutant sources.[123–125]

While this all-encompassing strategy has virtue, the problem with continuing ICE systems is that the devices used to lower pollutant emissions from ICEs result in the relative tailpipe concentrations of CO_2 per unit of fuel to increase. This is no longer an advantageous situation, as CO_2 has been legally defined as an air pollutant in the US and Canada. How can this be, since almost all atmospheric CO_2 comes from natural sources? These include ocean outgassing, biomass and decomposing vegetation, and volcanic eruptions, as well as humans and some other animals exhaling CO_2. On the surface, these legal decisions appear absurd given that CO_2 is vital to the survival of all life forms, but as typified by impositions of carbon taxes, only CO_2 emissions resulting from human activities associated with the combustion of fossil fuels are considered to be an air pollutant.

Apart from natural sources, which are significant in the case of nitrogen oxides, the main sources of both sulfur and nitrogen oxides have been the burning of fossil fuels. As previously discussed in Sections 1.3.1 to 1.3.2.2, in general, the eventual elimination of both industrial smog and photochemical smog, PM, and ground-level ozone to improve air quality will be heavily reliant on a transition from the use of fossil fuels to low or non-carbonaceous fuels in transportation, power generation, cooking, and building heating and cooling. Such a transition would also cut the levels of atmospheric CO concentrations, which are almost exclusively associated with the burning of carbon fuels. Although such energy transitions are underway, it is not clear when the use of fossil fuels will cease, despite a plethora of predictions ranging from the end of the decade, to the end of the century and beyond. To reduce pollutant emissions to more acceptable, if not completely safe levels, the use of carbon dioxide sequestration and storage and further improvements in the operational effectiveness of criteria pollutant removal devices could offer a stop-gap, at least until the energy transition is substantially accomplished, though the issue of the carbon-neutrality of biomass and biofuels will need to be revisited.

There is one other major air pollutant, ammonia (NH_3), that appears in Table 1.4 because it is recognized as such by Canada, but is now attracting increased political awareness and scientific attention in a number of countries, including the US, the UK, and the EU.[97,126] The major source of NH_3, now considered a significant component of secondary $PM_{2.5}$, is the agricultural sector, from both synthetic fertilizers and livestock waste.[127] In North America, because of the higher densities of livestock, the NH_3 emissions per hectare are three to five times those of Europe.[97] Globally, the agricultural sector contributes 80% to 90% of all ammonia emissions,[i] with Canada leading the way with 93% as of 2020, which possibly explains why Canada categorizes ammonia as an air contaminant. Although not yet attracting as much attention as other primary air pollutants, ammonia concentrations, sources, and adverse health effects appear to have been underestimated, especially in urban environments, as previously noted.[87]

As shown in Figure 1.6, many countries are improving their air quality towards meeting the WHO's AQGL standards, but sometimes such indicators can hide the success and scale of the measures being taken to

---

[i]Ammonia emissions are also associated with damaging soil acidification and eutrophication — the excess growth of plants and algae.

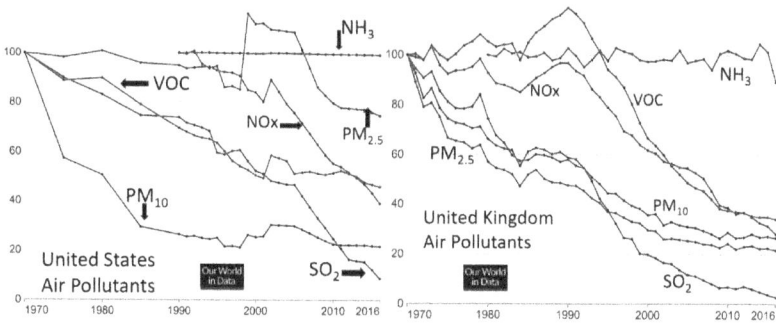

Figure 1.17.   USA and UK air pollutant levels 1970–2016.[128]

attain the AQGL benchmarks. For example, the scale of the UK's efforts is somewhat masked by the data used to produce Figure 1.6, but if the relative emission reductions achieved since the 1970s to 2016 are mapped, as seen in Figure 1.17, then air quality improvements can be readily appreciated.[128] Similarly, although the examples in Figure 1.6 indicate the great success the US has enjoyed in air quality improvements over a similar timeframe, particularly over the past decade in relative terms, the efforts to reduce ammonia and especially $PM_{2.5}$ have not been as successful, as shown in Figure 1.17. The arrival of commercially viable green ammonia, whereby a zero-carbon synthetic version can be produced by using the hydrogen produced by electrolysis together with nitrogen obtained by air separation in a Haber-Bosch process, may address some of the issues.[129] To be green, all steps in the processes use only electricity generated from renewable sources. In addition to existing uses such as fertilizer production, applications such as energy storage and as a transportation fuel are also being pursued. Research and development in Australia, Japan and, more recently, China, is gaining momentum but while it could offset some of the anthropogenic climate change concerns, its impact on air pollution has yet to be thoroughly investigated.

Overall, there are many reasons to be cautiously optimistic about the nourishing needs for air quality being achieved this century, but only if affordable energy transitions can be realized in a timely manner, especially for those populations having to use unclean cooking fuel. Nevertheless, coal will still be a major energy source to at least the mid-century and for much longer in India and China, largely because of the size of their populations. However, although coal production is likely to increase for some time in these countries, the percentage share of coal in their energy markets will likely decrease. Moreover, by the end of the century, the populations of these two countries are predicted to have peaked; by mid-century in India and earlier in China. Even so, by the end of the century, the combined populations of India, China, and Sub-Saharan Africa are forecasted by various models to be more than half of the total global population.[31,130] As many air contaminants are transboundary, the continued use of coal for electricity generation and building heating in India and China, and unclean cooking fuel in Sub-Saharan Africa over the next three decades and perhaps beyond, could cause the cautiously optimistic suggestion to be more akin to a proverbial forlorn hope.[j]

1.3.3. *Water, Sanitation, and Hygiene (WASH) needs*

The virtues of a clean water supply were recognized as soon as permanent sizeable human settlements were first established several millennia ago, but it was only in 2010 that a UN resolution was passed, recognizing that water and sanitation were human rights, and called upon member states to provide "safe, clean, accessible and affordable drinking water and sanitation for all."[131,132] It would be another five years before "sanitation was explicitly recognized as a distinct right by the UN General Assembly."[133] In the same year, the action plan for global sustainable development, "Transforming Our World", was adopted by the member countries and states of the UN. The laudable main aim of this transformation is to eliminate all global poverty by achieving the 17 Sustainable Developments Goals (SDGs) including SDG 6, "water and sanitation for all".[5,134] The connection between clean water and sanitation only became evident in the 18th and 19th centuries, as cities became larger and drinking water retrieval and sewage effluent occurred in close proximity to the water sources, usually rivers. Even so, the smell–taste–sight indicator, used since ancient times, was still the benchmark for assessing if water was drinkable by humans, until scientific investigations demonstrated that water that appeared drinkable could be a carrier of harmful diseases, such as cholera, typhoid, polio, and dysentery.[135] The main sources of these diseases and the pathogens (i.e., bacteria, viruses, and parasites) which cause them being human and animal waste such as faeces, and since the start of the Second Industrial Revolution (circa 1880s), industrial

---

[j]An extremely difficult initiative with little chance of success.

effluents and materials, such as lead piping, have played an increasing role in water contamination. Presently, the USEPA have named more than 90 such contaminants for which concentrations are legally limited.[136]

With a long history of water quality awareness, and the development of more modern techniques to identify unclean water, what was the reason for the UN declarations of human rights concerning water and sanitation, and if there are any nourishing needs? The answers are embedded in the stated aims of SDG 6 emphasizing the phrase "for all." Many people living in rich countries have long experienced clean water and good sanitation, albeit with the occasional disruptor events associated with temporary infrastructure failures caused by accidental harmful spills, pipeline breaks, and severe weather events like flooding. In some rich countries such as Canada, various indigenous and remote communities have experienced long-term clean water supply and sanitation problems, requiring so-called drinking water advisories.[137,138] Although 79% of these long-term advisories have been lifted since 2015, 33 communities still have such advisories in place as of April 2022.[139] So what about the global situation?

In 2017, an authoritative study found that 1.2 million people died prematurely from unsafe water, three times higher than from homicides, and a further 0.75 million people died from unsafe sanitation.[140] Although these data represent unsettling statistics if used in isolation, they concealed the real overall state of the global safe sanitation and drinking water needs. The key global facts are summarized in Table 1.6.

Table 1.6. Key facts: Sanitation and drinking water.[140–142]

Sanitation	Drinking water
4.2 billion people use a safely managed sanitation service (54% of the global population in 2020)	5.8 billion people used a safely managed drinking water service (74% of the global population in 2020)
More than 1.7 billion people do not have access to a basic sanitation service	2 billion people use a drinking water source contaminated with faeces (as of 2019)
673 million people defecated in the open in 2017, reduced to 494 million by 2020	785 million people lacked even a basic drinking water service in 2017
Can transmit diseases such as Hepatitis A, diarrhea, cholera, dysentery, typhoid, and polio	Can transmit diseases such as diarrhea, cholera, dysentery, typhoid, and polio
432,000 diarrheal deaths annually	485,000 diarrheal deaths annually
Unsafe sanitation is responsible for 775,000 deaths each year[143]	6% of deaths in low-income countries are the result of unsafe water sources

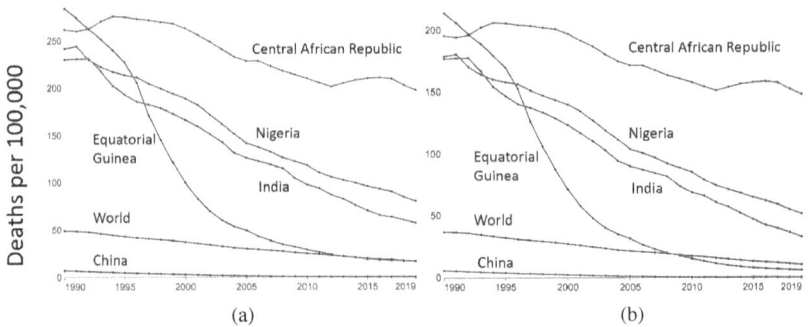

Figures 1.18. Death rates per 100,000 for unsafe water (a) and unsafe sanitation (b).[140,143]

Source: IHME Global Burden of Disease & Our World in Data.

However, as the examples in Figure 1.18 show, the death rates per 100,000 inhabitants caused by unsafe drinking water and unsafe sanitation vary significantly depending on the country. Exemplars for rich countries (GDP per capita) have not been shown in the figures because the death rates are usually less than 0.1 deaths per 100,000 inhabitants, for both unsafe drinking water and sanitation, although interestingly, the rates in many of these particular countries, e.g., Canada and the US, have increased over the last three decades, albeit staying less that 0.1. In some countries whose income (GDP) per capita has markedly improved since 1990, the deaths rates have been significantly reduced. For example, China, with its large population and now a World Bank upper medium income status, has reduced its death rates by more than 90%. Similarly, a small country in terms of population such as Equatorial Guinea, which is now a lower medium income country, has achieved similar reduction rates.[140,144]

The links between a country's wealth and its unsafe sanitation death rates are evident, as shown in Figure 1.19(a), but the relationship between access to improved drinking water sources and per capita wealth is not as obvious, as illustrated by Figure 1.19(b).[143,145] Should greater focus be directed to sanitation improvements compared with those of drinking water?

1.3.3.1. *Identifying the drinking water nourishing needs*

Achieving universal access to drinking water and sanitation by 2030 are two of eight key targets defined by the UN for SDG 6. Yet, despite the

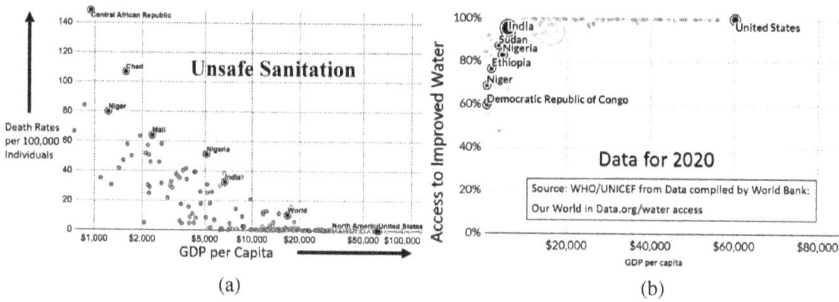

Figures 1.19. (a) Exemplars of unsafe sanitation, and (b) access to improved water based on GDP per capita.[143,146]

global progress that has been made in providing access to safe drinking water and safely managed sanitation facilities, a quarter of the world's population still does not have access to safe drinking water and almost half of the world's population does not have access to safe sanitation.[140] The 2030 timeline seems overly ambitious when based on the agreed tracking indicators used to determine the progress towards achieving targets. However, it has to be noted that the SDGs have only been in force since 2015 and during the first five years of their existence, access to safe drinking water has increased from 70% to 74% and access to adequate sanitation from 47% to 54%, which are measurable successes.[143,146] Nonetheless, if these rates of progress were to be maintained to 2030, then 18% of the global population would still not have access to drinking water and 32% would not enjoy access to adequate sanitation. Achieving the "for all" targets for water and sanitation by 2030 will require at least a three-fold or four-fold global increase from 2020, in terms of an average accelerated rate of progress, and for some regions, the required rate could be higher than 20-fold.[143,146] As the rates of progress in the pre-pandemic world have proved insufficient, it seems that the preparations required *today* to achieve the nourishing needs of *tomorrow* will likely be delayed, as the economics of a post-pandemic world struggle to recover. But what progress can be made and what could be the key factors?

In this discussion of drinking water and sanitation, terms such as "adequate," "safe," "safely managed," and "improved" are encountered, but these are not just interchangeable euphemisms for "clean," though all the terms refer to the definitions used in the measures that need to be taken to achieve universal access, as specified in SDG 6. To highlight the necessary measures, the current situation in the provision of drinking water and

Table 1.7.   Definitions of the key SDG 6 targets and indicators.

SDG	Definitions[147,148]
Target 6.1	"...achieve universal and equitable access to safe and affordable drinking water for all," by 2030
Indicator 6.1.1	"Proportion of population using safely managed drinking water services"
Target 6.2	"...achieve access to adequate and equitable sanitation and hygiene for all and end open defecation," by 2030
Indicator 6.2.1	"Proportion of population using (a) safely managed sanitation services and (b) a hand-washing facility with soap and water"

sanitation, locally, regionally, and nationally must first be established. The levels of improvements required can then be more readily identified and the progress made in achieving the SDG 6 goal and its associated targets ascertained. For the water and sanitation goals there are eight targets and eleven indicators and as given in Table 1.2, various agencies were tasked with tracking the progress towards these targets. For drinking water, the main target is SDG 6.1 and arguably the key indicator is 6.1.1, whereas for sanitation, the main target is SDG 6.2 and the associated indicator is 6.2.1. The slightly paraphrased definitions of SDG 6.1 and 6.2 and their indicators are given in Table 1.7.[147,148]

The indicators are then used as a measure of the improvements that have been or are being made as a percentage of the population. However, to benchmark and graphically report the progress being made, the WHO and UNICEF use definitions of five service levels, each being assigned a different color code for visual usefulness.[143,146,149] For drinking water, the service levels representation is termed a "ladder." For both drinking water and sanitation, the top three levels are considered to signify "improved" water sources and "improved" sanitation facilities respectively. These service level benchmarks are summarized in Table 1.8, with '1' to '3' being "improved".[143,146]

In many countries, treated drinking water is then provided directly to consumers using networks of pipelines especially, but not exclusively, for urban and suburban communities. If these communities have small populations, defined by the USEPA as between 25 and 10,000, which tend to be situated in more rural areas, so-called small water treatment plants will then provide the drinking water, once again, using a pipeline. However, the water supply may be by standpipe, where the water has to be collected by individuals in acceptable containers (hopefully). If it takes less than

Table 1.8.   Definitions of the drinking water ladder and access to sanitation.[143,146,149]

Service level	Drinking water	Sanitation
1. Safely managed	"Located on premises, available when needed, and free from contamination"	"Not shared with other households, excreta safely disposed of *in-situ* or transported for treatment off-site"
2. Basic	"Misses a 'Safely managed criteria', under 30 minutes roundtrip to collect"	"Misses a 'Safely managed criteria', not shared with other households"
3. Limited	"Misses a 'Safely managed criteria', over 30 minutes to collect"	"Misses a 'Safely managed criteria', are shared with other households"
Unimproved	"Water from an unprotected dug well or spring"	"Use of pit latrines without a slab or platform, hanging latrines or bucket latrines"
Surface water	"Water directly from a river, dam, lake pond, stream or canal"	
Open defecation		"Disposal of human faeces in fields, forests, bushes, open bodies of water, beaches and other open spaces"

30 minutes to collect water, but does not meet all the safety criteria, the service level is still considered to be "improved" but only "basic." Whereas, if similar conditions apply to an "improved" level but it takes more than 30 minutes to collect, i.e., to travel to the drinking water source and return to the point of departure, then the level is designated as "limited."

Government authorities identify contaminants, akin to those defined for air quality, and regulate their concentrations, at least for public water supply systems.[k,150] In most situations, ensuring compliance with regulations is a local or municipal responsibility. Thus, when problems are identified by regular testing, a variety of remedial actions are usually

---

[k] This is because there are legal loopholes in various systems. For example, in the US, the regulations, "do not apply to privately owned wells or other individual water systems. Owners of private wells are responsible for ensuring that their well water is safe from contaminants."

taken, including addressing any treatment plant and distribution system deficiencies, temporary or otherwise, and providing accompanying public announcements or advisories. In the case of a newly-identified permanent contaminant, the necessary corrective actions may not be timely. For example, in the US, if a new toxin is scientifically recognized as harmful to health, it could be three to five years before the treatment for its removal becomes compulsory, the timeline often depending upon the additional costs associated with the necessary corrective actions that have to be taken and the scale of actions.

1.3.3.2. *Treating water sources for consumption*

All water is classified by its salinity, i.e., the quantity of dissolved salts it contains measured in parts per thousand[l] (ppt), the most common compound being sodium chloride. For water to be considered fresh, its salinity must be 0.5 ppt or less. The salinity of drinking water is even lower, at a maximum of 0.1 ppt. However, for land irrigation purposes, water salinity up to 2 ppt can and is used. To put these levels in context, the average salinity of seawater is 34.7 ppt. On a mass basis, the absolute amount of Earth's water is sensibly constant,[m] but only about 2.5% of the total is freshwater and more than two-thirds of this freshwater is trapped in glaciers, ice caps, and permafrost. The main sources of freshwater are groundwater and surface water, but it should be noted that not all water from such sources is classified as fresh, e.g., water from global lakes is almost equally divided between fresh and saline.[151] Indeed, only a small amount of the Earth's freshwater, about 0.03%, is in lakes, rivers, and wetlands. Hence, although it can seem that there are endless sources of water, especially as the land is surrounded by massive oceans and seas, little of this water is available as freshwater and most certainly not as drinking water. However, processes such as desalination can be used to convert high salinity water sources, including sea water, to agricultural water, freshwater and, eventually, drinking water, but the technologies employed are energy-intensive.

The initial processes, i.e., sedimentation and filtration by which fresh water sources are treated to produce drinking water today, closely

---

[l]The mass of the dissolved material in a unit mass of solution.

[m]The ramifications of the 'Water Cycle' which readers will likely have studied at grade school or high school.

resemble those first used several millennia ago, albeit they have been improved by the addition of chemical coagulants to speed up the settling time during sedimentation and superior filter materials. Although these courses of action help water pass the human sight, taste, and smell criteria, as mentioned in Section 1.1, it was only with 19th-century scientific investigations of water and the identification of cholera as a waterborne disease that disinfection became the next step in water treatment, with the aim of removing harmful pathogens.[152,153] Rules were also developed in this era to ensure that raw sewage and other harmful effluent were not discharged into surface water sources such as rivers in close proximity to the locations where water was being extracted for drinking.

Today, modern water treatment plants employ a basic five-step process to make the resulting water safe for human consumption: (i) Water source protection, (ii) Sedimentation, (iii) Filtration, (iv) Disinfection, and (v) Safe water storage.[154] The storage facilities can be large, e.g., reservoirs, clear wells, but small household-scale storage is also possible using the type of containers recommended by the US Centers for Disease Control and Prevention (USCDC) and WHO.[155,156] Free-standing elevated water towers and rooftop water tanks are frequently assumed to be water storage devices, but their primary function is to provide sufficient water pressure to ensure the distribution of public water supplies if the main delivery pumps from the water treatment plants experience power reductions or complete electrical outages. However, during daily peak demand periods or in case of emergencies, e.g., firefighting, they may also be used to supplement the normal water supplies. Not all towers are connected to the drinking water or potable water supply systems.

Actual treatment plants, which are at the core of public water supply systems, are based on the five-step methodology, but they are technologically complex and come in a variety of forms. Operational and design improvements are invariably and continually pursued. For example, during the disinfection step, the plants may traditionally use chemical compounds such as chlorine, but ozone and ultraviolet light are increasingly utilized. Moreover, there are many different mechanical configurations used in the filtration procedures and additionally, separate filtration process may be employed at various stages of the water treatment. Once the drinking water has been produced, it is supplied to consumers. In domestic and some commercial situations, after consumption, e.g., cooking, washing, cleaning, and flushing of toilets, the water becomes wastewater. This is also true of water discharged from industrial processes and some agricultural activities. Before wastewater is returned to a surface water source, it will be treated in a similar manner to the production of drinking water. There are numerous public, educational, and industrial publications dealing with the variety of configurations of both drinking water and wastewater treatments plants, and for readers new to the topic and wishing to explore these in more detail, there are references that provide insightful further reading.[157–160]

Surface water is not the only water source used in water treatment plants. As mentioned earlier, the largest source of freshwater is groundwater, which is water found below the land's surface in sand, gravel, and creviced rocks; indeed, any permeable strata. The water descends through the cracks, spaces, and gaps until it encounters impermeable rocks. In this way, it is stored in layers called aquifers, which can be accessed through drilling and pumping, and occur in the so-called groundwater saturated zone, though there can also be a natural flow of water out of rock and soil materials.[161] Water from aquifers is usually cleaner than surface water because of the reduced concentrations of contaminants of all kinds. The main source of both groundwater and surface water is precipitation, usually rain. While the amount of surface water varies with seasonal weather, groundwater quantities tend to be more constant. In the US, about 40% of municipal water (i.e., public supply) comes from groundwater, with a similar amount going to agricultural water.[162,163]

In general, there are three basic uses of freshwater drawn from groundwater and surface water: (i) for public supply drinking water, (ii) agricultural, i.e., irrigation, aquaculture, and livestock, and (iii) a large variety of industrial processes. Of all the water withdrawn from the two sources globally, only about 11% is used for household drinking water. However, there are variations depending on the regional agricultural and industrial activities, annual precipitation rates and total amounts, population growth, and the natural characteristics of the inherent ecosystem.[164] As the exemplars show in Table 1.9, for some of the countries which have kept reliable, rigorously measured, and open access available data, the percentages of freshwater withdrawn for household use have remained sensibly the same since the start of this millennium.

Thus, the drinking water treatment systems described condition only a relatively small percentage of the withdrawn freshwater. Globally, about 71% of this freshwater is used for agricultural purposes with large regional and national variations, as illustrated in Figure 1.20.[165] Although municipally supplied drinking water can be used for agricultural

Table 1.9. Percentage of freshwater used for household use (exemplars).[164]

Country	2002	2007	2012	2017
India	7.0	7.2	7.4	7.4
China	11.2	12.4	12.0	13.3
United States	11.5	12.1	12.9	13.1
Germany	14.3	16.2	19.3	18.0
Australia	20.9	25.4	23.4	20.5
Brazil	25.3	26.5	24.0	25.5

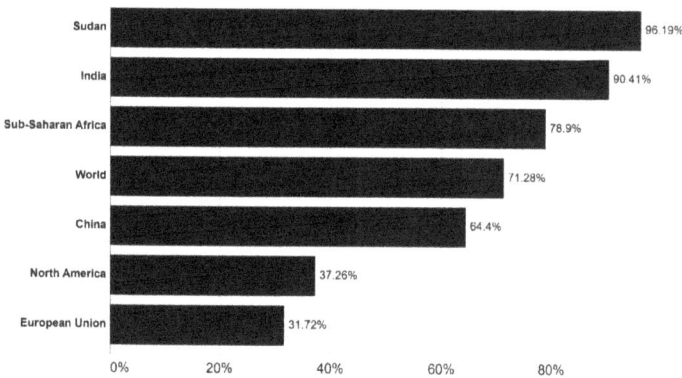

Figure 1.20. Share of agricultural water of total water withdrawal in 2017, exemplars.[165]

activities, and is often used for household gardens, the only sources used in addition to traditional groundwater and surface water freshwater are rainwater catchment systems e.g., rain barrels,[n] and lower quality non-conventional water (NCW) sources such as wastewater, saline groundwater, and surface water.

Agricultural and industrial water supplies undergo treatments not dissimilar to those used for drinking water.[157,158] Untreated irrigation water, especially if containing human and animal fecal matter, is responsible for harmful and sometimes deadly waterborne decreases. In the US, about 40% of agricultural water is used for crop irrigation, with a greater share coming from surface water than safer groundwater sources.[166] Wastewater

---

[n] Also used by some households.

from industrial and agricultural activities are frequently discharged into surface water sources and to avoid contaminating the source, there are a wide range of national regulations, regional standards, and global guidelines aimed at ensuring that the withdrawn water is of a suitable quality for its intended purpose, pending appropriate treatment. However, the inspection, testing, and monitoring frameworks for agricultural wastewater is rarely as timely or as rigorous as in drinking water treatment operations.

For industrial activities, the wastewater may not be discharged into a surface water source but it will need to undergo levels of treatment pertinent to the specific industrial process if it does, e.g., all forms of manufacturing from food to construction materials. In some instances, the chemicals used or chemical by-products formed in the processes will be extracted from the wastewater and recycled. The reduction of water contamination, enhanced conservation of water, and decreased generation of wastewater are common themes in many industrial efforts to address water quality issues, but the actual approaches vary depending on the industry. As a share of the total surface water and groundwater withdrawn, the use of water for industrial purposes is extremely low in low-income countries. However, in high-income countries, the share averages about 17%, almost an order of magnitude higher. Examples of specific industrial approaches in high income countries have been made publicly available in a USEPA database, which provides details of 41 national case studies from 1979 to 2016.[167]

However, in all cases, great care needs to be taken in the interpretation of published water data, such as in Figure 1.20, especially the use of terms such as "withdrawn" water. This is because not all withdrawn water is consumed in the processes for which it is withdrawn. For example, in terms of actual withdrawal amounts, the largest percentage is not for agricultural purposes but for thermoelectric power generation, i.e., steam-driven electrical generation and its subsequent cooling, then depending upon the actual cooling system, most of the unconsumed water is sent to a wastewater treatment system and subsequently to its original sources. This is known as return flow or non-consumptive water use. In the US, only about 3% to 4% of withdrawn thermoelectric water is actually consumed, and continuous improvements are constantly being sought to reduce even this consumption.[168,169]

As thermoelectric power generation consumes so little water, its use is not included in most datasets concerning the shares of total water withdrawals, which usually compare the amounts used for domestic, industrial, and agricultural purposes. Clearly then, water use and water consumption are quite different measures; the latter is based on the portion of water that is not returned to its original source, whereas water use can be presented as equivalent to water withdrawals. However, this approach can be misleading. Despite the idiosyncrasies in how freshwater from groundwater and surface water sources is used, measured, and consumed, in terms of the nourishing needs for safely managed drinking water, it is demonstrable that water and wastewater treatment needs to be vastly improved for *all* societal uses involving freshwater sources. However, it must be recognized that while there are clear connections between the sources and treatment of drinking, agricultural, and industrial water and their wastewater discharges, the UN human rights pronouncement covers only personal and domestic uses. In spite of these boundaries, the UN encourages the use of treated water if it is instrumental in preventing starvation and disease and especially with agricultural activities.[132] If treated wastewater of any kind is to be used for crop production irrigation schemes, then great care has to be taken in ensuring its quality.[170]

Most of the premature deaths associated with unclean or unimproved drinking water are also linked to unsafe sanitation so, regardless of the efforts to improve drinking water, without a rapid improvement in safely managed domestic and personal sanitation, the end results will be disappointing to say the least.

1.3.3.3. *Identifying the nourishing needs for safely managed sanitation and hygiene*

The infrastructure needs for improved sanitation services are outlined in Table 1.8 and are all indicative of what can be described as Community Congregate Settings.[171] These settings include permanently occupied single and multiple dwellings, but also other more temporary or shorter-term human gathering places such as workplaces, hotels, educational institutions, shopping centres, restaurants and bars, sports and entertainment venues, medical, travel and incarceration facilities, and so on. All these locations require improved sanitation services but they are not universally enforced or maintained although regulated by building standards in many countries. Once again, while the existence of such infrastructure appears strongly related to a country's wealth and political priorities, the major single factor which needs to be addressed is the elimination of open defecation, especially in close proximity to surface water sources. However, seepage into groundwater sources, e.g., freshwater wells, is also problematic. It is often assumed that open defecation is only associated with rural communities but as global homelessness in urban environments increases (as seen in Section 1.3.5), so does urban open defecation, even in high income countries.[172] Nevertheless, 91% of all people practicing open defecation live in rural areas.[173]

With the large number of attributed premature deaths in 2020, equivalent to almost the entire population of the US city of San Francisco, open defecation is a serious problem that should not be ignored since, according to the UN Children's Fund (UNICEF), "one gram of faeces can contain 10 million viruses, one million bacteria and one thousand parasite cysts."[174] Significant improvements have been made by many countries, such as India, to reduce the proportion of their population openly defecating. Over the past two decades, India has achieved a reduction of almost 60% in open defecation, but 15% of the population continues the practice, albeit much lower than many African countries percentage-wise. However, because of the scale of India's population, 15% of the population means that more than 206 million people continue to defecate in the open. If this situation existed in Europe, it would amount to the combined populations of France, Germany, and the UK. Moreover, while open defecation is now limited to 6% of the global population, there are still countries, especially in parts of sub-Saharan Africa, where almost 50% to 70% of the population are impacted by the practice. In the same region where many countries have made improvements, there are others, such as the Democratic Republic of Congo and the Central African Republic, where there have been no overall percentage reductions in the past 20 years, and their populations are increasing.

What actions can be taken to address the open defecation issues? The simple answer is safely managed working toilets for all, as seen in Table 1.8. According to the UN, 3.6 billion people do not enjoy such facilities but, by 2030, all could have access to a basic level of sanitation for an investment of US$195 billion, over a 10-year period.[173] This may seem to be a considerable financial commitment but, over the same period, at least US$1 trillion is slated to be spent on household pets in the US as well as some US$260 billion on toys and hobbies.[175,176] While pets, toys, and hobbies contribute to human well-being, it has also been estimated that a comparatively modest investment in global sanitation service could provide $4 to $5 of benefits, including saving medical costs, for every single dollar spent.[177,178] A depressing statistic — only three years ago, 21% of global healthcare facilities had no sanitation service, along with 620 million children at their schools.[179,180]

Providing basic toilet services alone will not ensure that the open defecation problems are resolved. If hands are not washed after open defecation, or indeed after any form of defecation, and before eating or handling food and especially if soap is not used when washing hands, then there is a high probability of harmful contact with viruses, bacteria, and parasites. Remarkably, almost one-third of the global population do not have access to soap and water handwashing services, and almost 10% have no access to any handwashing facilities. With access to drinking water, safe sanitation and effective water treatment improvements have been made and levels of hygiene have also increased, particularly over the past five years. However, if the same rate of progress is maintained up to 2030, 20% of the global population will still not be able to experience basic handwashing facilities. Moreover, providing more toilets and hygiene capabilities will prove ineffective if people do not use them. Ensuring awareness through education will then be a necessary adjunct to any infrastructure initiatives, and the UN also suggests that as open defecation is a cultural norm in some communities, "a sustained shift in the behaviour of whole communities" will be required.[173]

The challenges associated with the provision of drinking water, safely managed sanitation, and handwashing facilities now, and for the nourishing needs, are all the same and interrelated. They include the lack of sufficient capital funding to provide new infrastructure and a similarly insufficient operational budget to maintain existing facilities, including inadequate testing regimes for certain types of water and wastewater treatment plants. Increasing populations will exacerbate these problems but progress has been and continues to be made. It is the speed of progress that needs to be accelerated. In terms of the WASH targets for 2030, it seems unlikely that they will be achieved in its entirety until the latter quarter of this century.

1.3.4. *Food*

Food, or access to food, is essential for the survival of all living things. However, exactly how long humans can survive without food even with access to clean air and safe drinking water remains difficult to quantify or be determined through methodical investigation, as any such studies would breach scientific and ethical standards. The information that is available tends to be more circumstantial than statistical, although studies of voluntary hunger strikes are occasionally reported and death by starvation has been estimated by various UN agencies for circumstances involving conflict and crop failures.[26] In such cases, starvation is usually taken to mean "a severe lack of food for a prolonged period."[181] The available literature indicates that survival can be between 21 and 73 days.[182,183] Like sanitation and water, food was globally pronounced as a human right embedded in Article 25 of the 1948 UN Declaration of Human Rights, which states that "[e]veryone has the right to a standard of living adequate for the health and well-being of himself° and his family, including food...."[2] What constitutes an adequate living standard remains a matter of opinion and debate, especially with regards to food. For example, how much food does the average person require to satisfy their daily needs and what types of food should be consumed to remain healthy?

Food provides the energy for all our activities, whatever they may be. There are ways to quantify the energy content of food and but it is arguably harder to also determine how much energy the average person needs. Food, including restaurant menus and beverages in many countries are now required by regulation to list the calorie amounts of their ingredients. A calorie is defined as the amount of energy required to raise the temperature of one gram of water by 1°C. Historically, this was referred to as a "small" calorie (c), but the term is now rarely used except by some research scientists. Similarly, the term "large" calorie was also used but has been largely replaced by kilocalorie (kcal), the amount of energy required to raise the temperature of one kilogram of water by 1°C. In terms of food nutrition values, calories and kilocalories are used interchangeably with their specific use depending upon the country.[184] These energy values are used by national and international agencies to determine daily food requirements per person assessed to be sufficient to maintain a healthy body weight. Of course, age, gender, height, different personal activity levels, and country of residence will all influence the minimum daily per capita dietary energy needs. For 2020, the global average for the defined minimum need was 1,828 kcal, as shown in Figure 1.21, together with exemplars of countries above and below the global average.[185]

If these minimum needs are not met for all the people in a particular region or country, the UN regards the portion of the population below the

---

° Phraseology of the time.

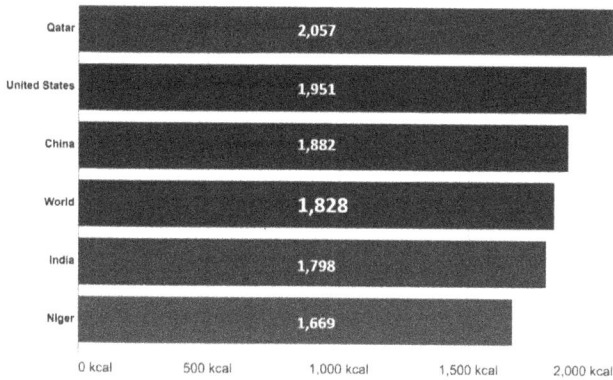

Figure 1.21.　Minimum defined daily Kilocalorie (kcal) amounts for 2020.[185]
Source: Food and Agriculture Organization of the United Nations

minimum as being undernourished. However, a universal abundance of food would not necessarily reduce the undernourished portion to zero. Indeed, some of those enjoying a sufficiency of food in calorific terms may still be suffering from malnutrition which is caused by a diet with an inadequate content of nutrients and proteins. However, the inadequacy may be the result of a lack of food, i.e., undernourishment. Undernourishment and malnutrition are not identical measures and should not be used interchangeably, as neither term captures both the wants and needs of the level of food quality and of food quantity. To address this issue, the UN Committee on World Food Security uses the term "food security," which is defined as "all people, at all times, have physical, social, and economic access to sufficient, safe and nutritious food that meets their food preferences and dietary needs for an active and healthy life".[186] Other agencies use somewhat plainer ways of expressing food security, or insecurity, such as the US Department of Agriculture (USDA) which defines food insecurity as "a household-level economic and social condition of limited or uncertain access to adequate food."[187] Like the other nourishing needs, issues connected with access to food also have a prescribed sustainable development goal, the SDG2, which has an overall aim to end hunger, or more specifically to "[e]nd hunger, achieve food security and improved nutrition and promote sustainable agriculture."[188] To achieve this admirable goal by 2030, the UN have defined eight targets and 13 indicators.

1.3.4.1. *Elements of sustainable development goal 2*

Without doubt, all eight targets of SDG 2 are important but Target 2.1, "the universal access to safe and nutritious food," and Target 2.2, "to end all forms of malnutrition," are arguably the most critical. Target 2.1 has two indicators, and Target 2.2 has three indicators to enable the tracking of progress towards these targets by 2030, with interim values by 2025 for indicators 2.2.1 and 2.2.2; and also by 2025 for the progress of specific items, namely "wasting" and "stunting" for children under five years old; the third indicator being "the prevalence of anaemia in women aged 15 to 49 years of age, by pregnancy status."[189] The two indicators for Target 2.1 are (i) the Prevalence of Undernourishment (PoU), and (ii) the prevalence of moderate or severe food insecurity, which is based on a relatively new measure called the Food Insecurity Experience Scale (FIES).[190] A single indicator now reflecting SDG Targets 2.1.1, 2.2.1, and 2.2.2 emerged in the middle of the first decade of this millennium called the Global Hunger Index (GHI), which also includes a measure of the mortality of children under the age of five, as seen in Section 1.3.4.2. All these indicators, especially GHI and FIES, provide not only insights into how well the global progress to achieve the goal of SDG2 is unfolding, but are also indicative of the nourishing needs of tomorrow and the preparations which need to be taken today.

1.3.4.2. *The global hunger index*

While national and global undernourishment can be assessed by comparing the data with pre-determined levels of food energy consumption, the determination of food quality is more problematic as is measuring the combined impact of food quality and quantity. To produce meaningful information regarding the combination, a Global Hunger Index (GHI) was developed in the early years of this century by non-governmental organizations (NGOs) based in three countries — the International Food Policy Research Institute in the US, the Deutsche Welthungerhilfe in Germany, and Concern Worldwide in Ireland.[191] The GHI has three dimensions measured by four indicators, which are then used to calculate the value of the index using an agreed numerical expression.[20] The values are expressed on a reversed 100-point scale where '0' is the best score and '100' is the worst. Within the overall scale, five categories are identified by their calculated values; if a country or region has a GHI of more than

'35,' that is considered to be in the "alarming" category in terms of food quantity and quality, whereas a score of '50' or higher is labelled "seriously alarming".[20] The latest global data shows the average GHI is '17.9,' based on information from 116 countries.[192]

Of the five categories, the global average is towards the upper boundary of the "moderate" band. This may seem to represent a reasonable foundation scenario, which could readily be improved upon by 2030, and so meet the nourishing needs of tomorrow, as defined at the start of this chapter. However, there are wide variations across major world regions and individual countries, as shown in the examples quoted in Table 1.10.

Reflecting the SDG Targets 2.1 and 2.2 identified earlier, the three dimensions of the GHI are (i) undernourishment for all, and for children under five years of age (ii) mortality, and (iii) undernutrition. For dimensions (i) and (ii), there is a single indicator, while for dimension (iii), there are two indicators known as "wasting," meaning low weight for height, and "stunting," meaning low height for age.[20] It is worth noting that the undernutrition indicator classifications of wasting and stunting are based on the 1970s research work of the British physiologist Dr. JC Waterlow, though in the 1930s, the Anglo-Jamaican physician Dr. CD Williams discovered a link between child wasting and a form of malnutrition caused

Table 1.10. GHI exemplars for regions and countries.[192]

Global geographical region	GHI	Country, *above world average* — Examples	GHI	Country, *at or below world average* — Examples	GHI
Africa South of the Sahara	27.1	Somalia	50.8	Gambia	17.6
South Asia	26.1	Yemen	45.1	Cambodia	17.0
World	**17.9**	Haiti	32.8	Philippines	16.8
West Asia and North Africa	12.7	Nigeria	28.3	Egypt	12.9
Latin America and the Caribbean	8.7	India	27.5	Mexico	8.5
East and South-East Asia	8.5	Venezuela	22.2	Saudi Arabia	6.8
Europe and Central Asia	6.5	Indonesia	18.0	Brazil, China, Chile, Cuba, Slovakia, Turkey + 12 others	< 5.0

by a lack of protein in their diet, referring to it as a disease called "kwashiorkor," a term used by African women.[193] However, while she eventually went on to have an internationally acknowledged stellar career in advancing pediatrics and child healthcare in developing countries, her male colonial colleagues in particular largely refused to acknowledge that kwashiorkor was a unique disease. Today, kwashiorkor is recognized as a form of severe protein malnutrition, though the exact cause is still unknown and it can occur even with a sufficient calorie intake.[194]

The work of Waterlow and Williams are examples of just how difficult it can be to quantify and qualify food. Nevertheless, the increasingly used measure, GHI, can be calculated using a simple arithmetic expression[192]:

$$GHI = \frac{1}{3} \times \text{Standardized PUN} + \frac{1}{6} \text{Standardized CWA} + \frac{1}{6} \text{Standardized CST} + \frac{1}{3} \text{Standardized CM}$$

Where PUN is the proportion (%) of the population that is undernourished[p].

CWA is the prevalence of child wasting under 5 years old (%).

CST is the prevalence of child stunting under 5 years old (%).

CM is the proportion (%) of children dying before the age of 5 and *Standardized* means the annual reported indicator value compared to its maximum observed value between 1988 and 2013 which for PUN = 80, CWA = 30, CST = 70 and CM = 35.

For example, the 2021 GHI for India, given in Table 8, is 27.5 which is calculated from India's data, i.e., PUN = 15.3, CWA = 17.3, CST = 34.7, CM = 3.4, so that the standardized indicators become, PUN = 15.3/80 × 100; CWA = 17.3/30 × 100; CST = 34.7/70 × 100; CM = 3.4/35 × 100 and thus the above expression, India's GHI is therefore:

$$GHI = 6.375 + 9.611 + 8.262 + 3.238 = 27.466 \ (27.5)$$

Ultimately, the aim should be to have a GHI of '0,' but an increasing number of countries and regions have GHIs of '10' and less which, in terms of categorization, is considered to be "low." However, while the GHI can be used to identify regional, national, and global hunger trends, it does not pinpoint the individuals who are experiencing undernourishment or exactly where they live. To address these issues, the UNFAO

---

[p]Derived from PoU data.

launched the "Voice of the Hungry Project" almost a decade ago, based on previously published research and developed in conjunction with other international agencies such as Gallup® polls and the Food Insecurity Experience Scale (FIES), which is now used by more than 150 countries.[195] The rationale behind the FIES is to ask individuals directly about their access to food and the severity of their food insecurity. The collection of data to determine the FIES is carried out by targeting selected individuals or groups, and asking them to answer a specific set of questions, i.e., conducting a traditional survey.

1.3.4.3. *The food insecurity experience scale*

As the global population continues to increase, it is reasonable to ask if there will be a sufficient amount of food, i.e., dietary energy, available for all? The answer appears to be yes, since about 30% more food in calorific terms is produced annually than is needed or consumed. This global abundance of food, even in times of famine, is not a new phenomenon, and yet 1.9 billion people are moderately or severely food insecure.[20] How can this be? More than four decades ago, the Nobel prize winner, Indian economist Dr. Amartya Sen, published a book on the causes of starvation and famine, arguing that poverty and the resulting inequalities in the distribution of food were prime factors.[196] This work and subsequent research would eventually lead to both the UN's FIES and its Human Development Index (HDI). Although Sen's original works contained some analysis, his assertions that (i) all the problems of hunger could not be solved by producing more food, and (ii) that food abundance did not preclude hunger for particular individuals were really hypotheses. To test such judgments using the principles of the scientific method, it would be necessary to collect empirical evidence to gauge the reliability and reproducibility of Sen's observations and his explanations.

Researchers worldwide, but especially in the US, acknowledged that the causes of hunger in times of ample food supplies needed to be studied and measured. A Cornell University team seeking to understand the experience of hunger by those who had experienced it selected a rural area of the New York State, where they spoke to women about the problem and the subsequent actions they had taken.[197] All the women reported that they had gone through a similar process at times when they were experiencing hunger, namely (a) worrying about not having enough food, (b) stretching their available food resources, (c) reducing the quality of food in their diet, (d) cutting down on portion sizes, (e) skipping meals entirely, and (f) not eating at all for at least a day or more, i.e., fasting. Similar studies were carried out throughout the US, with similar feedback and eventually resulted in the advent of the USDA's Household Food Security Survey from 1995 onwards.[198] Subsequently, this type of investigative survey work was carried out in South America, Asia, and Africa, and once again results were comparable. Armed with these findings, the Voices of the Hungry Project was launched.

The USDA's survey consists of 18 questions for households, 10 for adults and also has a short six-question version.[199] The FIES involves questionnaires focused on individuals and households seeking answers to eight questions which cover the (a) to (f) items stated in the previous paragraph, but asking two more questions about running out of food and being unable to eat healthy and nutritious food at any time during a 30-day or 12-month reference period.[200] In all cases, codicils are added to all the questions, specifically if individual or household food insecurity experiences are due to "not enough money or other resources" or "a lack of money or other resources".[201] For example, Question 7 of the survey module asks individuals: "You were hungry but did not eat because there was not enough money or other resources for food"? or, for households, the opening phrase of the question is: "Was there a time when you or others in your household…"[200,201] The questions are usually asked by a qualified interviewer and the responses to the questions noted and given a numerical value where '1' = "yes," '0' = "no," '98' = "don't know," and '99' = "refused to answer".[200]

Like other food security (insecurity) surveys, the FIES results are represented by a statistical scale similar to other widely-accepted scales based on what are known as Item Response Theories (IRT) or Latent Response Theories (LRT), which are designed to measure unobservable traits. These unobserved variables are used to try and explain correlations between measured parameters, especially when a direct causal relationship has not been fully established or universally accepted. The results of unobservable surveys then add self-judgments of a particular situation; in this case, the lack of access to adequate levels of quality food.[202] Although directly measured parameters may indicate that average household income is above a nationally defined poverty line and that there is an abundance of quality food available, the FIES surveys can indicate that these measured data mask the actual situation for many individuals and households, and they can provide the reasons why this is so.

Moreover, as the FIES is based on survey questions, the scale can be used to study the food insecurity features of global, national, or localized occurrences. For example, in the UNFAO 2020 survey for Iraq, 19 "governorates" were individually analyzed as well as the whole nation, whereas for Nigeria, FIES was used for the country and its 36 federated states in addition to its capital territory.[203] In this way, not only can FIES issues be identified by country, but also can be focused on specific areas within a country. But how can the accuracy, or at least the reliability, of the data be confirmed? Normally, the results from data FIES surveys are validated using an IRT–Rasch Model to make sure that they represent a reliable measure of food insecurity.[204,205]

The FIES methodology may appear to be somewhat of an esoteric and academic approach to the determination of the scale of food insecurity. This is accurate, to some extent, and to fully appreciate the system requires some post-secondary level knowledge of fundamental psychological principles, mathematical analysis, and the efficient use of sociological surveys. The UNFAO offer a detailed online course dealing with the FIES through its e-learning academy that enables professionals of various backgrounds to obtain an understanding of the system and to receive UN certification.[206] There are a number of such courses dealing with SDGs. But how does FIES help to identify the actions that need to be taken to meet the nourishing needs of tomorrow?

While the determination of FIES appears rather convoluted, the results can be classified into three more straightforward food insecurity categories — (i) "mild," (ii) "moderate," and (iii) "severe" — where "mild" is defined as worrying about the ability to obtain food; "moderate" means reducing quantities, skipping meals, compromising food quality and variety; and "severe" refers to experiencing hunger.[207] All these items come from particular answers to the eight standard questions asked in FIES surveys. The combination of the moderate and severe categories, $FI_{mod+serv}$, is used as the 2.1.2 indicator for SDG 2. Knowing the geographical locations where $FI_{mod+serv}$ is above predefined levels and, since the experience question codicils are "lack of money or other resources," then, arguably, the provision of greater funding and more resources should address the nourishing needs, especially if there is an abundance of food.

As discussed, there are measures to gauge the prevalence of undernourishment (PoU) and intuitively, it could be supposed that the number of people affected would likely be in harmony with those suffering moderate or severe food insecurity. However, as the data example for

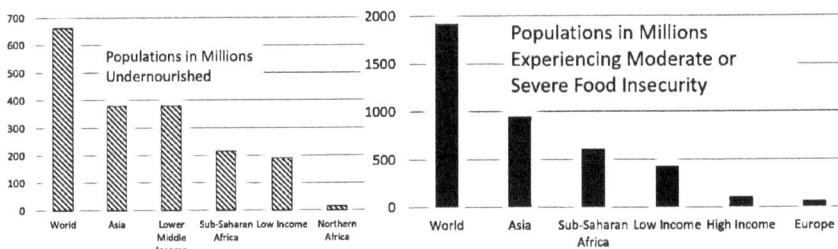

Figure 1.22.   2017 Populations undernourished or experiencing moderate or severe food insecurity.[185]

2017 shows, (Figure 1.22), food insecurity is far more dominant than undernourishment. This is underscored by the fact that while 663 million people were assessed as lacking nourishment, almost three times more, 1.92 billion, were found to be experiencing moderate or severe food insecurity.[26] The data values are somewhat in agreement in that Asia and Sub-Saharan Africa provide the bulk of the global deficiencies in both cases. It can also be seen on the exemplars that income, or the lack of it, does play a role but it does not provide a unique explanation for either undernourishment or food insecurity alone. There are other drivers involved. The key drivers, identified by the UNFAO and many others are: (i) the intensity and frequency of armed conflict,[q] (ii) climate variability (as opposed to anthropogenic climate change) and extreme weather events, and (iii) economic slowdowns and recessions. The latter obviously impacts income level, and this can be the result of many different factors such as the continuing COVID-19 pandemic. Depending on the region and country all these drivers, (i) to (iii) may be present. In terms of armed conflicts, considered to be the main driver, Africa and parts of Asia continue to be the areas of the highest occurrences which provides further insight into the results given in Figure 1.22. In 2020, conflict was the cause of hunger for 99 million people in 23 countries.[203] Moreover, as of March 2022, armed conflict spread to Eastern Europe and, combined with the lasting effect of the pandemic and the resulting changes in food insecurity and PoU, the ramifications are unlikely to help the global achievement of the SGD 2 targets by 2030, as more than 60 million people in Europe were already enduring food insecurity in 2017 pre-pandemic and conflict.[20]

---

[q]Wars, rebellions, skirmishes as defined by the International Committee of the Red Cross https://www.icrc.org/en/doc/assets/files/other/irrc-873-vite.pdf.

Natural events such as volcanic eruptions, earthquakes, tsunamis, hurricanes etc., are not only difficult to predict, they and their resulting weather events cannot be prevented. There are provisions which can be taken to lessen the negative effects on food production, and availability of seasonal weather events, such as drought and flooding, but there is great concern that climate change and global warming is exacerbating the magnitude of such events. Whether or not climate mitigation and adaptation prove ultimately successful, they likely will have little immediate impact on food issues. Moreover, there have been wars and conflicts globally for 90% of the time over the last three millennia. Indeed, up to June 2020, the US has been in wars and conflicts, with very few continuous periods of peace since 1776.[208] Some suggest that world peace is inevitable, while a large body of literature argues that it is possible but, given the facts, the realization of the concept seems to be wishful thinking. In courteous terms, the UN have stated with regard to SDG 2 that "[a]chieving this goal by the target date of 2030 will require a profound change of the global food and agriculture system."[209]

It appears that despite the horrors of armed conflict, civilization has yet to find wholly effective alternatives to resolve its problems. As the total elimination of armed conflicts by 2030 is not going to happen, if ever, and as the possibility of the human prevention of climate variabilities is nonsensical, then the economic factors alone, or the stated goal of SDG 1, i.e., the "end of poverty in all its forms everywhere," is the only practical approach to purge the world of global hunger. Unfortunately, after the success of reducing global undernourishment between 2001 and 2017, the percentage of people is now on the rise again, with the 2020 levels being comparable to those of 2009.[20,209] From 2019 to 2020, there was an increase of more than 160 million people experiencing undernourishment, equivalent to almost twice the population of Germany, the European Union's most populous country.[209] Given all the facts regarding the SDG 2 indicators, the probability of achieving the goal is remote.

1.3.5. *Housing, Shelter, and Homelessness*

Humans have always needed shelter to protect themselves from the vagaries of the prevailing weather and the threats of other animals. For nomadic hunter–gatherers, the shelter would be temporary and transportable in some instances, but the agricultural revolution enabled more permanent shelters to be constructed. The right to housing and the ownership of property is embedded in Articles 17 and 25 of the 1948 UN Universal Declaration of Human Rights.[2] Housing or shelter is also part of SDG 11, whose goal is to "[m]ake cities and human settlements inclusive, safe, resilient, and sustainable."[210] Of SDG 10's targets and 15 indicators, the one dealing specifically with housing is Target 11.1 which is to "ensure access for all adequate, safe and affordable housing and basic services and upgrade slums" by 2030, representing several interrelated factors.[211] To achieve this target is a much larger challenge than may be generally appreciated since, it was estimated in 2018 that 1.6 billion people globally live in inadequate housing, including the one billion living in slums and informal settlements.[212] And even these large numbers may be an underestimation.

The scale of this problem is then on par with the other identified nourishing needs — clean air, water, and sanitation (WASH), and food. Although the problem is frequently associated solely with urban environments, the lack of adequate housing and homelessness can also occur in rural settings. Nevertheless, despite the vagaries of the current pandemic, urbanization, i.e., the continued migration from rural areas to cities, continues and is compounded by the growing global population. This means that cities are largely the focus of SDG 11, and this is reflected in SDG Indicator 11.1.1: "[p]roportion of urban population living in slums, informal settlements or inadequate housing."[213] Obstacles arise in collecting the global data necessary to track this indicator because of the lack of universally accepted definitions concerning the terminology of Indicator 11.1.1. Indeed, several countries refuse to use the term "slum," while others use the term "slum" and "informal settlement" interchangeably.

Defining homelessness is also inconsistent globally. Although permanent shelters are now associated with buildings or homes, and those in many informal settlements are far from permanent, the current lexicon is to use the term "shelter" in the context of describing homelessness, e.g., sheltered, or unsheltered homelessness.[214] Moreover, the problems in collecting and collating Indicator 11.1.1 data are trifling in comparison to obtaining global measurements of homelessness. How then can the housing, shelter, and homelessness nourishing needs of tomorrow and the actions that ought to be taken by 2030 be satisfactorily established?

1.3.5.1. *Slums, informal settlements, and Inadequate housing*

At the start of this millennium, the UN in conjunction with other international agencies produced a series of Millennium Development Goals (MDGs), the forerunners of the recent SDGs. As part of the MDGs, the criteria for classifying housing into slums, informal settlements, and inadequate housing were developed and are still in use today, especially for tracking Indicator 11.1.1. These criteria, or the factors which are considered, e.g., access to water, in determining the MDG/SDG category are summarized in Table 1.9.[27] Care needs to be taken when using these factors to interpret exactly how they are used to classify the types of housing conurbations. For example, a slum is not defined by its access to water but by its lack of access to improved water and, as such, is related to Indicator 6.1.1 as given in Table 1.7. Similarly, it is the lack of access to improved sanitation, Indicator 6.2.1, that is a characteristic for a slum. The same codicils also apply when labeling housing as informal settlements or as inadequate. If there is a "lack of" for one or more of Criteria 1 to 5 in Table 1.9, a household will be identified as a slum dwelling. This approach also provides the benchmark for classifying informal settlements.

Despite the reluctance by some nations to use the term "slum" and instead preferring "informal settlement," others differentiate between the two by considering the latter to be, (i) housing which is built on land to which the occupants have no legal claim and are occupying illegally, (ii) unplanned areas of housing that are not in compliance with building or planning codes and regulations, and (iii) usually lack city infrastructure and formal basic services.[212,215] The third category of Indicator 11.1.1 is inadequate housing as shown in the right-hand column of Table 1.11 and it requires a further three criteria (numbered 6 to 8) to be met, compared to that for a slum classification. Of these, the measurement of Criteria 8, "cultural adequacy," is perhaps the most prone to subjective assessment, as housing is not considered adequate "if it does not respect and take into account the expression of cultural identity and ways of life."[212] But exactly what is "cultural identity" and how could this impact the design, construction, and planning of buildings? This is a field of sociological study that is beyond the scope of this chapter's discussion but, suffice to say that cultural identity involves religious beliefs, ethnicity, nationality and location, social class and status, gender, sexuality, food choice and preparation, and more.[216,217]

Table 1.11. Criteria (factors) involved in housing classifications.[27]

Criteria	Slums	Informal settlements	Inadequate housing
1. Access to water	X	X	X
2. Access to sanitation	X	X	X
3. Sufficient living space, overcrowding	X		X
4. Structural quality, durability, and location	X	X	X
5. Security of tenure	X	X	X
6. Affordability			X
7. Accessibility			X
8. Cultural adequacy			X

Accessibility is more easily defined as addressing the needs of disadvantaged and marginalized individuals, families, and groups, e.g., those with physical, mental health, and economic challenges. The latter may be due to general poverty, or the desire to live in close proximity to their workplaces or essential services and, even if they are considered to have incomes above the national, regional, or local poverty benchmarks, they cannot afford to own or rent adequate accommodation. The SDG indicator measurement of Criteria 6, "affordability," historically considered as a straightforward relationship between house prices and incomes, is now recognized as a far more challenging concept though less so than cultural identity. Housing is considered inadequate, or at least non-adequate, with regard to affordability "if its cost threatens or compromises the occupants' enjoyment of other human rights."[212] One dictionary definition of enjoyment is a feeling of happiness or pleasure,[218] once again a subjective and emotional self-assessment that has been studied by religious leaders, philosophers, and psychologists from Aristotle to Nietzsche to Maslow. But how can happiness be measured?

The measurement of happiness was suggested in the late-18th century largely as a means of determining how a government was performing, but it would be another two centuries before methods of measuring happiness, such as the Subjective Happiness Scale (SHS), were developed.[219,220] Very much like the approach used in the FIE scale (Section 1.3.4.2), happiness[r]

---

[r]The United States Declaration of Independence identified the 'pursuit of happiness' as an unalienable right of all humans.

would be measured by conducting surveys using formalized questionnaires. However, the seminal event which brought a global happiness measure to public and political attention was a UN Resolution 65/309, "Happiness: Towards a holistic approach to development", sponsored by Bhutan[s] and adopted by the UN General Assembly in 2011. The resolution urged governments to "give more importance to happiness and well-being in determining how to achieve and measure social and economic development."[221] Since then, an annual World Happiness Report has been published, currently under the joint auspices of the UN Sustainable Development Solutions Network (SDSN) and Columbia University in the US, with funding from several private, public, and government agencies and data from the Gallup® World Poll.[222]

Over the past decade, interest has rapidly increased in the concepts, science, and measurement methodology of happiness, which continues to be a term growing in popularity compared to the more traditional state of national development parameters such as GDP per capita. The results of the annual happiness reports are now widely covered by media outlets. The annual data are reported to three decimal places, as an index with a scale between '0' (the lowest) and '10' (the highest), based on a statistical analysis of the answers to the survey questions. Of the 146 countries ranked in the 2022 report, the index values vary between '7.821' (Finland) to '2.404' (Afghanistan). Of the 20 least happy nations, which includes India, more than 60% are countries in the African regions below the Sahara, whereas of the 20 happiest countries, more than 70% are European led by the Scandinavian nations, with the others being Israel, the US, and the British Commonwealth countries, Australia, Canada, and New Zealand.[222]

Is happiness then directly proportional to affordability, and is affordability similarly a measure of GDP per capita? Finland, for example, which is not ranked in the top 20 countries by GDP per capita, still has a GDP per capita more than 3.5 times the global average. Indeed, only just over a half of the top 20 happiest nations appear in the top 20 countries by GDP per capita. Thus, it could be argued that while measures such as happiness and cultural identity may seem eccentric in determining the adequacy of housing, they could provide useful insights into the nourishing needs of both today and tomorrow. However, although some data on these criteria are being collected, especially happiness, the tracking of the SDG 11.1.1 indicators, with regard to inadequate housing, still rely on the more

---

[s]Located in the Eastern Himalayas, between China and India.

traditional measures of affordability, as defined by the UN-Habitat organization, namely the capital and operating costs of a dwelling compared to the incomes of the occupants.[212,223]

The "costs versus income" measures have prescribed benchmarks to define affordability. A median-priced dwelling is deemed unaffordable if it costs more than three years of yearly median income. Additionally, if the monthly mortgage or rental and utility fees, i.e., the running costs of a house, are more than 30% of the household's monthly income, then that is also considered to indicate unaffordability. In a recent UN-Habitat publication, it was reported that the global capital cost ratio was estimated at 6.1 years and operating costs at 35%.[223] The least affordable regions were Eastern Asia/Pacific at 8.1 years but surprisingly while above the cost benchmark, the ratio in Sub-Saharan Africa was only four years, though more than 55% of people living in the region did not have access to affordable housing because of operating costs. On the other hand, the affordability timelines for the purchase of dwellings for those living in Western Asia and North Africa was greater than both Sub-Sahara Africa and the UN benchmark at 4.2 years, although only just. In North Africa, only 5.5% of the population did not have access, which is lower than North America, Europe, and Australia, which have the highest global happiness ratings. It seems that the data derived for measuring Criteria 6 to 8, to help identify inadequate housing, are not yet proving as helpful as expected by their protagonists, but any rush to judgment must be tempered by acknowledging there is still much room for improvements to emerge in definitions, measurement methodologies, and data gathering. So, what is known about slums and informal settlements?

As monitored by UN-Habitat and the World Bank, the 'World' trend in the share of urban populations living in slums, as defined by the lack of at least one of the five criteria given in Table 1.9, is downward since 1990 and in many countries, as shown in Figure 1.23.[224] The Central African Republic is an exception to this trend, as was Iraq between 2000 and 2005, as both are examples of conflict-affected countries. India also experienced a greater proportion of slum-dwellers since 2014 after reducing the share by a half from 1990 to 2014.[224] Unfortunately, while the general proportion of those living in slums has been decreasing, the actual number has been increasing due to population growth, often coupled with more people moving to urban areas, e.g., cities, from rural areas. Care has to be taken with the data, since different countries use diverse

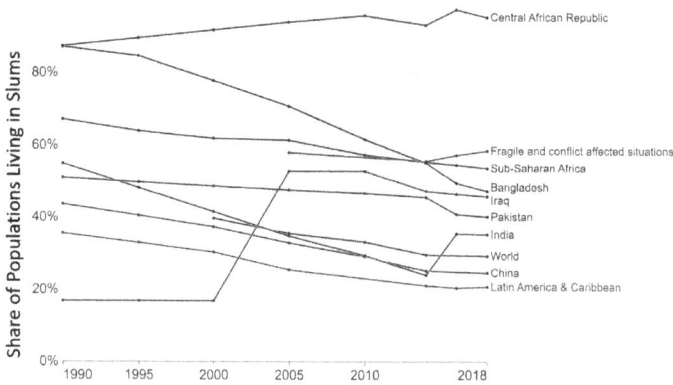

Figure 1.23.   Exemplars of Urban Slum Population Share Changes 1990–2018.[224]

Source: UN-HABITAT (via World Bank)

definitions[t] of "urban" in their censuses and similar national statistics in determining the urban versus rural populations. The urbanization of society has also gained rapid momentum since the start of the Industrial Revolution and it is forecasted to further increase, so that by 2050, more than two-thirds of the global population will live in urban settings.[225, 226]

In 2017, it was reported that if no improvements were made to reduce the proportion of the population living in slums, then those numbers could rise to 2.5 billion by 2030 and approach three billion by 2050.[227] However, the estimates for 2030 and 2050 could well be conservative since, by mid-century, it is projected that Japan and the US will have at least 89% of their populations living in cities, with urbanization in heavily populated countries like China reaching 80%, accompanied by large increases in urban dwellers in India and Sub-Saharan Africa, as seen in Figure 1.24.[224] Globally, by 2050, more than one in two people will be living in China, India, and Sub-Saharan Africa. Given these situations, it is very difficult to envision that the housing problems are going to be solved in time to meet tomorrow's nourishing needs. It can appear that some countries may not have nourishing needs, such as the US, Canada, and the UK because they do not publish data on slums. Perhaps this is not too surprising since the 2020 UN-Habitat World Cities Report indicates that these countries

---

[t]For example, Sweden and Denmark define the *urban threshold* as 200 inhabitants while Japan's threshold is 50,000.

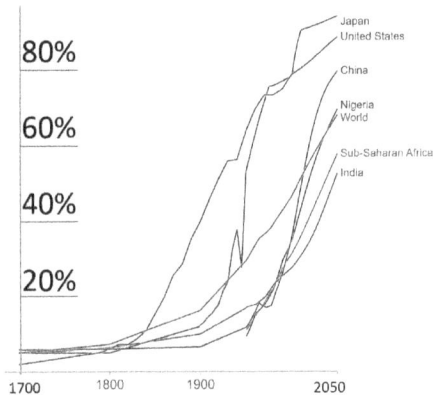

Figure 1.24.   Portion of Population living in Urban Centers 1700–2050.[224]

contribute less than 0.1% of the global urban population living in slums,[228] a statistical nicety likely to be of little comfort to those living in slums in those countries, as defined by the criteria of the 11.1.1 indicator. But what about those people who are homeless?

1.3.5.2. *Shelter and Homelessness*

There is a scarcity of rigorous data with respect to global homelessness. Much of the publicly available government information comes from medium- and high-income countries, although other sources such as the Millennium Alliance for Humanity and the Biosphere (MAHB) and the Ruff Institute of Global Homelessness (IGH) are attempting to quantify the global state of affairs.[229,230] Furthermore, it has been over a decade and a half since the UN last attempted a global survey of homelessness, when it was estimated that 100 million people were homeless, which was about 1.5% of the world population at the time. Countries that have reported their version of population homelessness rarely carry out annual assessments of the problem. A proverbial obstacle of acquiring an accurate worldwide assessment is the lack of a globally accepted definition of homelessness. For example, the IGH framework expresses it as "lacking access to minimally adequate housing," which, as discussed in Section 1.3.5.1, is more akin to the inadequate housing criteria provided in Table 1.9 rather than explicitly dealing with homelessness in the same way as some countries.

Perhaps the most useful definition of homeless is found in the US Code of its general and permanent laws, which simply states that the terms "homeless, homeless individual and homeless person" mean "an individual or family who lacks a fixed, regular and adequate nighttime residence."[231] This terminology comes from Title 42, item §11302 of the Code, which deals with public health and welfare, where further elaboration of what homelessness means for individuals and families is presented under six main legal characterizations, together with an acknowledgment that domestic violence and other threatening conditions can contribute to an individual's or a family's homelessness.[232] The US Code elements appears to have been very influential in more countries categorizing homelessness as being "sheltered" or "unsheltered," albeit the term "literally homeless" is also used to describe unsheltered homelessness, and "rough sleepers" to quantify the visible elements of the same category.

Moreover, the US has standard mechanisms and methodologies for on-street community counting of homelessness, the results of which are reported to their Congress. Arguably then, their definitions provide a helpful insight into homelessness. These can be paraphrased to (i) "sheltered homelessness" identifying those people who are staying in emergency shelters, transitional housing programs, or safe havens, whereas (ii) "unsheltered" means those with a primary nighttime residence, private or public, which are not, as a rule, used as regular sleeping accommodation such as a vehicle, a park, an abandoned building, open land areas, streets, and public spaces at bus and train stations or airports.[32,233] In 2007, the total number of the homeless in the US was almost 650,000 people which, a decade later, had been reduced by an overall 15% and by almost 30% in the case of unsheltered homelessness.[214] However, since 2014, the number of unsheltered people has started to rise and, in some jurisdictions such as Los Angeles County, both sheltered and unsheltered numbers continue to increase by 2020, with more than 63,000 to 66,000 homeless people, over 2.5 times more, were living under unsheltered conditions compared to sheltered.[234,235]

What insights do the more recent detailed data from LA County provide? With more than 10 million people, according to the most recent US census, the county's population is greater than that of 40 individual US States and with a GDP of more than US$1 trillion, it has the third-largest metropolitan economy in the world, even though almost 15% live below the national poverty line.[236] The county is a diverse multiracial society. As of 2020, the Hispanic/Latino community accounted for 48% of the population, Caucasians accounted for 25.6%, Asians accounted for 14.7%, and Black/African Americans accounted for 7.6%. These ethnic proportions

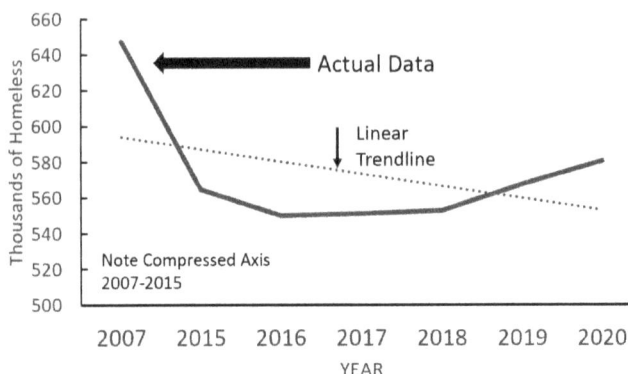

Figure 1.25.   Homeless numbers in thousands USA 2007–2020.[32]

are not reflected in the relative numbers of the homeless, with Black/ African Americans being 33.8% (39% in LA City itself), only slightly less than the Hispanic/Latino with 36% (34.6% in the city) and the Asians with 1.2%.[234] Other characteristics of the homeless in the LA County include more than 58% being in the 25–54 age group, and for those aged 18 or over, 36.2% are "chronically homeless" individuals, with domestic violence, mental illness, and substance abuse accounting for significant causes of homelessness. Cities like Los Angeles use only homelessness as the measure of Indicator 11.1.1 but do so in great detail, and the information is made public via a series of websites.[237-239]

A snapshot of the 2015 homeless rate in the US was provided in Figure 1.1. Between 2007 and 2016, the total number of the homeless did decrease from 647,258, and the overall trend is downwards. However, since 2017, the actual number has started to increase being 580,466 in 2020, as shown in Figure 1.25.[240] Media headlines dealing with US homelessness have proclaimed that there is a homeless crisis or catastrophe, and the National Homelessness Law Center,[u] a US advocacy group, have since 2015 suggested that the reported numbers by the US Department of Housing and Urban Development (USHUD) are too low and in reality, the homeless numbers are more likely to be between 2.5 and 3.5 million people.[241] These are significant differences but come about largely because of differing methodologies for determining homelessness. Until USHUD's 2018 report, the homeless count in the US was based solely on what is known as a Point-in-Time (PIT) methodology, which involves

---

[u] Formerly known as the National Law Center.

data on a single night in the latter part of the month of January for both sheltered and unsheltered homeless collected by official local, regional, and state agencies from about 3,000 national-wide sources.[242] So, if an individual is homeless in June, or any time other than the chosen date in January, they would not be included in the annual homeless count.

However, the provision of more housing especially in parts of California became a political issue and, along with the continuing media and advocacy groups, the criticisms of the under-reported nature of homelessness levels and the publication of a less-than-flattering report on human rights, including issues regarding the homeless by a UN Special Rapporteur likely led to the USHUD adopting Longitudinal Systems Analysis (LSA) as a homeless count methodology.[242,243] LSA involves the same data collecting agencies as used in the PIT approach but now, any individual experiencing homelessness at "some time during in the year" is counted. In 2018, USHUD reported that 1,446,000 individuals had faced homelessness, less than suggested by non-government sources, but more than 2.5 times greater that the reported PIT levels. This also implies that homelessness affects more people in the US than slum dwelling. If homelessness in a rich country like the US has been considerably under-estimated, it is reasonable to conclude that this problem is replicated in the other few countries who attempt to measure homelessness.

1.3.5.3. *Is housing an intractable problem?*

In 2018, with more than one billion slum dwellers globally, and homeless-ness, as far as it is known, on the increase, what actions need to be taken to meet Indicator 11.1.1? Certainly, housing, or more accurately, the lack of housing worldwide and especially in expanding urban centers and cities requires government intervention. There are a few jurisdictions where all levels of government are starting to make efforts to address the issues, but these usually focus on residential building programs aimed at increasing the supply of affordable housing.[244–246] It can appear that these initiatives are often political gestures, rather concerted plans, given that there was already a serious shortage of housing, whether to own or rent, even before the pandemic and the more recent Ukraine–Russia conflict.[245] If afforda-bility is a key factor in all housing issues, then the situation seems to have worsened since the arrival of the pandemic, as illustrated by the exemplars shown in Figure 1.26 for national housing price changes in 2020 and large city increases in 2021.[247,248]

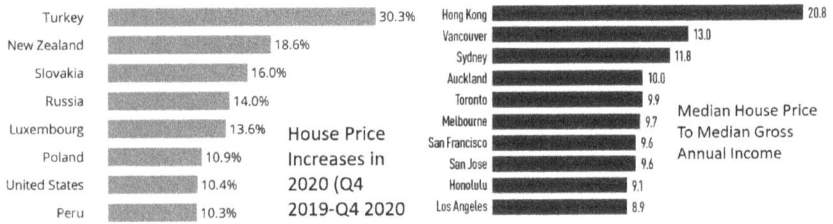

Figure 1.26.   National housing price increase 2020 and city house prices 2021.[247–248]

The measures advocated to address the global housing crises and in some instances the actions taken vary, which include (i) increasing housing densities by amending space regulations, (ii) using cheaper constructional materials and labor, (iii) providing financial incentives for mainly private sector builders, and (iv) removing more income-related obstacles to acquiring a mortgage.[249] These suggestions have not been greeted with universal or societal enthusiasm and, where implemented, they have shown little success. If today's needs cannot be achieved, it is hard to imagine that the SDG 11 target timeline of 2030 will be satisfied, especially with continually growing populations. This does not bode well for the housing nourishing needs of tomorrow. Moreover, as homelessness increases, studies have indicated that the probability of more open defecation and urination will not only impact the health and wellbeing of those personally affected, but also the general attempts to address global WASH problems (see Section 1.3.3).

1.3.6. *Access to affordable energy*

As with the other nourishing needs identified in Table 1.1, affordable energy is another key factor in the well-being of *all* people, albeit not required for basic survival in the same way as air and water. Population growth and increased life expectancy, for example, are both the result of more people having access to energy derived from the Earth's internal natural resources, covering biomass to carbonaceous fuels, and external sources, such solar radiation and gravitational forces. Humans have also harnessed the power of the atom, i.e., nuclear energy, over the past seven decades. While the exploitation of the internal resources provided many advantages, there are growing concerns that the continued use of carbon-containing fuels is not only unsustainable, but also responsible for adversely affecting the global climate and weather patterns, mainly by warming both the atmosphere and the oceans. To limit and perhaps reverse these damaging effects, there is a strong political and scientific impetus to use only renewable and sustainable sources, largely those externally derived, i.e., a major transition in the types of source energy used to meet human needs. Historically, such energy source transitions were driven by availability and reliability of supply and cost.[250]

All the many facets involving global energy consumption are embedded in the goals, targets, and indicators of SDG 7.[35] The goal of SDG 7 was given in Section 1.1, but is repeated here for convenience, namely, "[to] ensure access to affordable, reliable, sustainable, and modern energy for all." To achieve this goal, there are four indicators associated with three national targets, and two international targets with two indicators, as given in Table 1.12.

Table 1.12.   SDG 7 targets and indicators.[251]

SDG 7 target	Target description, by 2030	SDG 7 indicator	Indicator description
7.1	"Ensure universal access to affordable, reliable, and modern energy services"	7.1.1	"Proportion of population with access to electricity"
		7.1.2	"Proportion of population with primary reliance on clean fuels and technology"
7.2	"Increase substantially the share of renewable energy in the global energy mix"	7.2.1	"Renewable energy share in the global final energy consumption"
7.3	"Double the rate of improvement in energy efficiency"	7.3.1	"Energy intensity in terms of primary energy and GDP"
7.a	"Enhance international cooperation to facilitate access to clean energy research and technology including renewable energy, energy efficient and advanced and cleaner fossil fuel technology, and promote investment in energy infrastructure and clean technology"	7.a.1	"International financial flows to developing countries[i] in support of clean energy research and development and renewable energy production, including hybrid systems"

(*Continued*)

Table 1.12.   (*Continued*)

SDG 7 target	Target description, by 2030	SDG 7 indicator	Indicator description
7.b	"Expand infrastructure and upgrade technology for supplying modern and sustainable energy services for all in developing countries, in particular least developed countries, small island developing States, and land-locked developing countries, in accordance with their respective programmes of support"	7.b.1	"Installed renewable energy generating capacity in developing countries (in watts per capita)"

Note: [i] The term "developing country" is an historical artefact of UN's nomenclature but continues to be used for consistency across is myriad of publications since 1948.

Essentially, Targets 7.1 to 7.3 are benchmarks for the countries with the means to realize them, while Targets 7.a and 7.b can be viewed as UN overtures for those with the means to share their resources and technology with everyone else.

Targets 7.1 and 7.2 refer to modern energy services and renewables. Often, by inference, modern energy services are taken to be solar, wind, and water power, but as usable energy sources, there is nothing new about these. Rather, it is the energy convertors associated with these sources that can be labelled as modern. The only modern energy source is nuclear energy, though the electricity produced by nuclear power stations uses energy convertors developed since the 18th century, i.e., steam engines. Renewable energy sources such as biomass, as delineated presently, are also not new, and indeed biomass dominated global energy consumption until the advent of the Age of Coal in the 19th century.[252] By the 20th century, coal as the primary energy source was increasingly replaced by fossil oil, and as the new millennium approached, so began the Age of Electricity. The history of energy, and energy transition pathways, is really the history of civilization.

The only energy source exploited by our very early ancestors was their muscle power, which was eventually boosted by similarly sourced power from tamed animals. Hunter–gatherers learned how to harness the properties of fire for cooking and heating likely by observing lightening strikes, and then how to purposely self-start fires using wood and other combustible vegetation such as straw. These traditional biomass fuels dominated energy consumption until the 19th and 20th centuries. After the advent of fire and biomass combustion, humans discovered additional energy sources, such as wind and waterpower, and learned how to construct more efficient methods of exploiting muscle power, e.g., the pulley mechanism. Moreover, the history of solar power utilization did not start with solar photovoltaic (PV) panels, as often imagined, but at least two millennia ago when the Greeks and Romans used reflective materials — so called burning mirrors[253] — to light fires and for rudimentary magnifying lenses for the space heating of buildings. Geothermal energy, in the form of natural pools and hot springs, is known to have been used for cooking, space heating, and bathing for at least 10 millennia.[254] All these energy sources are still available today and likely will be for the rest of this century and probably beyond, as they are considered renewable.

However, while an energy source may be considered to be renewable, depending upon the conversion technique used, the source may be identified as environmentally unfriendly, such as large-scale hydroelectric production. This is mainly because of the possibly significant emission levels of carbon dioxide and methane, and habitat degradation associated with hydropower reservoirs and dams. There is then a large suite of technological proven energy sources in use now and even more that could be available in the future. Will all be needed? In this millennium, the global population has increased annually by about 80 million people and, although since 2001, the rate of yearly population rises has and continues to diminish, it is expected that the total population will reach 11.0 billion by 2100.[29,30] It is found that there is not sufficient primary energy to satisfy present needs and eventually those of an additional three billion people, not all will be able to afford reliable and modern energy services, regardless of the availability of numerous energy sources, the existing populations cannot all afford the current energy mixes. According to a recent World Bank analysis, which estimated that 3.4 billion people, 44% of the then global population, "still struggle[s] to meet basic needs," a situation echoed in a Stanford Social Innovation Review report about "the true extent of global energy poverty."[255,256] At the moment, many individuals cannot afford sufficient quantities of quality food, as manifest in the Global Hunger Index, or adequate housing, and the same is true of energy, regardless of the level of access. Is SDG Indicator 7.1.1, access to electricity, then a suitable measure of energy affordability?

1.3.6.1. *Access to electricity*

In 2019, about 10% of the global population or 760 million people did not have access to electricity, the main proxy indicator for SDG 7.1, access to affordable energy.[28] While this may seem high, in 1998, there were 1.6 billion people, about 26% of the population, without such access.[28,] Without doubt, the present situation represents a significant achievement, but it must be realized that defined access levels are only 250 kWh per year for rural households and 500 kWh per year for urban households. [28,257] The IEA have recommended that 1250 kWh per year should be the bench-mark, which would enable a household to power four lightbulbs for five hours per day, a cell phone charger, a radio or television for four hours daily, together with a fan for six hours daily.[258] Even this is a very modest target when compared to the average annual usage of electricity in Canada and the US of about 4.5 MWh. This is three times more than most European households, and approximately 11 times more than Chinese households.[259] Given that Canadians and Americans have 100% access to electricity, according to pre-pandemic data from the World Bank, does it mean that despite the high level of electricity use, all people in Canada and the US can afford to pay for this access?

1.3.6.2. *Energy poverty*

Canada is the fourth largest country in terms of land area alone[v] and it has a dispersed population living in its cold climate. As a rich, high-income country, most of its people enjoy a high standard of living. It is also one of the largest global energy producers and consumers of energy. Yet, according to the Federal Government, between 6% and 13% of regional populations also experience what they term as "fuel poverty," defined as households having to spend more than 10% of their income on energy.[260] A 2021 study suggests that the government data may be underestimating the issue and in fact, as many as 19% of Canadian households are living with energy poverty.[261] Globally, there are several other terms in use to describe fuel/energy poverty, such as energy insecurity or energy bur-den.[262, 263] Congressional politicians in the US favor the latter term. The threshold for defining their energy burden is lower than Canada's at 6%,

---

[v]Not including water bodies such as lakes, rivers, reservoirs, etc.

i.e., if a household needs to spend more than 6% of its income on energy, then the burden is deemed to be unaffordable.[264]

According to their Federal data, Canada's national average for fuel poverty is 8%, which appears to be much higher than the comparable US data, which indicates an average energy burden of 2.7% with variations in individual states from 2% to 3%, except Maine and the territory of Puerto Rico at 4%.[265] It could be concluded that there are only minor energy affordability problems in the US. However, an analysis of the 2015 USEIA's Residential Energy Consumption Survey (RECS) indicated that a third of US households have faced a financial challenge in meeting their energy needs, as illustrated in Figure 1.27.[266] The average energy burden data masks the experiences of low-income households, which constitute 44% or 50 million of the national US housing stock,[267] which were found to have an energy burden three times the national average, i.e., 8.6%, but for those living on the US Federal poverty line, this rises to 17%.[268] Just prior to the RECS survey, a White Paper for the US Congress estimated that if household energy costs rose by 10%, a further 840,000 would be pushed into poverty.[269] Nevertheless, after the congressional White Paper, the US Census Bureau reported five consecutive years of poverty proportion declines. In 2021 however, this Bureau found that the poverty rate increased to 11.4% in the 2020 pandemic year, almost a full percentage point higher than 2019, with 37.2 million people in poverty.[270] Hardly an encouraging trend for the laudable SDG Goal 1 to end poverty in the world's largest economy by 2030, let alone in less well-off countries.

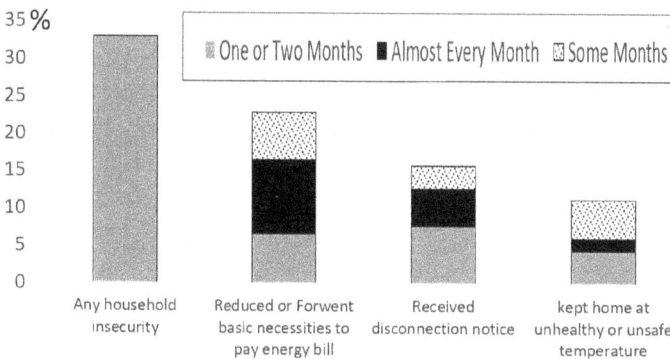

Figure 1.27.   US households experiencing energy insecurity in 2015.[266,271]

Global data shows that in many countries which have a GDP per capita of less than $25,000, large swathes of their populations do not have access either to electricity or clean cooking fuels. Ironically, these countries do not significantly contribute to global GHG emissions. However, there are a growing number of 70 and more countries, including the major emitters, who have plans in place to achieve zero GHG emissions by 2050 or soon after. Once again, paradoxically, the very poorest countries in Africa with large populations such Burundi and the Democratic Republic of Congo are already close to this target as are other less populous countries, but invariably the majority of their populations suffer from energy poverty.[272] How can energy poverty be reduced while simultaneously eliminating anthropogenic GHG emissions? It can appear that the two targets are in opposition, in that if energy is not consumed, then no emissions result, but that situation is not helpful to those suffering energy poverty. Arguably, the issue is the use and affordability of energy sources which do not generate GHG emissions. Consequently, the solution, perhaps somewhat utopian, is the 100% use of affordable renewable energy sources accompanied by the complete rejection of the use of carbonaceous fuels, especially the traditional fossil fuels, coal, oil, and natural gas, as part of the global energy mix.

With the renewable scenario in mind, legislation to abolish the manufacture of energy convertors requiring the combustion of fossil fuels during this decade or soon after was passed by several countries, ostensibly to accelerate the timelines for the total replacement of fossil energy and, subsequently combat anthropogenic climate change in sufficient time to ensure that terrestrial and ocean surface temperature increases do not exceed 1.5°C over pre-industrial levels. It is unlikely, however, that these actions can be achieved by 2030 or even 2050, according to the USEIA's international energy outlook report, as illustrated in Figure 1.28.[273] Nevertheless, the indications are that there will be a significant increase in the use of renewable energy by 2050, almost double the 2018 proportion, if biofuels are considered renewable and this increase is accompanied by a decreasing share of fossil fuels.

1.3.6.3. *Renewables, Fossil Fuels, and Nuclear*

There are several energy sources which are considered to be renewable and sustainable, but the favored or at least the most popular are solar, wind, and some forms of water power. Although it seems unlikely that

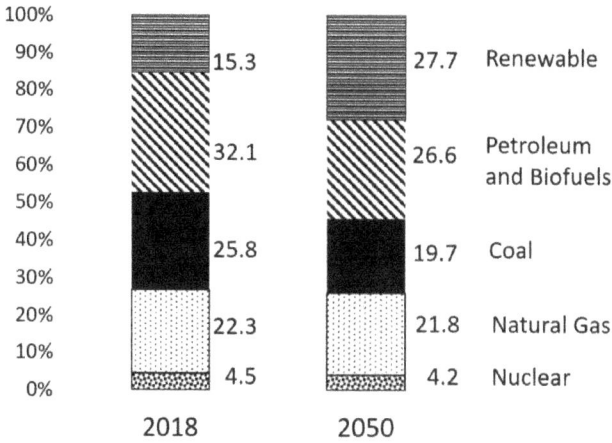

Figure 1.28.   Energy source transition forecast globally from 2018 to 2050 using USEIA data.[273]

these energy power forms could provide 100% of global primary energy consumption, not all agree. Wide-ranging and detailed papers published by members of the Stanford University Atmosphere/Energy Program demonstrated that 100% renewable energy could be achieved in 139 countries by 2050 and that by 2030, the SDG 7 target date, 85% of energy supplies in 50 US states could be renewable.[274,275] The major concerns about solar and wind power are intermittency and variability coupled with insufficient global electrochemical battery storage capacity to continuously support baseload electricity demands and the lack of country-wide electricity transmission grid networks. The recent announcements by national governments such as the French, Canadian, and the UK on renewing and expanding their nuclear power sector would appear to suggest that lawmakers are not as convinced as the Stanford group regarding the solar-wind-water scenarios.[276–278] However, there are other renewable energy sources such as geothermal, green hydrogen, some forms of hydropower, various ocean energy harvesting techniques, and biomass which could be used to support solar and wind power.

Of all these other renewables, the most contentious is the use of biomass, largely wood, since burning biomass releases carbon dioxide and PM, impacting anthropogenic climate change and air quality respectively. The incongruity is that there is staunch political enthusiasm for using carbon-neutral biomass, especially in the European Union.[279] But how can a carbonaceous fuel be considered carbon neutral? The protagonists argue that the emitted carbon dioxide will be sequestrated eventually by new growth. However, estimates for eventually achieving 100% sequestration vary from eight years to centuries in some boreal regions.[280,281] This is hardly an effective situation to meet the SDG target deadlines and, at the same time, increasing PM emissions will not ameliorate hazardous air pollution. If carbon dioxide from wood burning can be sequestrated, If carbon dioxide from wood burning can be sequestrated why not from the combustion of other carbonaceous sources such as fossil fuels? The answer is that with fossil fuels the sequestration would need to be in real-time, but this is technically feasible using carbon capture, storage, and utilization (CCUS).[282] Indeed, many international organizations have also stated that fossil fuel usage will remain part of the global energy mixes well into the second half of this century, especially if CCUS becomes commercially and economically viable.[283] Consequently, the 2050 forecast energy mix, as shown in Figure 1.28, is likely to prove to be a reasonably realistic representation. If this is the case, then SDG Target 7.2 is unlikely to be achieved until much later in the 21st century.

1.3.6.4. *The cost factor*

Using the historical rule of thumb regarding energy transitions, availability or security of supply and cost are key factors. Since before the 14th century's bubonic plague pandemic in Europe and the UK, the growing scarcity and the rising cost of wood precluded its use for heating and cooking for all but the nobility. The impoverished classes, i.e., the majority of people, were then allowed to use cheaper and more freely available coal. The aftermath of the horrific death toll from the various waves of plague pandemic ironically meant that with fewer people, the supply of wood was greater than the demand and its cost significantly reduced, so biomass again became the energy source of choice. It would be another three centuries before the general use of coal became prevalent again, especially in the UK. If the cost of renewable energy and the electricity generated from renewable sources is not affordable, then their use is highly unlikely to be universally popular with consumers. Convincing individuals in low-income situations probably living below national poverty lines that it is a good idea to pay more for energy is probably an insurmountable challenge, even if the rationale is that such a transition could result in the control of anthropogenic climate change and keep surface temperature increases to the Paris Agreement's 2°C or, ideally, the IPCC's 1.5°C requirement. However, if the alternatives to renewables start to become comparatively more expensive, then attitudes will no doubt change, but if the price of all forms of energy increases, then there could be even more people suffering energy poverty. One way to combat this dilemma is to improve the efficiency of energy conversion devices, which would result in less primary energy being consumed. If sufficient, the increase in efficiency would not only offset cost increases but could reduce energy poverty for some. However, there are a lot of "ifs" and "buts" associated with pitching energy costs such that energy poverty can be reduced and eventually eliminated.

Not surprisingly, several models using computer-based numerical analyses have been developed to explore the possible future economic viability and potential advantages of global energy transitions to renewable sources.[284] The models, wholly or partially, usually involve some form of life-cycle analysis (LCA), such as the Energy Return on Investment (EROI), i.e., the ratio of how much energy is produced in relation to the amount of energy used to create it.[285] Thus, for example, if the EROI is '1,' there is no return of the investment. According to the World Nuclear Association, the breakeven value of EROI is '7' but for the US the average is '40' for all generating technologies.[286] However, although EROI may seem a straightforward approach, its application is not as simple as may be imagined, and published efforts to produce definitive lists of EROIs for various energy sources have attracted academic discussion and rebuttals among researchers about modeling quality, definitions, and "confusions."[287]

The issue appears to be centered around claims that nuclear, hydropower, and fossil fuel energy systems are much more effective, i.e., higher EROIs, than solar photovoltaics and wind power.[288] While the claim of higher EROIs for nuclear, hydro, and some fossil fuel systems are still valid, the differences between these and non-hydro renewables such as wind power are shrinking, especially with respect to fossil fuel EROIs. While EROIs are used by investors in general and venture capitalists in particular, because of the frequent vagaries of energy source costs, e.g., impacts of conflicts, perhaps a more useful planning tool especially for making decisions about future infrastructure is the Levelized Cost of

Energy (LCOE) approach, which is used by both governments and energy consultants. Using a stock market analogy, EROI is more about day trading while LCOE is associated with much longer-term investment strategies, but that is just the author's opinion.

The LCOE is a measure of the actual costs of energy production calculated for a particular energy source for a generating plant over its remaining or complete lifetime.[289] The US Department of Energy (USDOE) has identified two key uses of the LCOE, namely that it (a) can be used to "calculate the present value of the total cost of building and operating a power plant over an assumed lifetime", and (b) "allows the comparison of different technologies (e.g., wind, solar, natural gas) of unequal life spans, project size, different capital cost, risk, return, and capacities."[289] In conjunction with the USDOE's National Renewable Energy Laboratory (NREL), an Excel™-based LCOE model was developed for analyzing (a) and (b) mentioned above, but for renewables, it is called the Cost of Renewable Energy Spreadsheet Tool (CREST). Originally developed between 2010 and 2014 to help and guide policymakers "in the evaluation and development of cost-based incentives: to support renewable energy technologies", this tool and its accompanying user manual are now available for public download and use.[290] While useful for determining the costs of various forms of renewable energy, the proprietary software, developed by one of the world's leading financial advisory and assessment energy companies Lazard®, a publicly traded company,[w] considers all forms of energy, a wider range of factors than CREST, and is constantly updated. They released Version 15 of their LCOE model analysis results in 2021.[291]

The latest Lazard analyses indicate that some renewable LCOE rates, in particular onshore wind and utility scale solar PV, are already competitive with conventional energy supply in the US provided that the renewable systems receive tax incentives and subsidies. The pricing of carbon (also known as carbon tax) will no doubt adversely impact the cost to the consumers of fossil fuel energy whether for electricity generation, basic domestic applications, or societal needs, such as health services and infrastructure construction. Moreover, as the majority of foodstuffs are transported by diesel trucks, locomotives, and shipping, higher fuel costs for transportation and distribution will lead to elevated food prices, which in

---

[w] NYSE: LAZ

turn will affect all people. The Canadian Federal government has imposed a carbon tax regime which is set to rise to C$170 per tonne by 2030. The carbon tax is part of their strategy to combat the impact and cost of climate change and the accompanying extreme weather events. It is considered that these issues can be effectively addressed by reducing GHG emissions. Whether or not this is an accurate depiction remains to seen, but the government's imposition of the tax, and the reasons for having to do so, convinced the Supreme Court of Canada to declare it legal and constitutional.

Since the inception of the tax, the government has claimed that most households ended up better off financially from the carbon tax and rebate system, whereby 90% of the tax is returned to families.[292] Until recently, there was some justification for this claim, but Canada's independent Parliamentary Budget Officer (PBO) has now challenged this assertion because of planned increases in the carbon tax. The rate is set to double in 2022, and increase to 8.5 times the original rate by 2030. The PBO has calculated that Canadians will now start paying more than they get back in carbon tax rebates and the country's GDP will diminish by 1.4% by 2030, and total labor income will reduce by 2.3%.[293,294] The cost of tackling climate change and modifying weather patterns may be acceptable to all Canadians, but for those suffering a lack of nourishing needs such as energy poverty for example, and with little potential for improvements over the next decade, it can hardly be an attractive proposition.

Globally, in 2021, the World Bank identified over 60 jurisdictions of different levels with official carbon pricing policies and have recommended even higher carbon taxes if mitigation of climate change is to be successful.[295] However, subjecting the burning of biomass to carbon pricing is a contentious matter. Some countries, such as the US and the UK, do not tax biomass; in fact, the UK provides significant subsidies to encourage biomass use.[296] Other countries do have carbon pricing policies for biomass, but only under certain conditions and for certain uses. As more than three billion people use unclean cooking fuels, largely wood, with many living in low-income countries, how will imposing a carbon tax alleviate not just energy poverty but general poverty? Indeed, how could a carbon tax be imposed in many countries where using biomass fuels for cooking is prevalent because the fuel has little or no cost in monetary terms as it is collected by waste scavenging or cutting off low-hanging branches. If carbon pricing is the most effective way of mitigating anthropogenic climate change, then the GHG emissions from utility scale hydropower and large water reservoirs and dams should also be subject to the tax. These are politically challenging questions, especially since the LCOEs for particular hydropower installations, reported by such agencies as IRENA, can be lower than all forms of solar PV and wind power.[297]

Notwithstanding the relative simplicity of energy source comparators, they are notoriously difficult to apply in practice, and there is considerable variation in the results produced from such analyses.[298,299] Nevertheless, the general trend of the calculated LCOEs for utility-scale renewable wind and solar have dramatically decreased over the past 10 years and both are becoming increasingly cost competitive. However, the global economic bases for a complete energy transition to renewables, considered to be necessary to address the concerns linked to potentially catastrophic anthropogenic climate change, has still to be fully established. What is clear is that financial incentives and penalties applied to the various energy source mixes have significant impacts, especially on LCOE calculations.[300]

4.0. Concluding Remarks

The UN 2030 Agenda "Transforming Our World" lists 17 Sustainable Development Goals which will need to be fulfilled if Goal 1 to "[e]nd poverty in all its forms everywhere" is to be achieved. Five of these SDGs, as given in Table 1.1, are closely associated with the nourishing elements which humans require for (i) survival, air, water, sanitation, and food, and (ii) equitable standards of living for all humanity, adequate and affordable housing, and universal access to affordable energy. It is not clear how much global investment will be required to realize the nourishing needs identified in this chapter, but it has been estimated that it will take US$4 to US$5 trillion per year between 2021 and 2030 to achieve all 17 SDGs.[301] This figure is equivalent to about 6% of the world's total GDP, a seemingly modest amount, but where will the 6% come from? The total GDP of global low-income people is less than US$0.5 trillion so, even in the unlikely scenario that all of it was used in pursuit of the SDGs, the financial shortfall would be 90% of that required. On the basis of "who can, pays," i.e., the richest nations, the US alone would need to contribute 25% of its annual GDP every year for a decade, whereas for China, the

Table 1.13. Possible achievement of nourishing needs and current actions.

Nourishing element	Associated SDG #	SDG Tracking benchmarks met by 2030	Improvements by 2030	Tomorrow's nourishing needs comments
Clean air	Elements of 3, 7 and 11	No	Yes	Continued population growth and enthusiastic promotion of biomass use will blunt the desired rate of improvement but reductions in fossil fuel usage, especially coal, should help
Clean water	6	No	Yes	Better use of appropriate water treatment technologies especially with respect to sewage and agricultural practices. Realization that not all clean water needs to meet drinking water standards
Safe sanitation	6	No	Maybe	More public education should help increase awareness but cultural and shelter challenges (homelessness) will negate some of the measures
Adequate food	2	No	Maybe, but conflict and climate variability occurrences dependent	Better performance by end of century but conflict, poverty, and climate change and variability will continue to challenge the quality of food supplies and their distribution
Adequate shelter	11	No	No	Supply and affordability issues unlikely to be meaningfully resolved by political action largely because of the scale of the problem and the continued increases in global urbanization
Access to affordable energy	7	No	Access — Yes Affordability — No	Greater access to electricity dependent on capital and operating costs but access measurement could be adversely affected if global basic needs benchmarks are changed

contribution of the total amount would represent about 34% of its GDP, and for Japan 100%.[302] These financial scenarios are as unrealistic as the low-income scenario.

However, if the 38 countries who are members of the Organization for Economic Co-operation and Development (OECD) whose mantra is "to build better policies for better lives," were collectively to contribute just over 9% of their GDPs to the SDG and nourishing need objectives, then they could be achieved if the financial estimates are accurate. Almost 18% of the global population live in OECD countries; so, in essence, the contributions of one in five people would enable the other four to achieve the goal of ending poverty, a commitment adopted by 193 member states of the UN. Apart from general economic growth, the intentional redistribution of resources from the rich to the poor is obviously fundamental in addressing the nourishing needs of tomorrow. However, can money alone solve the existing problems? For example, how will resource recipients prioritize their needs in a transparent and acceptable manner to the aid donors? Moreover, how would the collection of aid and its redistribution be organized and by whom? A central international banking agency would likely be necessary with the International Monetary Agency (IMF) being an obvious candidate.

Despite the myriad of issues and obstacles involved in the practical achievement of the admirable SDGs, there have still been global improvements in most aspects over the past three decades, e.g., reductions in extreme poverty. However, there remains significant concerns about the rate of improvements and the adverse impact of the global pandemic, and continuing instances of armed conflicts on global finances, i.e., the ability to contribute. The financial commitments surrounding combating anthropogenic climate change concerns add to the complexity of solving humanity's nourishing needs mainly because energy transitions are never easy, never swift, and always expensive, at least initially. In all aspects of securing a better life for everyone, it seems affordability will be the yardstick whether it be for clean air, clean water, sanitation, food, shelter, and/or access to energy.

At the start of the chapter, Table 1.1 presented the author's perspective of the nourishing needs for a better tomorrow. To complete this chapter, the prospects for achieving these needs by the latter part of this century and the probable progress in the tracking of the associated SDGs by 2030 are summarized in the final table, Table 1.13, once again from the author's perspective.

References

1. The Gifford Foundation (n.d.). Nourishing Tomorrow's Leaders. Retrieved from: https://giffordfoundation.org/capacity-building/ntl/.
2. United Nations (1948). *The Universal Declaration of Human Rights.* Retrieved from: https://www.un.org/sites/un2.un.org/files/2021/03/udhr.pdf.
3. Green, C. (2018). 70 Years of Impact: Insights on The Universal Declaration of Human Rights. Retrieved from: https://unfoundation.org/blog/post/70-years-of-impact-insights-on-the-universal-declaration-of-human-rights/.
4. Bachelet, M. (2018). Remarks to High-level Event Marking the 70th anniversary of The Universal Declaration of Human Rights: A Prevention Tool to Achieve Peace and Sustainable Development. Retrieved from: https://www.ohchr.org/sites/default/files/Documents/Events/StatementsChronologicalOrderUDHR70Event.pdf.
5. United Nations (2015). Transforming Our World: The 2030 Agenda for Sustainable Development. Retrieved from: https://sdgs.un.org/publications/transforming-our-world-2030-agenda-sustainable-development-17981.
6. Vaughan, A. (2021, November 2). COP26: 105 Countries Pledge to Cut Methane Emissions by 30 per cent. *New Scientist Environment.* Retrieved from: https://www.newscientist.com/article/2295810-cop26-105-countries-pledge-to-cut-methane-emissions-by-30-per-cent/.
7. GOV.UK. (2021). Net Zero Strategy: Build Back Greener. Retrieved from: https://assets.publishing.service.gov.uk/government/uploads/system/uploads/attachment_data/file/1033990/net-zero-strategy-beis.pdf.
8. Scott, J. (2021, October 27). Australia Rejects U.S., EU Call for Methane Agreement. Retrieved from: https://www.bloomberg.com/news/articles/2021-10-27/australia-rejects-u-s-eu-call-for-global-methane-agreement.
9. Brimblecombe, P. (1976). Attitudes and Responses Towards Air Pollution in Medieval England. *Journal of the Air Pollution Control Association.* *26*(10), 941–945. DOI: 10.1080/00022470.1976.10470341.
10. Stets, E. (n.d.). Water Quality Evolution: From Industrialization to the age of the internet. [Powerpoint Presentation]. Retrieved from: https://acwi.gov/monitoring/webinars/industrial_internet_11242015.pdf.
11. Sun, J., Chen, X., Yu, J., et al. (2021). Deciphering Historical Water Quality Changes Recorded in Sediments Using eDNA. *Frontiers in Environmental Science.* https://doi.org/10.3389/fenvs.2021.669582.
12. United States Environmental Protection Agency (2022). History of Air Pollution. Retrieved from: https://www.epa.gov/air-research/history-air-pollution.

13. Hasell, J. (2018, March 22). Famine mortality over the long run. *Our World in Data*. Retrieved from: https://ourworldindata.org/famine-mortality-over-the-long-run.
14. Lowcock, M. and Trotsenburg, A. V. (2021, February 2). Prevent the next food crises now. *World Bank Blogs*. Retrieved from: https://blogs.worldbank.org/voices/prevent-next-food-crisis-now.
15. Charpentier, A. (2017). The U.S. has been at war 222 out of 239 years. Retrieved from: https://www.greynun.org/gn2/wp-content/uploads/2018/10/US-Involvement-in-War.pdf.
16. Ortiz-Ospina, E. and Roser, M. (n.d.). Homelessness — Evidence from cross-country studies. *Our World in Data*. Retrieved from: https://ourworldindata.org/homelessness#evidence-from-cross-country-studies.
17. Global SDG Indicator Platform (2018). 7.1.1. Proportion of the Population with Access to Electricity. Retrieved from: https://sdg.tracking-progress.org/indicator/7-1-1-proportion-of-the-population-with-access-to-electricity/.
18. The World Bank (n.d.). Access to Electricity (% of Population). Retrieved from: https://data.worldbank.org/indicator/EG.ELC.ACCS.ZS.
19. Ritchie, H., Roser, M. and Rosado, P. (2020). Access to Electricity part of Energy article. *Our World in Data*. Retrieved from: https://ourworldindata.org/energy-access#what-share-of-people-have-access-to-electricity.
20. Roser, M. and Ritchie, H. (2019). Hunger and Undernourishment. *Our World in Data*. Retrieved from: https://ourworldindata.org/hunger-and-undernourishment.
21. New Zealand PM Arden Says COVID-19 Has Been 'Eliminated' in Her Country (2020, June 8). *Voice of America*. Retrieved from: https://www.voanews.com/a/covid-19-pandemic_new-zealand-pm-ardern-says-covid-19-has-been-eliminated-her-country/6190694.html.
22. Roser, M. (2021, November 25). Data Review: How Many People Die From Air Pollution? *Our World in Data*. Retrieved from: https://ourworldindata.org/data-review-air-pollution-deaths.
23. UNICEF (2019). Progress on Household Drinking Water, Sanitation, and Hygiene 2000–2017: Special Focus on Inequalities. Retrieved from: https://data.unicef.org/resources/progress-drinking-water-sanitation-hygiene-2019/.
24. Gleick, P. H. (2002). *Dirty Water: Estimated Deaths from Water-Related Disease 2000–2020*. Pacific Institute Research Report. Retrieved from: https://pacinst.org/wp-content/uploads/2013/02/water_related_deaths_report3.pdf.
25. Ritchie, H. and Roser, M. (2021). Clean Water and Sanitation. *Our World in Data*. Retrieved from: https://ourworldindata.org/clean-water-sanitation.
26. FAO, IFAD, UNICEF, WFP, WHO (2021). *The State of Food Security and Nutrition in the World 2021.Transforming Food Systems for Food Security*,

Improved Nutrition and Affordable Healthy Diets for All. Rome, Italy: Food and Agriculture Organization of the United Nations. Retrieved from: https://doi.org/10.4060/cb4474en.

27. UN-Habitat (2021). SDG Indicator Metadata. Retrieved from: https://unstats.un.org/sdgs/metadata/files/Metadata-11-01-01.pdf.

28. Ritchie, H., Roser, M. and Rosado, P. (2021). Energy. *Our World in Data*. https://ourworldindata.org/energy.

29. United Nations Department of Economic and Social Affairs (n.d.). World Population to reach 9.8 billion in 2050, and 11.2 billion in 2100. Retrieved from: https://www.un.org/en/desa/world-population-projected-reach-98-billion-2050-and-112-billion-2100.

30. United Nations (2017). World Population Prospects: The 2017 Revision — Key Findings and Advance Tables. Retrieved from: https://population.un.org/wpp/publications/files/wpp2017_keyfindings.pdf.

31. Roser, M. (2019). Future Population Growth. *Our World in Data*. Retrieved from: https://ourworldindata.org/future-population-growth.

32. The U.S. Department of Housing and Urban Development (2022). The 2021 Annual Homeless Assessment Report (AHAR) to Congress. Retrieved from: https://www.huduser.gov/portal/sites/default/files/pdf/2021-AHAR-Part-1.pdf.

33. United Nations Economic and Social Council (2009). Enumeration of Homeless People. (Report ECE/CES/GE.41/2009/7). Retrieved from: https://unece.org/fileadmin/DAM/stats/documents/ece/ces/ge.41/2009/7.e.pdf.

34. United Nations Economic and Social Council (2018). Understanding the system of custodian agencies for Sustainable Development Indicators. (Report ECE/CES/2018/39). Retrieved from: https://unece.org/fileadmin/DAM/stats/documents/ece/ces/2018/CES_39.pdf.

35. IEA, IRENA, UNSD, World Bank, WHO (2021). Tracking SDG 7: The Energy Progress Report. The World Bank. Retrieved from: https://trackingsdg7.esmap.org/data/files/download-documents/2021_tracking_sdg7_report.pdf.

36. Mulholland, E., Dimitrova, A., and Hametner, M. (2018). SDG Indicators and Monitoring: Systems and Processes at the Global, European, and National Level. *ESDN Quarterly Report 48*. Vienna, Austria: ESDN Office. Retrieved from: https://www.esdn.eu/fileadmin/ESDN_Reports/QR_48_Final_Final.pdf.

37. United Nations Human Rights Council (2017). Special Rapporteur on the Rights of Indigenous Peoples. (Report A/HRC/36/46). Retrieved from: https://www.ohchr.org/EN/HRBodies/HRC/RegularSessions/Session36/Documents/A_HRC_36_46.docx.

38. United Nations, Human Rights — Office of the High Commissioner. (n.d.). About the Mandate — Special Rapporteur on Climate Change. Retrieved from: https://www.ohchr.org/en/specialprocedures/sr-climate-change/about-mandate.

39. WHA (n.d.). What is an oxygen-enriched atmosphere? Retrieved from: https://wha-international.com/what-is-an-oxygen-enriched-atmosphere/.

40. Oschlies, A. A. (2021). Committed fourfold increase in ocean oxygen loss. *Nature Communications, 12*(2307), https://doi.org/10.1038/s41467-021-22584-4.

41. Oschlies, A., Brant, P., Stramma, L., and Schmidtko, S. (2018). Drivers and Mechanisms of Oxygen Deoxygenation. *Nature Geoscience.* 11: 467–474. URL: www.nature.com/naturegeoscience. Available at https://doi.org/10.1038/s41561-018-0152-2.

42. Mosley, S. (2009). A Network of Trust: Measuring and Monitoring Air Pollution in British Cities, 1912–1960. *Environment and History. 15*(3), 273–302. DOI: 10.3197/096734009X12474738131074.

43. United States Environmental Protection Agency (1996). *EPA Takes Final Step in Phaseout of Leaded Gasoline.* Retrieved from: https://archive.epa.gov/epa/aboutepa/epa-takes-final-step-phaseout-leaded-gasoline.html.

44. WHO (2017). Evolution of WHO air quality guidelines: past, present, and future. Copenhagen: WHO Regional Office for Europe. ISBN 978289052306. Retrieved from: https://www.euro.who.int/__data/assets/pdf_file/0019/331660/Evolution-air-quality.pdf.

45. Air Pollution Control Association (1968). The Air Quality Act of 1967. *Journal of the Air Pollution Control Association. 18*(2), 62–71. DOI: 10.1080/00022470.1968.10469096.

46. United States Environmental Protection Agency (2021). Evolution of the Clean Air Act. Retrieved from: https://www.epa.gov/clean-air-act-overview/evolution-clean-air-act.

47. Hatton, T. (2017, November 14). Air Pollution in Victoria Era Britain — Its effect on health now revealed. *The Conversation.* Retrieved from: https://theconversation.com/air-pollution-in-victorian-era-britain-its-effects-on-health-now-revealed-87208.

48. Farrington, D. and Bogle, E. (2016, May 6). In photos: Canada's Devastating Ford McMurray Wildfires. *NPR.* Retrieved from: https://www.npr.org/sections/thetwo-way/2016/05/06/477014138/in-photos-canadas-devastating-fort-mcmurray-wildfires.

49. United States Environmental Protection Agency (2022). Hazardous Air Pollutants. Retrieved from: https://www.epa.gov/haps.

50. Government of Canada (2017). Criteria Air Contaminants. Retrieved from: http://www.ec.gc.ca/air/default.asp?162lang=En&n=7C43740B-1.

51. United States Environmental Protection Agency (2021). Criteria Air Pollutants. Retrieved from: https://www.epa.gov/criteria-air-pollutants.
52. WHO (n.d.). Types of Pollutants. Retrieved from: https://www.who.int/teams/environment-climate-change-and-health/air-quality-and-health/health-impacts/types-of-pollutants.
53. Schraufnagel, D. E. (2020). The health effects of ultrafine particles. *Experimental and Molecular Medicine.* 52: 311–317. https://doi.org/10.1038/s12275-020-0403-3.
54. United States Environmental Protection Agency (2022). What is PM, and how does it get in the air? Retrieved from: https://www.epa.gov/pm-pollution/particulate-matter-pm-basics.
55. United States Environmental Protection Agency (2014). Air Quality Index, A Guide to Air Quality and Your Health. Retrieved from: https://www.airnow.gov/sites/default/files/2018-04/aqi_brochure_02_14_0.pdf.
56. United States Environmental Protection Agency (2018). Technical Assistance Document for the Reporting of Daily Air Quality — The Air Quality Index (AQI). (Report EPA 454/B-18-007). Retrieved from: https://www.airnow.gov/sites/default/files/2020-05/aqi-technical-assistance-document-sept2018.pdf.
57. World Health Organization (2021). WHO Air Quality Guidelines: Particulate matter ($PM_{2.5}$ and PM_{10}), ozone, nitrogen dioxide, sulfur dioxide and carbon monoxide. Retrieved from: https://apps.who.int/iris/bitstream/handle/10665/345329/9789240034228-eng.pdf.
58. Government of Canada (2022). Greenhouse Gas Polluting Price Act, S.C. 2018, c. 12, s. 186. Retrieved from: https://laws-lois.justice.gc.ca/PDF/G-11.55.pdf.
59. Supreme Court of Canada (SCC) (2021). Case in Brief: References re Greenhouse Gas Polluting Price Act. Retrieved from: https://www.scc-csc.ca/case-dossier/cb/2021/38663-38781-39116-eng.aspx.
60. United States Environmental Protection Agency. Ground-level Ozone Basics (2022). Retrieved from: https://www.epa.gov/ground-level-ozone-pollution/ground-level-ozone-basics.
61. Sukumara-Pillai Krishnamohan, K.-P., Bala, G., Cao, L., Dua, L. and Caldeira, K. (2019). Climate system response to stratospheric sulfate aerosols: Sensitivity to altitude of aerosol layer. *Earth System Dynamics.* *10*(4), 885–900. https//doi.org/10.5194/esd-10-2019.
62. Takemura, T. (2020). Return to different climate states by reducing sulphate aerosols under future CO_2 concentrations. *Scientific Reports.* 10(21748). https://doi.org/10.1038/s41598-020-78805-1.
63. World Data Centre for Greenhouse Gases (2018). Retrieved from: https://gaw.kishou.go.jp/about_wdcgg/wdcgg.
64. Humidex (2022). Retrieved from: https://en.wikipedia.org/wiki/Humidex

65. Tegtmeier, S., Hegglin, M. I., Anderson, J., Funke, B., Gille J., et al. (2016). The SPARC Data Initiative: Comparisons of CFC-11, CFC-12, HF and SF_6 climatologies from international satellite limb sounders. *Earth System Science Data. 8*(1), 61–78. DOI:10.5194/essd-8-61-2016. Retrieved from: http://www.ace.uwaterloo.ca/publications/essd-8-61-2016.pdf.

66. United States Environmental Protection Agency (2022). Cost Reports and Guidance for Air Pollution Regulations. Retrieved from: https://www.epa.gov/economic-and-cost-analysis-air-pollution-regulations/cost-reports-and-guidance-air-pollution

67. International Energy Agency (2022). SDG7: Data and Projections — Access to Affordable, Reliable, Sustainable and Modern Energy. Retrieved from: https://www.iea.org/reports/sdg7-data-and-projections.

68. United Nations Department of Economic and Social Affairs (2022). World Population Prospects: The 2019 Revision. Retrieved from: https://population.un.org/wpp/Publications/.

69. Li, N., Georas, S., Alexis, N., Fritz, P., et al. (2016). A Work Group Report on Ultrafine Particles (American Academy of Allergy, Asthma & Immunology): Why Ambient Ultrafine and Engineered Nanoparticles Should Receive Special Attention for Possible Adverse Health Outcomes in Human Subjects. *Journal of Allergy and Clinical Immunology, 138*(2), 386–396. DOI: 10.1016/j.jaci.2016.02.023.

70. Donaldson, K., Stone, V., Clouter, A., Renwick, L., and MacNee, W. (2001). Ultrafine Particles. *Occupational and Environmental Medicine. 58*(3), 211–216. http://dx.doi.org/10.1136/oem.58.3.211. Retrieved from: https://oem.bmj.com/content/oemed/58/3/211.full.pdf.

71. Green Facts™ (2005). Air Pollution Particulate Matter. *Green Facts: Facts of Health and The Environment*. Retrieved from: https://www.greenfacts.org/en/particulate-matter-pm/index.htm.

72. Ritchie, H. and Roser, M. (2022). Outdoor Air Pollution. *Our World in Data*. Retrieved from: https://ourworldindata.org/outdoor-air-pollution#share-exposed-to-air-pollution-above-who-limits.

73. World Health Organization (2016). Burning Opportunity: Clean Household Energy for Health, Sustainable Development, and Wellbeing of Women and Children. Retrieved from: https://apps.who.int/iris/handle/10665/204717.

74. Speake-Cole, R. (2021, October 5). Biomass is promoted as a carbon neutral fuel. But is burning wood a step in the wrong direction? *The Guardian*. Retrieved from: https://www.theguardian.com/environment/2021/oct/04/biomass-plants-us-south-carbon-neutral.

75. Mueller, W., Loh, M., Vardoulakis, S., Johnston, HJ., Steinle S., et al. (2020). Ambient Particulate Matter and Biomass Burning: An Ecological Time Series Study of Respiratory and Cardiovascular Hospital Visits in

Northern Thailand. *Environmental Health. 19*(77), https://doi.org/10.1186/s12940-020-00629-3. Retrieved from: https://ehjournal.biomedcentral.com/articles/10.1186/s12940-020-00629-3.

76. U.S. Energy Information Administration (2020). Country Analysis Executive Summary: India. Retrieved from: https://www.eia.gov/beta/international/analysis.php?iso=IND and https://www.eia.gov/international/content/analysis/countries_long/India/india.pdf.

77. McKenna, P. and Lavelle, M. (2021, November 2). Over 100 Nations at COP26 Pledge to Cut Global Methane Emissions by 30 Percent in Less Than a Decade. *Inside Climate News.* Retrieved from: https://insideclimatenews.org/news/02112021/global-methane-pledge-cop26/.

78. Borunda, A. (2020, February 20). Natural Gas is a Much 'Dirtier' Energy Source than We Thought. *National Geographic.* Retrieved from: https://www.nationalgeographic.com/science/article/super-potent-methane-in-atmosphere-oil-gas-drilling-ice-cores.

79. Climate and Clean Air Coalition (2021). Global Methane Assessment: Benefits and Costs of Mitigating Methane Emissions. Retrieved from: https://www.ccacoalition.org/en/resources/global-methane-assessment-full-report.

80. Doniger, D. (2020, September 15). NRDC and Partners Go to Court on EPA's Methane Surrender. *NRDC.* Retrieved from: https://www.nrdc.org/experts/david-doniger/nrdc-and-partners-go-court-epas-methane-surrender.

81. Carrington, D. (2021, May 6). Cutting Methane Emissions is Quickest Way to Slow Global Heating — UN Report. *The Guardian.* Retrieved from: https://www.theguardian.com/environment/2021/may/06/cut-methane-emissions-rapidly-fight-climate-disasters-un-report-greenhouse-gas-global-heating.

82. National Research Council (2010). *Global Sources of Local Pollution: An Assessment of Long-Range Transportation of Key Air Pollutants to and from the United States.* Washington D.C., U.S.A.: The National Academies Press. Retrieved from: https://doi.org/10.17226/12743 and https://nap.nationalacademies.org/read/12743/chapter/1.

83. Khan, R. K. and Strand, M. A. (2018). Road Dust and Its Effect on Human Health: A Literature Review. *Epidemiol Health.* DOI: https://doi.org/10.4178/epih.e2018013. Retrieved from: https://www.e-epih.org/journal/view.php?doi=10.4178/epih.e2018013 and https://www.ncbi.nlm.nih.gov/pmc/articles/PMC5968206/pdf/epih-40-e2018013.pdf.

84. Bosch® (2014). *Automotive Handbook* (9th Ed.). Chichester, England: John Wiley.

85. Catalytic Converter (2022). Retrieved from: https://en.wikipedia.org/wiki/Catalytic_converter#Construction.

86. Plautz, J. (2018). Ammonia, a Poorly Understood Smog Ingredient, Could be Key to Limiting Deadly Pollution. *Science.* Retrieved from: https://www.science.org/content/article/ammonia-poorly-understood-smog-ingredient-could-be-key-limiting-deadly-pollution.

87. Cao, H., Henze, D. K., Cady-Pereira, K., McDonald, B. C., et al. (2022). COVID-19 Lockdowns Afford the First Satellite-Based Confirmation that Vehicles are an Under-recognized Source of Urban NH_3 Pollution in Los Angeles. *Environmental Science & Technology Letters. 9*(1), 3–9. DOI: https://doi.org/10.1021/acs.estlett.1c00730. Retrieved from: https://pubs.acs.org/doi/abs/10.1021/acs.estlett.1c00730.

88. IPCC (2012). Bioenergy. *Special Report on Renewable Energy Sources and Climate Change Mitigation.* Cambridge University Press; pp. 209–332.

89. Verma, P., Stevanovic, S., Zare, A., Dwivedi, G., et al. (2019). An Overview of the Influence of Biodiesel, Alcohols, and Various Oxygenated Additives on the Particulate Matter Emissions from Diesel Engines. *Energies. 12*(25), DOI: 10.3390/en12101987.

90. Jääskeläinen, H. and Majewski, W. A. (2021). Effects of Biodiesel on Emissions. *DieselNet.* Retrieved from: https://dieselnet.com/tech/fuel_biodiesel_emissions.php.

91. Haines, D. and Van Gerpen, J. (2019). Biodiesel and the Food vs. Fuel Debate. *Farm Energy.* Retrieved from: https://farm-energy.extension.org/biodiesel-and-the-food-vs-fuel-debate/.

92. Ruzic, D. (2021, July 20). Decarbonizing Air Travel [Video file]. Retrieved from: https://www.youtube.com/watch?v=02cCzdFBMCY&t=137s.

93. Randall, C. (2022, January 4). Denmark aims for fossil-fuel free inland flights by 2030. *Electrive.com.* Retrieved from: https://www.electrive.com/2022/01/04/denmark-aims-for-fossil-fuel-free-inland-flights-by-2030/.

94. Federal Aviation Administration Office of Environment and Energy (2015). *Aviation Emissions, Impacts & Mitigation: A Primer.* Retrieved from: http://fortbertholdplan.org/wp-content/uploads/2017/11/Primer_Jan2015.pdf.

95. Ritchie, H., Roser M. and Rosado P. (2020). Energy. *Our World in Data.* Retrieved from: https://ourworldindata.org/energy-mix.

96. Jacobson, M. Z., Delucchi, M. A., Bauer, Z. A. R., Goodman, S. C., et al. (2017). 100% Clean and Renewable Wind, Water and Sunlight All-Sector Energy Roadmaps for 139 Countries of the World. *Joule. 1*(1): 108–121. Retrieved from: https://web.stanford.edu/group/efmh/jacobson/Articles/I/CountriesWWS.pdf.

97. United Nations Economic Commission for Europe (2021). Assessment Report on Ammonia. Retrieved from: https://unece.org/sites/default/files/2021-03/ECE_EB.AIR_WG.5_2021_7-2102624E.pdf.

98. Wolf, M. J., Esty, D. C., Kim, H., Bell, ML., et al. (2022). New Insights for Tracking Global and Local Trends in Exposure to Air Pollutants. *Environmental Science & Technology. 56*(7): 3,984–3,996. DOI: https://doi.org/10.1021/acs.est.1c08080

99. New World Encyclopedia™. Sub-Saharan Africa. Retrieved from: https://www.newworldencyclopedia.org/entry/Sub-Saharan_Africa.

100. Sub-Saharan Africa. Retrieved from: https://en.wikipedia.org/wiki/Sub-Saharan_Africa.

101. Selassie, A. A. and Hakobyan, S. (2021, April 15). Six Charts Show the Challenges Faced by Sub-Saharan Africa. *International Monetary Fund (IMF) Africa Department.* Retrieved from: https://www.imf.org/en/News/Articles/2021/04/12/na041521-six-charts-show-the-challenges-faced-by-sub-saharan-africa

102. Cook, N., Arieff, A., Blanchard, L.P., et al. (2017). Sub-Saharan Africa: Key Issues, Challenges, and U.S. Responses. *Congressional Research Service.* Retrieved from: https://sgp.fas.org/crs/row/R44793.pdf.

103. United States Environmental Protection Agency (2021). Basic Ozone Layer Science. Retrieved from: https://www.epa.gov/ozone-layer-protection/basic-ozone-layer-science.

104. Donev, J. M. K. C, et al. (2018). Energy Education — Photochemical smog. Retrieved from: https://energyeducation.ca/encyclopedia/Photochemical_smog.

105. Royal College of Physicians (2016). *Every Breath We Take: The Lifelong Impact of Air Pollution.* London: RCP, 2016. Retrieved from: https://www.rcplondon.ac.uk/projects/outputs/every-breath-we-take-lifelong-impact-air-pollution.

106. UK Met Office (n.d.). The Great Smog of 1952. Retrieved from: https://www.metoffice.gov.uk/weather/learn-about/weather/case-studies/great-smog.

107. University Corporation for Atmospheric Research (2017). Nitrogen Oxides. *Centre for Science Education.* Retrieved from: https://scied.ucar.edu/learning-zone/air-quality/nitrogen-oxides.

108. Thompson, T. M. (2019). Background Ozone: Challenges in Science and Policy. *Congressional Research Service.* Retrieved from: https://crsreports.congress.gov/product/pdf/R/R45482.

109. United States Environmental Protection Agency (2022). Volatile Organic Compounds: Impact on Indoor Air Quality. Retrieved from: https://www.epa.gov/indoor-air-quality-iaq/volatile-organic-compounds-impact-indoor-air-quality.

110. United States Environmental Protection Agency (2015). Indoor Air Pollution, An Introduction for Health Professionals. Retrieved from: https://www.epa.gov/sites/default/files/2015-01/documents/indoor_air_pollution.pdf.

111. Greenstone, M., Lee, K., and Sahai, H. (2021). Indoor Air Quality, Information, and Socioeconomic Status: Evidence from Delhi. *AEA Papers and Proceedings*. 111: 420–424. DOI: 10.1257/pandp.20211006; retrieved from: https://www.aeaweb.org/articles?id=10.1257%2Fpandp.20211006.

112. United States Environmental Protection Agency (2022). Sources of Hydrocarbon and NOx emissions in New England. Retrieved from: https://www3.epa.gov/region1/airquality/piechart.html.

113. United States Environmental Protection Agency (2022). Nitrogen Oxides (NOx) Control Regulations. Retrieved from: https://www3.epa.gov/region1/airquality/nox.html.

114. Hedges & Company (2022). How Many Cars are There in the US? How Many Vehicles in the US and Other Vehicle Registration Statistics? Retrieved from: https://hedgescompany.com/automotive-market-research-statistics/auto-mailing-lists-and-marketing/.

115. Scott, M. (2019, June 10). Electric Models to Dominate Car Sales by 2040, Wiping Out 13m Barrels a Day of Oil Demand. *Forbes Sustainability.* Retrieved from: https://www.forbes.com/sites/mikescott/2019/06/10/electric-models-to-dominate-car-sales-by-2040-wiping-out-13m-barrels-a-day-of-oil-demand/?sh=2db48556342e.

116. United States Department of Energy: Office of Energy Efficiency & Renewable Energy. (2017, January 30). Fact #962: January 30, 2017 Vehicles per capita: Other Regions/Countries Compared to the United States. Retrived from: https://www.energy.gov/eere/vehicles/fact-962-january-30-2017-vehicles-capita-other-regionscountries-compared-united-states.

117. Tetraethyllead. Retrieved from: https://en.wikipedia.org/wiki/Tetraethyllead.

118. Lobet, I. (2021, September 2). Finally, the End of Leaded Gas. *National Geographic*. Retrieved from: https://www.nationalgeographic.com/environment/article/finally-the-end-of-leaded-gas.

119. United States Environmental Protection Agency (2022). Lead Trends. Retrieved from: https://www.epa.gov/air-trends/lead-trends.

120. Global Alliance on Health and Population. Its time to put pollution on the map. Retrieved from: https://lead.pollution.org/.

121. National Instruments (2021). Subsystems Required to Control Low Temperature Combustion Engine. Retrieved from: https://www.ni.com/en-ca/innovations/white-papers/11/subsystems-required-to-control-low-temperature-combustion-engine.html.

122. Kumar, M., Jain, A., and Chhaganlal Vora, K. (2019). Combustion Optimization and In-Cylinder NO_x and PM Reduction by Using EGR and Split Injection Techniques (SAE Technical Paper 2019-28-2560). *SAE Mobilus*. DOI: https://doi.org/10.4271/2019-28-2560.

123. Ledna, C., Muratori, M., Yip, A., et al. (2022). Decarbonizing Medium- & Heavy-Duty On-Road Vehicles: Zero-Emission Vehicles Cost Analysis [PowerPoint slides]. Retrieved from: https://www.nrel.gov/docs/fy22osti/82081.pdf.

124. Continental Automotive GmbH (2019). Worldwide Emission Standards and Regulations: Passenger Cars/Light and Medium Duty Vehicles. Retrieved from: https://www.borgwarner.com/technologies/emissions-standards.

125. Winkler, S. L., Anderson, J. E., Garza, L. et al. (2018). Vehicle Criteria Pollutant (PM, NO_x, CO, HCs) Emissions: How Low Should We Go? *NPJ: Climate and Atmospheric Science.* 1(26). https://www.nature.com/articles/s41612-018-0037-5.

126. United States Environmental Protection Agency (2022). Ammonia. Retrieved from: https://www.epa.gov/caddis-vol2/ammonia.

127. Guthrie, S., Giles, S., Dunkerley, F., Tabaqchali, H., et al. (2018). *The Impact of Ammonia Emissions from Agriculture on Biodiversity: An Evidence Synthesis.* Santa Monica, CA: RAND Corporation. Retrieved from: https://royalsociety.org/~/media/policy/projects/evidence-synthesis/Ammonia/Ammonia-report.pdf.

128. Ritchie, H. and Roser, M. (2022). Outdoor Air Pollution. *Our World in Data.* Retrieved from: https://ourworldindata.org/outdoor-air-pollution.

129. The Royal Society (2020). *Ammonia: Zero-carbon Fertiliser, Fuel and Energy Store. Policy Briefing.* February 2020. Retrieved from: https://royalsociety.org/-/media/policy/projects/green-ammonia/green-ammonia-policy-briefing.pdf.

130. Ghosh, I. (2020, September 8). These Countries Will Have the Largest Populations — By the End of the Century. *World Economic Forum.* Retrieved from: https://www.weforum.org/agenda/2020/09/the-world-population-in-2100-by-country/.

131. UN (2010). The Human Right to Water and Sanitation (A/RES/64/292). Retrieved from: https://www.un.org/en/ga/search/view_doc.asp?symbol=A/RES/64/292.

132. UN Human Rights, Office of the High Commissioner (2010). *Fact Sheet No. 35: The Right to Water.* Retrieved from: https://www.ohchr.org/sites/default/files/Documents/Publications/FactSheet35en.pdf.

133. UN Water (2018). Human Rights to Water and Sanitation. Retrieved from: https://documents-dds-ny.un.org/doc/UNDOC/GEN/N09/479/35/PDF/N0947935.pdf?OpenElement.

134. UN Department of Economic and Social Affairs: Sustainable Development. (n.d.). SDG 6: Ensure Availability and Sustainable Management of Water and Sanitation for All. Retrieved from: https://sdgs.un.org/goals/goal6.

135. WHO (2022). Drinking-Water Key Facts. Retrieved from: https://www.who.int/news-room/fact-sheets/detail/drinking-water.

136. United States Environmental Protection Agency. (2022). National Primary Drinking Water Regulations. Retrieved from: https://www.epa.gov/ground-water-and-drinking-water/national-primary-drinking-water-regulations.

137. Luo, C. X. (2021, November 15). The Water Crisis in Canada's First Nations Communities: Examining the progress towards eliminating long-term drinking water advisories in Canada. University of Windsor, Ontario, Canada: Academic Data Centre, Leddy Library. Retrieved from: https://storymaps.arcgis.com/stories/52a5610cca604175b8fb35bccf165f96.

138. Lucier, K. J., Schuster-Wallace, C. J., Skead, D., et al. (2020). "Is There Anything Good about a Water Advisory?": An Exploration of the Consequences of Drinking Water Advisories in an Indigenous Community. *BMC Public Health.* 20(1704). DOI: https://doi.org/10.1186/s12889-020-09825-9.

139. Government of Canada (2022). Remaining Long-term Drinking Water Advisories. Retrieved from: https://www.sac-isc.gc.ca/eng/1614387410146/1614387435325.

140. Ritchie, H. and Roser, M. (2021). Clean Water. *Our World in Data.* Retrieved from: https://ourworldindata.org/water-access#unsafe-water-is-a-leading-risk-factor-for-death.

141. World Health Organization (2022). Sanitation Key Facts. Retrieved from: https://www.who.int/news-room/fact-sheets/detail/sanitation.

142. World Health Organization (2022). Drinking-water Key Facts. Retrieved from: https://www.who.int/news-room/fact-sheets/detail/drinking-water.

143. Ritchie, H. and Roser, M. (2021). Sanitation. *Our World in Data.* Retrieved from: https://ourworldindata.org/sanitation.

144. Hamadeh, N., Van Rompaey, C. and Metreau, E. (2021, July 1). New World Bank country classifications by income level: 2021–2022. Retrieved from: https://blogs.worldbank.org/opendata/new-world-bank-country-classifications-income-level-2021-2022.

145. Ritchie, H. and Roser, M. (2021). Access to improved water sources increases with income. *Our World in Data.* Retrieved from: https://ourworldindata.org/water-access#access-to-improved-water-sources-increases-with-income.

146. Ritchie, H. and Roser, M. (2021). Clean Water. *Our World in Data.* Retrieved from: https://ourworldindata.org/water-access#access-to-safe-drinking-water.

147. United Nations Department of Economic and Social Affairs (n.d.) Ensure availability and sustainable management of water and sanitation for all. Retrieved from: https://sdgs.un.org/goals/goal6.

148. United Nations Sustainable Development Goals (n.d.). Goal 6: Clean Water and Sanitation. Retrieved from: https://www.un.org/sustainabledevelopment/water-and-sanitation/.

149. UNICEF and World Health Organization (2017). *Safely managed drinking water — Thematic report on drinking water 2017.* Geneva, Switzerland: World Health Organization. Retrieved from: https://apps.who.int/iris/handle/10665/325897.

150. Centre for Disease Control and Prevention (CDC) (2020). Drinking Water Standards and Regulations. Retrieved from: https://www.cdc.gov/healthywater/drinking/public/regulations.html.

151. United States Geological Survey (2018). Where is Earth's Water? *Water Science School.* Retrieved from: https://www.usgs.gov/special-topics/water-science-school/science/where-earths-water.

152. Frerichs, R. R. (2021). Who First Discovered *Vibrio Cholera*? UCLA Department of Epidemiology. Retrieved from: https://www.ph.ucla.edu/epi/snow/firstdiscoveredcholera.html.

153. History of Water Supply and Sanitation (2021). Retrieved from: https://en.wikipedia.org/wiki/History_of_water_supply_and_sanitation#Understanding_of_health_aspects.

154. Safe Water Drinking Foundation (2017). Conventional Water Treatment: Coagulation and Filtration. Retrieved from: https://www.safewater.org/fact-sheets-1/2017/1/23/conventional-water-treatment.

155. United States Centers for Disease Control and Prevention (YEAR). Safe Water Storage. Retrieved from: https://www.cdc.gov/healthywater/global/safe-water-storage.html.

156. United States Centers for Disease Control and Prevention (2021). Creating and Storing an Emergency Water Supply. Retrieved from: https://www.cdc.gov/healthywater/emergency/creating-storing-emergency-water-supply.html.

157. Kemira® (n.d.). *About Water Treatment: A Water Handbook for Water Treatment Industry.* Retrieved from: https://www.kemira.com/insights/water-handbook-2020.

158. Aquatech® (2019). Industrial Water: Our Essential Guide to Pollution, Treatment and Solutions. Retrieved from: https://www.aquatechtrade.com/news/industrial-water/industrial-water-essential-guide/.

159. O'Neal, J. (2021, September 10). How City Water Purification Works: Drinking and Wastewater [video file]. Retrieved from: https://www.youtube.com/watch?v=KsVfshmK0Akk.

160. Samco (2017). What is an Industrial Water Treatment System and How Does It Work? Retrieved from: https://www.samcotech.com/what-is-an-industrial-water-treatment-system-process/.

161. USGS (2019). Aquifers and Groundwater. *Water Science School*. Retrieved from: https://www.usgs.gov/special-topics/water-science-school/science/aquifers-and-groundwater.

162. United States Department of Agriculture (2022). Irrigation and Water Use. *Economic Research Service*. Retrieved from: https://www.ers.usda.gov/topics/farm-practices-management/irrigation-water-use/.

163. Johnson, T. D., Belitz, K., Kauffman, L. J., Watson, E., et al. (2022). Populations Using Public-supply Groundwater in the Conterminous U.S. 2010; Identifying the Wells, Hydrogeologic Regions, and Hydrogeologic Mapping Units. *Science of the Total Environment. 806*(Part 2), 150618. DOI: https://doi.org/10.1016/j.scitotenv.2021.150618.

164. Ritchie, H. and Roser, M. (2018). Water Use and Stress. *Our World in Data*. Retrieved from: https://ourworldindata.org/water-use-stress#freshwater-use-for-households-and-public-services.

165. Ritchie, H. and Roser, M. (2018). Share of Freshwater Withdrawals used in Agriculture. *Our World in Data*. Retrieved from: https://ourworldindata.org/water-use-stress#share-of-freshwater-withdrawals-used-in-agriculture.

166. USGS (2018). Groundwater Use in the United States, *Water Science School*. Retrieved from: https://www.usgs.gov/special-topics/water-science-school/science/groundwater-use-united-states.

167. United States Environmental Protection Agency (2021). Pollution Prevention Case Studies. Retrieved from: https://www.epa.gov/p2/pollution-prevention-case-studies.

168. USGS (2019). Thermoelectric Power Water Use. *Water Resources*. Retrieved from: https://www.usgs.gov/mission-areas/water-resources/science/thermoelectric-power-water-use.

169. Jin, Y., Behrens, P., Tukker, A., and Scherer, L. (2019). Water Use of Electricity Technologies: A Global Meta-analysis. *Renewable and Sustainable Energy Reviews. 115*(109391). DOI: https://doi.org/10.1016/j.rser.2019.109391.

170. Allende, A. and Monaghan, J. (2015). Irrigation Water Quality for Leafy Crops: A Perspective of Risks and Potential Solutions. *International Journal of Environmental Research and Public Health. 12*(7): 7,457–7,477. DOI: https://doi.org/10.3390/ijerph120707457.

171. Uzicanin, A. and Gaines, J. (2018). Community Congregate Settings. In Rasmussen, S. A. and Goodman, R. A. (Eds.), *CDC Field Epidemiology Manual*. Retrieved from: https://www.cdc.gov/eis/field-epi-manual/chapters/community-settings.html.

172. Fyre, E. A., Capone, D., and Evans, D. P. (2019). Open Defecation in the United States: Perspectives from the Streets. *Environmental Justice. 12*(5). DOI: https://doi.org/10.1089/env.2018.0030.

173. United Nations Department of Economic and Social Affairs (2019, November 18). Transformation Benefits' of Ending Outdoor Open Defecation: Why Toilets Matter. *UNESA News*. Retrieved from: https://www.un.org/development/desa/en/news/sustainable/world-toilet-day2019.html.

174. C-Circle (2020, November 22). World Toilet Day: Open Defecation. Retrieved from: https://c-circle.org/blog/2020/11/22/world-toilet-day-open-defecation/.

175. Mitic, I. (2022, January 6). 20 Insightful Pet Spending Statistics: Americans are Spending More on Pets than Ever. *Fortunly*. Retrieved from: https://fortunly.com/statistics/pet-spending-statistics/.

176. U.S. Bureau of Labor Statistics (2021, September 9). Consumer Expenditures — 2020. Retrieved from: https://www.bls.gov/news.release/cesan.nr0.htm.

177. United Nations (2014, November 19). Every dollar invested in water, sanitation brings four-fold return in costs. *UN News*. Retrieved from: https://news.un.org/en/story/2014/11/484032-every-dollar-invested-water-sanitation-brings-four-fold-return-costs-un.

178. WaterAid (n.d.). Mission Critical: Invest in water, sanitation and hygiene for a healthy and green economic recovery. Retrieved from: https://washmatters.wateraid.org/sites/g/files/jkxoof256/files/mission-critical-invest-in-water-sanitation-and-hygiene-for-a-healthy-and-green-economic-recovery_2.pdf.

179. UNICEF and World Health Organization (2019). Water, Sanitation and Hygiene in Health Care Facilities: Global baseline report 2019. Retrieved from: https://data.unicef.org/resources/wash-in-health-care-facilities/.

180. Steele, R., et al. (2018). The Crisis in the Classroom, The State of the World's Toilets 2018. *WaterAid* report. Retrieved from: https://washmatters.wateraid.org/publications/the-crisis-in-the-classroom-the-state-of-the-worlds-toilets-2018.

181. Brink, S. (2016, January 20). What Happens to the Body and Mind When Starvation Sets In? *WAMU 88.5 — American University Radio*. Retrieved from: https://wamu.org/story/16/01/20/what_happens_to_the_body_and_mind_when_starvation_sets_in/.

182. Morrow, A. (2022, July 28). How Long Can You Live Without Food? *Verywellhealth*. Retrieved from: https://www.verywellhealth.com/the-decision-to-stop-eating-at-the-end-of-life-1132033.

183. Burch, K. (2021, September 3). How many days a person can survive without food and water. *Insider Health*. Retrieved from: https://www.insider.com/how-long-can-you-go-without-food.

184. Mona, B. (2019). Kcal vs. Calories: What's the Diff? *Greatist*. Retrieved from: https://greatist.com/health/kcal-vs-calories.

185. Roser, M. and Ritchie, H. (2019). Hunger and Undernourishment. *Our World in Data.* Retrieved from: https://ourworldindata.org/hunger-and-undernourishment#severe-food-insecurity.

186. International Food Policy Research Institute (n.d.). Food Security. Retrieved from: https://www.ifpri.org/topic/food-security.

187. USDA (2022). Definitions of Food Security. *Economic Research Service.* Retrieved from: https://www.ers.usda.gov/topics/food-nutrition-assistance/food-security-in-the-u-s/definitions-of-food-security/.

188. United Nations Department of Economic and Social Affairs (n.d.). Goals 2: End hunger, achieve food security and improved nutrition and promote sustainable agriculture. Retrieved from: https://sdgs.un.org/goals/goal2.

189. World Health Organization (n.d.). Prevalence of anaemia in women aged 15 to 49 years of age, by pregnancy status (%). Retrieved from: https://www.who.int/data/gho/indicator-metadata-registry/imr-details/4552.

190. United Nations Food and Agricultural Organization (2018). Food Insecurity Experience Scale (FIES). Retrieved from: https://www.fao.org/policy-support/tools-and-publications/resources-details/en/c/1236494/.

191. Global Hunger Index. Retrieved from: https://en.wikipedia.org/wiki/Global_Hunger_Index.

192. von Grebmer, K., Bernstein, J., Wiemers, M., Schiffer, T., et al. (2021). *Global Hunger Index: Global Hunger and Food Systems in Conflict Settings.* Bonn/Dublin, Germany: Welthungerhilfe/Concern Worldwide. Retrieved from: https://www.globalhungerindex.org/pdf/en/2021.pdf.

193. Williams, C. D. (1983). Fifty years ago. Archives of Diseases in Childhood 1933. A nutritional disease of childhood associated with a maize diet. Reprint from *Archives of Disease in Childhood* 1933; 8: 423–433. *Archives of Disease in Childhood.* *58*(7): 550–560. DOI: 10.1136/adc.58.7.550. PMC1628206.PMID 6347092.

194. Kwashiorkor. Retrieved from: https://en.wikipedia.org/wiki/Kwashiorkor.

195. United Nations Food and Agricultural Organization (2021). Voices of the Hungry: Bringing Experienced-Based Food Insecurity Measurement to the Global Level. Retrieved from: https://www.fao.org/in-action/voices-of-the-hungry/background/en/.

196. Sen, A. (1983). *Poverty and Famines: An Essay on Entitlement and Deprivation.* Oxford Academic. Retrieved from: https://oxford.universitypressscholarship.com/view/10.1093/0198284632.001.0001/acprof-9780198284635.

197. Kendall, A., Olson, C. M., and Frongillo, E. A. Jr. (1996). Relationship of Hunger and Food Insecurity to Food Availability and Consumption. *Journal of the American Dietetic Association.* 96(10): 1,019–1,024; quiz 1025-6. DOI: 10.1016/S0002-8223(96)00271-4.

198. United States Department of Agriculture (2022). Food Security in the U.S. History & Background. *Economic Research Service.* Retrieved from: https://www.ers.usda.gov/topics/food-nutrition-assistance/food-security-in-the-u-s/history-background/#history.

199. Bickel, G., Nord, M., Price, C., et al. (2000). *USDA Guide to Measuring Household Food Security.* Revised 2000. Retrieved from: https://fns-prod.azureedge.us/sites/default/files/FSGuide.pdf.

200. United Nations Food and Agricultural Organization (n.d.). Global Food Insecurity Experience Scale Survey Modules. *Voices of the Hungry.* Retrieved from: https://www.fao.org/3/bl404e/bl404e.pdf.

201. United Nations Food and Agricultural Organization (n.d.). The Food Insecurity Experience Scale: Measuring food insecurity through people's experience (Report i7835EN/1/09.17). *Voices of the Hungry.* Retrieved from: https://www.fao.org/3/i7835e/i7835e.pdf.

202. United Nations Food and Agricultural Organization (2022). The Food Insecurity Experience Scale: 8 Key Questions. *Voices of Hungry.* Retrieved from: https://www.fao.org/in-action/voices-of-the-hungry/fies/zh/.

203. Boero, V., Cafiero, C., Gheri, F., Kepple, A. W., et al. (2021). *Access to Food in 2020, Results of 20 National Surveys Using the Food Insecurity Experience Scale (FIES).* Rome, Italy: Food and Agricultural Organization. Retrieved from: https://doi.org/10.4060/cb5623en.

204. United Nations Food and Agricultural Organization (2020). *2.0 Hunger: Using the FIES App: A Simple Tool for the Analysis of Food Insecurity Experience Scale data.* Rome, Italy: Food and Agricultural Organization. Retrieved from: http://www.fao.org/3/ca9318en/ca9318en.pdf.

205. Erdenesan, E. (2020). Online tool for FIES Data Validation and Indicator Compilation: Experience of Mongolia. *28th Session of Asia and Pacific Commission on Agricultural Statistics.* Retrieved from: http://www.fao.org/3/cb3333en/cb3333en.pdf.

206. United Nations Food and Agricultural Organization (2018). SDG Indicator 2.1.2 — Using the Food Insecurity Experience Scale (FIES). Retrieved from: https://elearning.fao.org/course/view.php?id=360.

207. FAO, IFAD, UNICEF, WFP, WHO (2017). *The State of Food Security and Nutrition in the World 2017: Building Resilience for Peace and Food Security.* Rome, Italy: Food and Agricultural Organization. Retrieved from: https://www.fao.org/3/I7695e/I7695e.pdf.

208. Congressional Research Service (2020). U.S. Periods of War and Dates of Recent Conflicts. Retrieved from: https://sgp.fas.org/crs/natsec/RS21405.pdf.

209. United Nations (n.d.). Peace, Dignity and Equality on a Healthy Planet: Global Issue Food. Retrieved from: https://www.un.org/en/global-issues/food.

210. United Nations Department of Economics and Social Affairs (n.d.). Sustainable Development, Goal 11. Retrieved from: https://sdgs.un.org/goals/goal11.

211. Shulla, K. and Köszeghy, L. (2021). Progress Report: Sustainable Development Goal 11: Target 11.1. *Habitat for Humanity International®*. Retrieved from: https://www.habitat.org/sites/default/files/documents/SDG%20Progress%20Report_0.pdf and https://www.habitat.org/sites/default/files/documents/Solid-Ground-SDG_booklet-update-2021.pdf.

212. UN-Habitat (2018). Metadata on SDGs Indicator 11.1.1. Retrieved from: https://unhabitat.org/sites/default/files/2020/06/metadata_on_sdg_indicator_11.1.1.pdf.

213. United Nations (2018). Global Indicator Framework for the Sustainable Development Goals and Targets of the 2030 Agenda for Sustainable Development (A/RES/71/313 E/CN.3/2018/2). Retrieved from: https://unstats.un.org/sdgs/indicators/Global%20Indicator%20Framework%20after%20refinement_Eng.pdf.

214. Ortiz-Ospina, E. and Roser, M. (2017). Homelessness. *Our World in Data.* Retrieved from: https://ourworldindata.org/homelessness.

215. Law Insider (n.d.). Informal Settlement Definition. Retrieved from: https://www.lawinsider.com/dictionary/informal-settlement.

216. Wilson, V. (2022). What Is Cultural Identity and Why Is It Important? *Exceptional Futures.* Retrieved from: https://www.exceptionalfutures.com/cultural-identity/.

217. Serai, Y. (2018). What Is Cultural Identity? *Classroom.* Retrieved from: https://classroom.synonym.com/what-is-culture-identity-12082328.html.

218. Cambridge English Dictionary (n.d.). Enjoyment. Retrieved from: https://dictionary.cambridge.org/us/dictionary/english/enjoyment.

219. Happiness Economics. Retrieved from: https://en.wikipedia.org/wiki/Happiness_economics.

220. Lyubomirsky, S. and Lepper, H. S. (1999). A Measure of Subjective Happiness: Preliminary Reliability and Construct Validation. *Social Indicators Research.* 146, 137–155. DOI: https://doi.org/10.1023/A:1006824100041.

221. United Nations (2011, July 19). Happiness should have a great role in development policy — UN member States. *UN News.* Retrieved from: https://news.un.org/en/story/2011/07/382052.

222. Helliwell, J. F., Layard, R., Sachs, J. D., et al. (2022). *World Happiness Report.* Retrieved from: https://happiness-report.s3.amazonaws.com/2022/WHR+22.pdf.

223. UN-Habitat (2019). The Global Housing Affordability Challenge: A more comprehensive understanding of the housing sector. *Urban Data Digest.*

Retrieved from: https://unhabitat.org/sites/default/files/2020/06/urban_data_digest_the_global_housing_affordability_challenge.pdf.

224. Ritchie, H. and Roser, M. (2019). Urbanization. *Our World in Data.* Retrieved from: https://ourworldindata.org/urbanization.

225. Ritchie, H. and Roser, M. (2019). What share of people will live in urban areas in the future? *Our World in Data.* Retrieved from: https://ourworldindata.org/urbanization#what-share-of-people-will-live-in-urban-areas-in-the-future.

226. UNDESA (2018). Population Division. World Urbanization Prospects: The 2018 Revision. Retrieved from: https://population.un.org/wup/Download/.

227. Rai, S. S. (2017, April 22). Global Explosion of Slums: The Next Biggest Planetary Health Challenge. *Amplify.* Retrieved from: https://medium.com/amplify/global-explosion-of-slums-the-next-biggest-planetary-health-challenge-49424f27ba16.

228. UN-Habitat (2020). World Cities Report 2020: The Value of Sustainable Urbanization. *UN-Habitat.* Retrieved from: https://unhabitat.org/sites/default/files/2020/10/wcr_2020_report.pdf.

229. Kuo, G. (2019, September 3). Yet Another Emerging Global Crisis — Homelessness. The Millennium Alliance for Humanity and the Biosphere (MAHB). Retrieved from: https://mahb.stanford.edu/library-item/yet-another-emerging-global-crisis-homelessness/.

230. Ruff Institute of Global Homelessness (2019). *What is IGH?* Retrieved from: https://ighomelessness.org/about-us/.

231. Cornell Law School (n.d.). 42 U.S. Code § 11302 — General Definition of Homeless Individual. *Legal Information Institute.* Retrieved from: https://www.law.cornell.edu/uscode/text/42/11302.

232. United States Code (n.d.). General Definition of Homeless Individual. Retrieved from: https://www.govinfo.gov/content/pkg/USCODE-2010-title42/pdf/USCODE-2010-title42-chap119-subchapI-sec11302.pdf.

233. Henry, M., Mahathey, A., Takashima, M., et al. (2020). *The 2018 Annual Homeless Assessment Report to Congress Part 2: Estimates of homelessness in the United States.* Retrieved from: https://www.huduser.gov/portal/sites/default/files/pdf/2018-AHAR-Part-2.pdf.

234. Los Angeles Homeless Services Authority (n.d.). 2020 Homeless Count by Supervisorial District. Retrieved from: https://www.lahsa.org/data?id=43-2020-homeless-count-by-supervisorial-district.

235. Los Angeles Almanac™ (n.d.). Homelessness in Los Angeles County 2020. Retrieved from: http://www.laalmanac.com/social/so14.php.

236. Los Angeles County, California (n.d.). Retrieved from: https://en.wikipedia.org/wiki/Los_Angeles_County,_California.

237. Los Angeles Sustainable Development Goals (n.d.). City of Los Angeles data for Sustainable Development Goal Indicators. Retrieved from: https://sdgdata.lamayor.org/.

238. Los Angeles Sustainable Development Goals (n.d.). Data Reporting Platform. 2022. Retrieved from: https://sdg.lamayor.org/our-work/data-reporting-platform.

239. Los Angeles Sustainable Development Goals (2021). 2021 Voluntary Local Review of Progress Towards the Sustainable Development Goals in Los Angeles. Retrieved from: https://sdg.lamayor.org/2021VLR.

240. National Alliance to End Homelessness (2022). State of Homelessness: 2021 Edition. Retrieved from: https://endhomelessness.org/homelessness-in-america/homelessness-statistics/state-of-homelessness-2021/.

241. National Law Center on Homelessness and Poverty (2015). Homelessness in America: Overview of Data and Causes. Retrieved from: https://homelesslaw.org/wp-content/uploads/2018/10/Homeless_Stats_Fact_Sheet.pdf.

242. U.S. Department of Housing and Urban Development (2018). The 2018 Annual Homeless Assessment Report (AHAR) to Congress: Part 1: Point-in-Time Estimates of Homelessness. Retrieved from: https://www.huduser.gov/portal/sites/default/files/pdf/2018-AHAR-Part-1.pdf.

243. UN (2018). Report of the Special Rapporteur on extreme poverty and human rights on his mission to the United States of America (A/HRC/38/33/Add.1). *Human Rights Council.* Retrieved from: https://documents-dds-ny.un.org/doc/UNDOC/GEN/G18/125/30/PDF/G1812530.pdf?OpenElement.

244. Housing Forward Virginia (n.d.). Affordable Housing 101: Who Are the Major Players and What are Their Roles? Retrieved from: https://housingforwardva.org/toolkits/affordable-housing-101/who-are-the-major-players-and-what-are-their-roles/.

245. *Around the House: Federal, State and Local Government Actions Affect Affordable Housing* (2021, December 11). Daily Commercial. Retrieved from: https://www.dailycommercial.com/story/lifestyle/2021/12/11/federal-state-and-local-government-actions-affect-affordable-housing/6425995001.

246. Arnold, C. (2021, November 6). Democrats are Seeking Largest Ever Investment in Affordable Housing. *NPR.* Retrieved from: https://www.npr.org/2021/11/06/1052876271/democrats-are-seeking-largest-ever-investment-in-affordable-housing.

247. Everett-Allen, K. and Harvey, M. (2020). Global House Price Index. *Knight Frank Research Reports.* Retrieved from: https://content.knightfrank.com/research/84/documents/en/global-house-price-index-q4-2020-7884.pdf.

248. Bucholz, K. (2022, March 21). Where Is the Hardest Place to Afford a Home? *Statista*. Retrieved from: https://www.statista.com/chart/16902/places-where-its-hardest-to-afford-a-home/.

249. Schuetz, J. (2019, October 15). How Can Government Make Housing More Affordable? Brookings Policy 2020. Retrieved from: https://www.brookings.edu/policy2020/votervital/how-can-government-make-housing-more-affordable/.

250. Smil, V. (2016). Examining Energy Transitions: A Dozen Insights Based on Performance. *Energy Research & Social Science.* 22: 194–197. DOI: https://dx.doi.org/10.1016/j.erss.2016.2016.08.017.

251. United Nations Environmental Programme (n.d.). GOAL 7: Affordable and Clean Energy. Retrieved from: https://www.unep.org/explore-topics/sustainable-development-goals/why-do-sustainable-development-goals-matter/goal-7.

252. Smil, V. (2018). *Energy and Civilization: A History*. Reprint edition. Cambridge, Mass: MIT Press; p. 562.

253. Rashed, R. (1990). A Pioneer in Anaclastics: Ibn Sahl on Burning Mirrors and Lenses. *Isis — A Journal of the History of Science Society.* 81(3): 464–491. DOI: https://doi.org/10.1086/355456. Retrieved from: https://www.journals.uchicago.edu/doi/abs/10.1086/355456?journalCode=isis.

254. Lund, J. W. (2018). Geothermal Energy. *Encyclopedia Britannica*. Retrieved from: https://www.britannica.com/science/geothermal-energy.

255. World Bank (2018). Nearly Half the World Lives on Less than $5.50 a Day. *World Bank Press Release.* Retrieved from: https://www.worldbank.org/en/news/press-release/2018/10/17/nearly-half-the-world-lives-on-less-than-550-a-day.

256. Herrling, S. and Moss, T. (2019). Scaling Power for Global Prosperity. *Stanford Social Innovation Review: Informing and Inspiring Leaders of Social Change.* 17(2), 59–60, DOI: https://doi.org/10.48558/YJFW-2K53. Retrieved from: https://ssir.org/articles/entry/scaling_power_for_global_prosperity.

257. IEA (2019). World Energy Outlook 2019. Retrieved from: https://www.iea.org/reports/world-energy-outlook-2019 and https://iea.blob.core.windows.net/assets/98909c1b-aabc-4797-9926-35307b418cdb/WEO2019-free.pdf.

258. Clemente, J. (2020, June 14). The World's Biggest Energy Problem: Access. *Forbes.* Retrieved from: https://www.forbes.com/sites/judeclemente/2020/06/14/the-biggest-energy-problem-in-the-world/?sh=107c9b8f4e29.

259. Wilson, L. (2022). Average Household Electricity Use Around the World. *Shrink That Footprint Research Group.* Retrieved from: http://shrinkthatfootprint.com/average-household-electricity-consumption.

260. Canada Energy Regulator (n.d.). Market Snapshot: Fuel Poverty across Canada — Lower Energy Efficiency in Lower Income Households.

Retrieved from: https://www.cer-rec.gc.ca/en/data-analysis/energy-markets/market-snapshots/2017/market-snapshot-fuel-poverty-across-canada-lower-energy-efficiency-in-lower-income-households.html.

261. Riva, M., et al. (2021). Energy poverty in Canada: Prevalence, Social and Spatial Distribution, and Implications for Research and Policy. *Energy and Social Science. 81*(102237). DOI: https://doi.org/10.1016/j.erss.2021.102237.

262. U.S. Energy Information Administration (2018). One in Three U.S. Households Faces a Challenge in Meeting Energy Needs. Retrieved from: https://www.eia.gov/todayinenergy/detail.php?id=37072.

263. American Council for an Energy-Efficiency Economy (2019). Understanding Energy Affordability. Retrieved from: https://www.aceee.org/sites/default/files/energy-affordability.pdf.

264. Drehobl, A., Ross, L., and Ayala, R. (2020). *How High are Household Energy Burdens? An Assessment of National and Metropolitan Energy Burden across the United States.* American Council for an Energy-Efficient Economy. Retrieved from: https://www.aceee.org/sites/default/files/pdfs/u2006.pdf.

265. USDOE (n.d.). Low-Income Energy Affordability Data (LEAD) Tool. *Office of Energy Efficiency and Renewable Energy.* Retrieved from: https://www.energy.gov/eere/slsc/maps/lead-tool.

266. U.S. Energy Information Administration (2018). Residential Energy Consumption Survey — Comparing the 2015 RECS with Previous RECS and Other Studies. Retrieved from: https://www.eia.gov/consumption/residential/reports/2015/comparison/.

267. USDOE (n.d.). Low-Income Community Energy Solutions. *Office of Energy Efficiency and Renewable Energy.* Retrieved from: https://www.energy.gov/eere/slsc/low-income-community-energy-solutions.

268. US Department of Health & Human Services (2022). HHS Poverty Guidelines for 2022. Retrieved from: https://aspe.hhs.gov/poverty-guidelines.

269. Murkowski, L. and Scott, T. (2014). *Plenty at Stake: Indicators of American Energy Insecurity.* 113th Congress, An *Energy 20/20* White Paper. Retrieved from: https://www.energy.senate.gov/public/index.cfm/files/serve?File_id=075f393e-3789-4ffe-ab76-025976ef4954.

270. Shrider, E. A. Kollar, M. Chen, F. Semega, J. (2021). *Income and Poverty in the United States: 2020 — Current Population Reports.* U.S. Census Bureau. Washington, D.C.: U.S. Government Publishing Office. Retrieved from: https://www.census.gov/content/dam/Census/library/publications/2021/demo/p60-273.pdf.

271. USEIA. One in three U.S. households faces a challenge in meeting energy needs. Last accessed April 2022. https://www.eia.gov/todayinenergy/detail.php?id=37072.

272. World Population Review. (n.d.). Greenhouse Gas Emissions by Country 2022. Retrieved from: https://worldpopulationreview.com/country-rankings/greenhouse-gas-emissions-by-country.
273. USEIA (2019). International Energy Outlook 2019. Retrieved from: https://www.eia.gov/outlooks/ieo/pdf/ieo2019.pdf.
274. Jacobson, M. Z. et al. (2017). 100% Clean and Renewable Wind, Water and Sunlight All Sector Energy Roadmaps for 139 Countries of the World. *Joule* 1, pp. 108–121. Retrieved from: https://web.stanford.edu/group/efmh/jacobson/Articles/I/CountriesWWS.pdf.
275. Jacobson, M. Z., Delucchi, M. A., Bazouin, G., Bauer, Z. A. F., et al. (2015). 100% clean and renewable wind, water, and sunlight (WWS) all-sector energy roadmaps for the 50 United States. *Energy & Environmental Science.* 8: 2,093–2,117. DOI: https://doi.org/10.1039/C5EE01283J.
276. NRCan (2019). Canadian Small Modular Reactor (SMR) Roadmap. Retrieved from: https://www.nrcan.gc.ca/our-natural-resources/energy-sources-distribution/nuclear-energy-uranium/canadian-small-modular-reactor-roadmap/21183.
277. UK planning for rapid nuclear expansion (2022, April 7). *World Nuclear News.* Retrieved from: https://www.world-nuclear-news.org/Articles/UK-planning-for-rapid-nuclear-expansion.
278. Macron sets out plan for French nuclear renaissance (2022, February 11). *World Nuclear News.* Retrieved from: https://www.world-nuclear-news.org/Articles/Macron-announces-French-nuclear-renaissance.
279. European Community (n.d.). Energy Biomass. Retrieved from: https://energy.ec.europa.eu/topics/renewable-energy/bioenergy/biomass_en.
280. Mitchell, S. R., Harmon, M. E., and O'Connell, K. E. B. (2012). Carbon debt and carbon sequestration parity in forest bioenergy production. *GCB-Bioenergy — Bioproducts for a Sustainable Bioeconomy,* 4: 818–827. DOI: https://doi.org/10.1111/j.1757-1707.2012.01173.x. Retrieved from: https://onlinelibrary.wiley.com/doi/full/10.1111/j.1757-1707.2012.01173.x.
281. Holtsmark, B. (2010). Use of wood fuels from boreal forests will create a biofuel carbon debt with a long payback time. *Statistics Norway.*
282. IEA (2020). Energy Technology Perspectives 2020: Special Report on Carbon Capture Utilization and Storage. CCUS in clean energy transitions. Retrieved from: https://iea.blob.core.windows.net/assets/181b48b4-323f-454d-96fb-0bb1889d96a9/CCUS_in_clean_energy_transitions.pdf.
283. Koh, W. (2022). CCUS Technologies — Can they mitigate climate change? Environmental, Social & Governance (ESG) Risk Briefing. *Allianz.* Retrieved from: https://www.agcs.allianz.com/news-and-insights/expert-risk-articles/ccus-technologies.html.
284. Renewable Energy Transition. Retrieved from: https://en.wikipedia.org/wiki/Renewable_energy_transition.

285. Hayes, A. (2022, June 6). Energy Return on Investment (EROI). *Investopedia.* Retrieved from: https://www.investopedia.com/terms/e/energy-return-on-investment.asp.

286. World Nuclear Association (2020). Energy Return on Investment. Retrieved from: https://world-nuclear.org/information-library/energy-and-the-environment/energy-return-on-investment.aspx.

287. Raugei, M., Carbajales-Dale, M., Barnhard, C., and Fthenakis, V. (2015). Rebuttal: "Comments on 'Energy intensities, EROIs (energy returned on invested), and energy payback times of electricity generating power plants' — Making clear of quite some confusion". *Energy.* 82: 1,088–1,091. DOI: 10.1016/j.energy.2014.12.060.

288. Weißbach, D., Ruprecht, G., Huke, A., Czerski, K., et al. (2013). Energy intensities, EROIs (energy returned on invested), and energy payback times of electricity generating power plants. *Energy.* 52: 210–221. DOI: https://doi.org.1016/j.energy.2013.2013.01.029.

289. USDOE, Office of Indian Energy (2013). Levelized Cost of Energy (LCOE). Retrieved from: https://www.energy.gov/sites/prod/files/2015/08/f25/LCOE.pdf.

290. National Renewable Energy Laboratory (2013). CREST: Cost of Renewable Energy Spreadsheet Tool: A Model for Developing Cost-Based Incentives in the United States — User Manual Version 4. Subcontract Report NREL/SR-6A20-50374. Retrieved from: https://www.nrel.gov/docs/fy13osti/50374.pdf.

291. Lazard® (2021). Lazard's Cost of Energy Analysis — Version 15. Retrieved from: https://www.lazard.com/media/451881/lazards-levelized-cost-of-energy-version-150-vf.pdf.

292. Nuccitelli, D. (2018, October 26). Canada Passed a Carbon Tax that will give most Canadians more Money. *The Guardian.* Retrieved from: https://www.theguardian.com/environment/climate-consensus-97-per-cent/2018/oct/26/canada-passed-a-carbon-tax-that-will-give-most-canadians-more-money.

293. Giroux, Y. (2022). A Distributional Analysis of Federal Carbon Pricing Under a Healthy Environment and a Healthy Economy (Report RP-2122-032-S_e). Ottawa, Canada: Office of the Parliamentary Budget Officer. Retrieved from: https://distribution-a617274656661637473.pbo-dpb.ca/6399abff7887b53208a1e97cfb397801ea9f4e729c15dfb85998d1eb359ea5c7.

294. Goldstein, L. (2022, March 26). Trudeau government lowballed cost of carbon tax. *Toronto Sun.* Retrieved from: https://torontosun.com/opinion/columnists/goldstein-trudeau-government-lowballed-cost-of-carbon-tax.

295. World Bank (2021). *State and Trends of Carbon Pricing 2021.* Washington, D.C.: World Bank, DOI: 10.1596/978-1-4648-1728-1. Retrieved from: https://openknowledge.worldbank.org/handle/10986/35620.

296. UK Government (2021). £4 million funding to boost UK biomass production. *Department of Business, Energy & Industrial Strategy.* Retrieved from: https://www.gov.uk/government/news/4-million-funding-to-boost-uk-biomass-production.
297. IRENA (2021). Renewable Power Generation Costs in 2020. *International Renewable Energy Agency.* Retrieved from: https://www.irena.org/publications/2021/Jun/Renewable-Power-Costs-in-2020.
298. USDOE, NREL (n.d.). Life Cycle Assessment Harmonization. Retrieved from: https://www.nrel.gov/analysis/life-cycle-assessment.html.
299. USEIA (2022). Levelized Cost and Levelized Avoided Cost of New Generation. Retrieved from: https://www.eia.gov/outlooks/aeo/pdf/electricity_generation.pdf.
300. Lazard® (2020). Levelized Cost of Energy, Levelized Cost of Storage and Levelized Cost of Hydrogen 2020. Retrieved from: https://www.lazard.com/perspective/lcoe2020.
301. Vorisek, D. and Yu, S. (2020). Understanding the Cost of Achieving the Sustainable Development Goals. World Bank Policy Research Working Paper No. 9146. *World Bank Group.* Retrieved from: https://documents1.worldbank.org/curated/en/744701582827333101/pdf/Understanding-the-Cost-of-Achieving-the-Sustainable-Development-Goals.pdf.
302. World Bank (n.d.). GDP (current US$). Retrieved from: https://data.worldbank.org/indicator/NY.GDP.MKTP.CD.

Chapter 2

Exploring the Link Between Irrigated Agriculture and Food Security: Evidence from the Case of Spain

Alfonso Expósito

Department of Applied Economics, University of Malaga, Ejido 6, 29071 Malaga, Spain

aexposito@uma.es

Abstract

Irrigation plays a fundamental role in guaranteeing food security. Increasing food production to meet the growing future demands of the world population will require not only the expansion of the irrigated area (i.e., through the conversion of rainfed land) and better water availability (both in space and time), but also greater efficiency in the use of resources. Both drivers of growth require complex institutional and policy reforms, especially in those regions of the world with water scarcity problems. In the particular case of Spain, the growth of agricultural production as a guarantor of food security appears to lack the support of an increase in resources (i.e., land and water), but not that of an increase in agricultural productivity. The analysis carried out in this chapter has shown that the achievement of a higher production per unit of productive

116 of A. Expósito

factor used (especially per unit of water used) has been decisive in maintaining the production of the Spanish agricultural sector.

Keywords: Irrigated agriculture, food security, productivity, Spain.

2.1. Introduction

When discussing the concept of food security, we usually ask ourselves the following question: Will we be able to produce sufficient food to meet the future needs of a constantly growing population in a context of climate change? The World Food Summit defined 'food security' as "when all people at all times have access to sufficient, safe, and nutritious food to maintain a healthy and active life."[1] This means that a quantitatively and qualitatively adequate supply of food must be maintained in a sustained manner over time, and that all members of society have ensured access to this food. In a global context of uncertainty due to factors such as climate change, exponential population growth, and variability in agricultural prices, a major challenge for agriculture involves the provision of a sustainable and secure supply of food. The world population is forecast to grow to more than nine billion by 2050, and demand for food will increase by 50% by 2030 and by 80% to 100% by 2050.[2] In this context, greater world food security will be impossible if the use of agricultural resources is unsustainable, especially if it is necessary to do so using fewer resources, such as farmland and water.[3] The phenomenon of climate change will affect the scope and productivity of irrigated and rainfed agriculture around the world, thereby increasing the demand for irrigation water and decreasing the productivity of crops in many regions. Furthermore, more than half of the world's population is forecast to live in water-scarce regions by 2050, which will exacerbate the agricultural sector's ability to supply food.[4]

In this challenging context, irrigation plays a major role in guaranteeing food security and in increasing agricultural production across the globe.[5] Worldwide, more than 40% of total food production comes from irrigated agriculture.[6] Numerous empirical studies have shown that irrigation exerts a positive impact both on food security and on reducing poverty across the world.[7–11] Given the fact that expansion of cultivated land will become limited in many regions of the world, the question arises as to whether increased productivity (i.e., yield per cultivated hectare) will be able to meet future demands for food. Agricultural yields have increased steadily in most of the world's major agricultural regions thanks to the conversion from rainfed to irrigated land, thus enhancing the ability to provide global food security.[12]

In the case of Spain, a more productive and competitive agriculture has been achieved, thanks to the development of irrigation. In fact, Spain has the largest irrigated area in the European Union (EU), with 3.63 million hectares irrigated in 2015, which represents 15% of the useful agricultural area and approximately 67% of the final crop production.[a] Furthermore, this sector consumes approximately 70% of the total water resources and uses 50% of the water stored in Spanish reservoirs in a regular year.[13] As can be observed throughout this chapter, the Spanish irrigation sector has undergone intense modernization with the aim of increasing the efficiency of use of the productive resources, especially that of water. The irrigated area that has been modernized in recent decades has reached the figure of 1.47 million hectares, with an investment close to 3,000 million euros, which has meant an estimated saving of almost 3,100 cubic hectometers per year. All this has increased the productivity obtained per hectare and per cubic meter of water used. Productivity per irrigated hectare is therefore six times higher than that in rainfed agriculture, and obtains an income four times higher than in rainfed agriculture.

The answer to the initial question posed, "Will we be able to produce sufficient food to meet the future needs of a population in constant growth and in a context of climate change?" can be approached using any of a variety of strategies. These will be analyzed in the following sections of this chapter, with a special analysis of the specific case of Spain. The first way to raise food production from irrigated land would be based on increasing available resources, that is, the supply of water and land above current levels. The second strategy would be focused on increasing irrigation productivity, either by improving yield or by improving the efficiency of water and/or land use.

2.2. Increase of Resource Availability

Between 1950 and 2000, the world's population grew more than during the previous four million years, from 2.5 billion to six billion inhabitants.

---

[a] 1 hectare = 2.47 acres.

In economic terms, the world economy has observed a remarkable growth. During the last half of the 20th century, the world economy expanded sevenfold. The demands from economic growth now exceed the natural capacity of the planet. In fact, Earth's natural life support systems have remained essentially the same, while demands on resources have risen dramatically to sustain economic growth. Thus, the use of water tripled, but the capacity of the hydrological system to produce fresh water by evaporation changed little.[14] The world population is projected to reach 9.8 billion in 2050,[15] with a per capita income that will double by that year. This higher income will also result in higher calorie intake and dietary changes with an increase in the proportion of meat and dairy consumed, especially in developing countries.[5] This combination of higher intake and changes in diet will generate an increase in the demand for primary food production and, therefore, for water resources.[16] In this context, the development of new water sources and/or the reallocation of water from other sectors to increase food production suffers from limited potential in many regions of the world.

In the case of Spain, final crop production (FCP) has undergone a very positive evolution in the last 50 years — a period in which the expansion of irrigation has played a highly significant role in increasing crop production. Figure 2.1 shows this evolution for the period of 1964–2016,

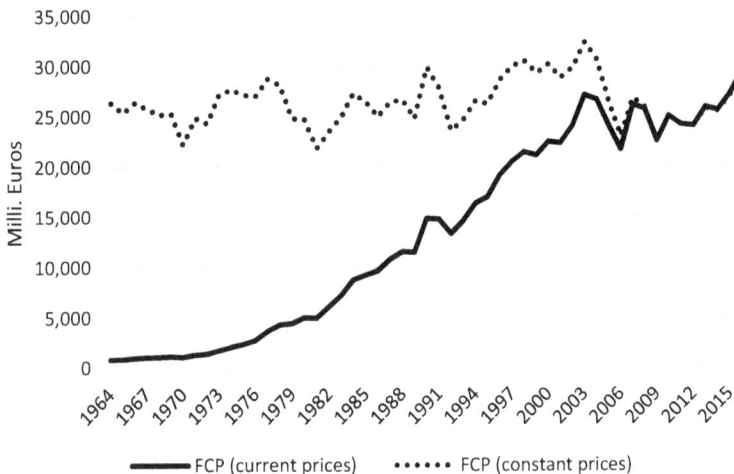

Figure 2.1.   Final crop production in Spain (in millions of euros).

Source: Spanish Ministry of Agriculture.

both at current prices (solid line) and at constant prices (dotted line), that is, without the effect of price increases. Currently, the FCP stands at 29 billion euros, which is more than 34 times that of the production recorded during the 1960s. The FCP represents 60% of the total agricultural final production, and its participation has not changed significantly in the last 50 years. As shown in the figure, this growth in FCP has been negatively affected by the impacts of periods of drought, especially in the 1990s and between 2003 and 2006. This reduction of the FCP can be more clearly observed in the series at constant prices (dotted line) since it eliminates the effect of increase in agricultural prices during periods of drought. Following the financial crisis in 2008, the FCP stagnated again, although it has been rising in recent years.

On a global scale, the expansion of cultivated land explains 20% of the increase in food production in recent decades. However, its growth potential has been restrained by urbanization expansion and land degradation.[17] In most cases, the increase in irrigated land has come at the cost of converting rainfed land into irrigation, although it is worth noting that irrigation and rainfed agriculture are complementary and not mutually exclusive.[18] Due to this conversion process, the share of rainfed production in the world food supply is expected to decline from 65% (2010) to 48% (2050).[16] The Food and Agriculture Organization (FAO) of the United Nations has made several projections of the demand and supply of irrigation water for 2025 and 2030. In developing countries in particular, it is expected that the area equipped for irrigation will have expanded by 20% (40 million hectares) by 2030. This suggests that 20% of the total land with irrigation potential but are not yet operational will be irrigated, and that 60% of all land with potential of irrigation (402 million hectares) will be in use by 2030. Thus, the net increase in irrigated land (40 million hectares) will be concentrated in the group of developing countries, where population growth is higher.[1]

In the EU, the relevance of irrigation in agriculture varies depending on the agronomic and climatic characteristics of each country. Most of the irrigated areas in the EU are concentrated in the Mediterranean region, where France, Greece, Italy, Portugal, and Spain represent 12 million hectares, corresponding to 75% of the total area equipped for irrigation in the entire EU-27. The irrigated area in Spain has registered a clearly increasing trend in the last 50 years. Figure 2.2 shows the evolution of the cultivated agricultural area in Spain, as well as its division between irrigated and rainfed land for the period of 1964–2016. As observed, the total

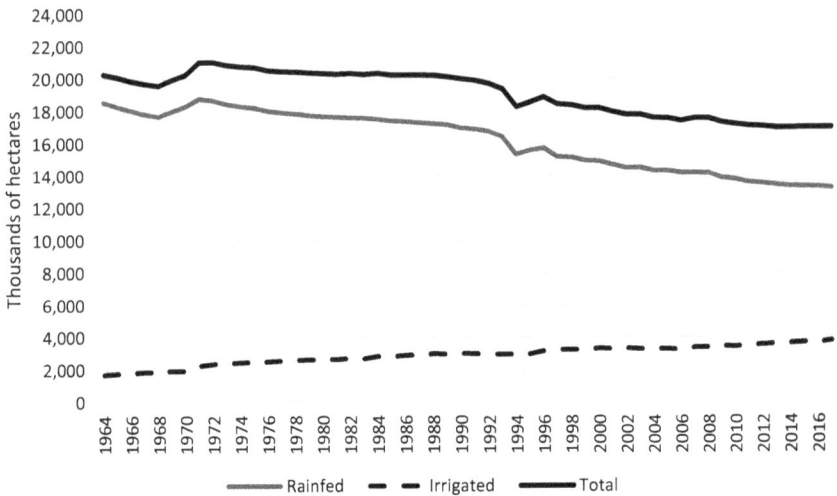

Figure 2.2.   Cultivated land in Spain (thousands of hectares).

Source: Spanish Ministry of Agriculture.

cultivated area has undergone a clear trend of reduction during the period analyzed, with the exception of the 1960s. A similar trend is observed in the cultivated area of rainfed land, which has been partly converted into irrigated land. Currently, the irrigated area remains stabilized at approximately 3.7 million hectares, although growth from 1.7 million hectares in 1964 has been highly significant. However, the growth of irrigated land in Spain has been insufficient to compensate for the loss of total cultivated land as a result of rural depopulation due to low profitability of crops, especially in the case of rainfed land in the central area of the country. This loss of cultivated land has been partly absorbed by urban expansion, forestry, and other uses.

As previously mentioned, food security in Spain is closely related to irrigation activity and availability of water. As shown in Figure 2.3, the water abstracted for irrigation in Spain has increased from 10,548 cubic hectometers in 1964 to the current 14,900 cubic hectometers per year, and reached a maximum of 17,800 cubic hectometers in 2004.[b] As of this year, the water abstracted for irrigation has experienced a decline and

[b]$1 \text{ Hm}^3 = 1 \text{ GL} = 810.71$ acre feet.

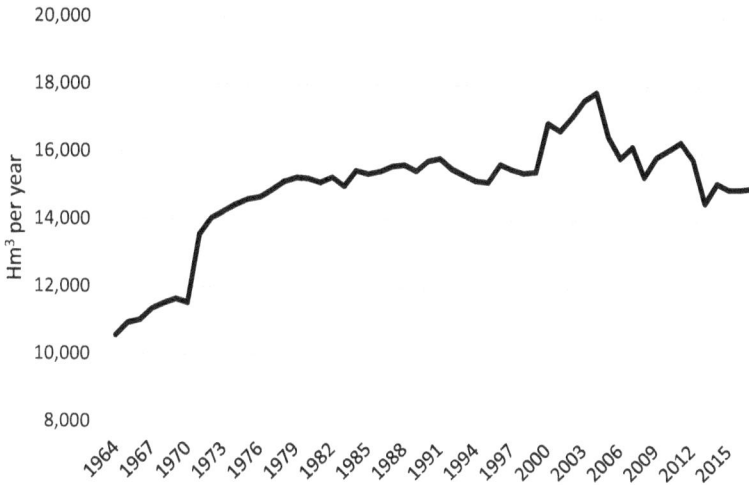

Figure 2.3.   Water withdrawals for irrigation in Spain (Hm^3 per year).
Source: Spanish Ministry of Agriculture.

stagnation of approximately 15,000 cubic hectometers per year. It should be noted that abstracted water, irrigated land, and FCP all grew in parallel until 2004, thus showing the major role that water availability plays in the expansion of crop production and food security in Spain. Likewise, the reduction in water withdrawals for irrigation in Spain since 2004 could be explained by the intense process of modernization observed in the last 20 years through more efficiency in irrigation systems in the use of water and the increase in productivity per hectare (defined as the value generated per cultivated hectare) thanks to irrigation expansion, the change of crops, and the use of new agronomic techniques.[19] These aspects are discussed in the next section of this chapter.

As a conclusion of this section, the analyzed data shows that the strategy based on the greater availability of resources (land and/or water) to increase crop production and thus improve Spain's capacity to face the challenges to food security appears to be infeasible due to the resource limits, especially that of water in a context of climate change. In the specific case of Spain, the growing scarcity of water has become a serious limiting factor for the growth of agricultural production, as well as for the cultivation of more irrigated hectares, thus compromising the future need to increase food production.

2.3. Increase of Agricultural Productivity and Resource-use Efficiency

At the beginning of the 21st century, the contribution of irrigated land to Spanish FCP was approximately 70%, despite it occupying only 13% of the usable agricultural area.[20] As mentioned at the beginning of this chapter, irrigated production registers a much higher productivity than rainfed agriculture, despite the difficulties in the availability of resources (especially water). According to the US Department of Agriculture's Economic Research Service, agricultural Total Factor Productivity (TFP) has raised by an average annual rate of 1.51% in the last decade.[21] Nevertheless, and following the Global Agricultural Productivity (GAP) Index,[22] the required rate of TFP growth should be 1.75% to double output by 2050 (as required by the population and economic growth). Additionally, and following the same GAP report, the TFP growth rate in low-income countries is too low (around 0.96%) and registering a negative trend, due to extreme weather and climate change impacts as one of the main explanatory factors. This fact poses a significant challenge to food security at the global scale since developing regions, such as the sub-Saharan Africa, are reporting higher growth rates of food demand together with decreasing agricultural productivity. Nevertheless, the FAO estimates that only 10% of food production comes from irrigated lands in Africa (30% of a total of 42.5 million hectares with irrigation potential), thus showing a significant potential growth in food production due to irrigation. In fact, the World Bank and the United Nations Development Programme (UNDP) calculates that irrigation could be expanded up to 110 million of additional hectares in developing countries in order to produce enough food for 2,000 million people.[23]

In the case of Spain, according to the estimations made by Grindlay et al.,[20] the strategy of increasing productivity has been the main source of FCP growth. Thus, the productivity of the agricultural sector can be measured in relation to the different productive factors used by irrigation, which are fundamentally water, land, and labor. As shown in the previous section, the number of cultivated hectares in Spain has experienced a decrease in the last 50 years, with a continuous conversion of land from dry land to irrigated land. This fact has led to the agricultural productivity of the land having increased continuously in this period. In relation to water productivity, its growth can be largely explained by a consequence of increased crop yields[24] and their greater added value,[19] and to a lesser extent by a reduction in the water used. This fact was especially obvious until 2004, a period in which water withdrawals for irrigation grew continuously (Figure 2.3). This increase in abstracted water stopped in 2004, when a period of trend change (decline) began as a consequence of the greater efficiency achieved by the modernization of irrigation infrastructures and the adoption of new agronomic techniques (e.g., deficit irrigation).

Table 2.1 shows the evolution of agricultural productivity per unit of resource used (water, land, and labor) both at constant and current prices for the period of 1985–2016 in Spain. As shown, the evolution of agricultural productivity measured as final crop production (in millions of euros) divided by abstracted water (in cubic hectometers, hm³), cultivated land (in thousands of hectares), and labor (in thousands of workers), has experienced a positive evolution in the analyzed period, both at constant and current prices. The highest rates of productivity growth have been observed for the labor factor, due to the intense reduction of the labor force employed in the agricultural sector and to its mechanization and modernization processes. Land productivity also experienced a significant increase between 1985 and 2016 — 31% at constant prices and 281% at current prices. Finally, water productivity (or value generated per unit of abstracted water) increased by 13% at constant prices and by 228% at current prices. The growth of prices (and of the crops with higher prices) might explain why the evolution followed by the productivity at current prices is greater than that at constant prices. Consequently, the observed increases in productivity would not have been possible without the expansion of irrigated agriculture.

Table 2.1. Agricultural productivities (at constant and current prices).

Year	Water (euro/m³) Constant	Water (euro/m³) Current	Land (10^3 euro/ha) Constant	Land (10^3 euro/ha) Current	Labor (10^3 euro/worker) Constant	Labor (10^3 euro/worker) Current
1985	1.72	0.60	1.31	0.45	13.62	4.74
2000	1.79	1.33	1.66	1.23	29.33	21.84
2016	1.95	1.97	1.72	1.73	38.22	38.63
Increase	13.1%	228.1%	31.4%	228.1%	180.7%	714.6%

Source: Author's own upon data gathered from the Spanish Ministry of Agriculture.

The agricultural use of water has two basic components — water consumed in the evapotranspiration (ET) process and water lost as run-off, infiltration, and percolation. Water-use efficiency has significantly increased by reducing water losses (and part of the water consumed in evaporation from the soil) through the implementation of agronomic and irrigation techniques of greater efficiency. The irrigated area that has been modernized in Spain, whose main objective is to maximize FCP per hectare and achieve a more efficient use of water, has reached the figure of 1.47 million hectares, with an investment close to 3,000 million euros.[18] Each irrigation method presents a different level of efficiency in the use of water (measured by the relationship between water applied and water actually used by the crop), and these vary from the range of 0.60–0.65 for furrow irrigation, 0.75–0.80 for sprinkler irrigation, and 0.90–0.95 for the case of localized drip irrigation.[25] As a consequence of the changes in irrigation methods, the modernization of irrigation systems has favored a decreasing trend in the total consumption of water for agricultural use in the last decade, with a reduction of 14%, while the irrigated area simultaneously increased by 7%. This saving has been made possible thanks to the investment in irrigation systems of greater efficiency in the use of water, among which localized drip irrigation systems deserve a special mention, since they represent 51% of the total irrigated area in Spain (see Table 2.2).

Figure 2.4 shows the increasing trend of the two most efficient irrigation techniques (drip and sprinkler) and the continued decrease of furrow irrigation in the period of 1964–2016. In the case of sprinkler irrigation, the continuous increase registered during the second half of the 20th century was not reflected in the first years of the 21st century. This stagnation can be explained by the significant development observed in drip irrigation, which is much more efficient in the use of water and has a high capacity to raise the FCP per irrigated hectare. The determined

Table 2.2.   Rate of use of irrigation techniques (% of irrigated hectares).

	1964	1980	2000	2010	2017
Furrow	100%	79.1%	43%	31%	25%
Sprinkler	0%	19.9%	28%	22%	24%
Drip	0%	1.0%	29%	48%	51%

Source: Spanish National Institute of Statistics.

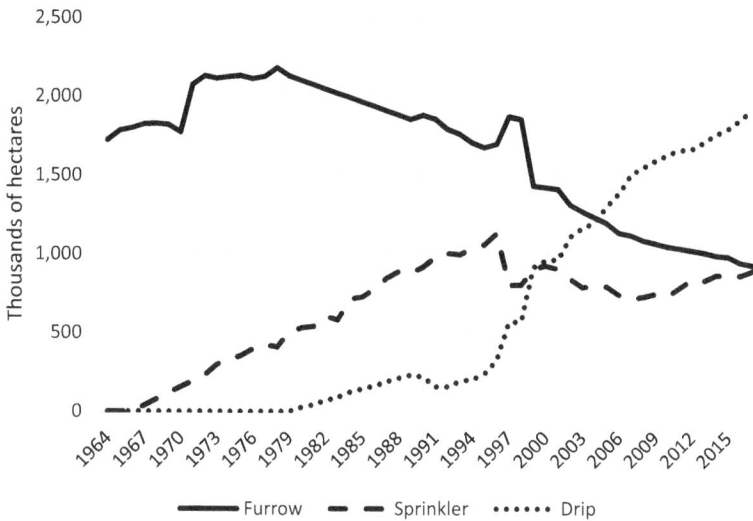

Figure 2.4.   Use of irrigation techniques (over thousands of hectares) from 1964 to 2015.
Source: Spanish National Institute of Statistics.

commitment of the different administrations and farmers to ensure a more efficient use of water resources can be clearly seen in the exponential increase in the number of hectares irrigated by drip techniques from the end of the 1990s to the present day. Table 2.2 also shows how furrow irrigation has come to represent only 25% of the irrigated area nowadays, having been greatly surpassed by drip irrigation, which represents 51% of the total irrigated area. Sprinkler irrigation is used in 24% of the irrigated area in Spain. This increase in irrigation efficiency has led to a higher estimated average irrigation efficiency from 0.5 in the 1960s to 0.7 in 2017 on a national level (including transport losses).[26]

Finally, food security is increasingly linked to the availability of energy in a highly technological agricultural sector. Although it is true that the irrigation modernization process has promoted the transformation of traditional surface irrigation into drip systems in the last 15 years, thus allowing more production with up to 25% less water, it has also exponentially skyrocketed energy consumption. In the case of Spain, electricity consumption associated to irrigation with energy-intensive methods (e.g., sprinkling and drip irrigation) has continuously grown during the period of 1964–2017, with the exception of the period of

1996–2003, thereby quadrupling the 1964 levels. Regarding the intensity of electrical energy use per cubic meter of abstracted water, this has risen from 0.04 GWh/m^3 in 1964 to the current 0.12 GWh/m^3.[26] This increase in irrigation production in a context with less water and land, poses the challenge of achieving double efficiency in modernized irrigation systems to save not only water, but also electricity. Therefore, continuing to advance in the modernization of irrigation implies adapting new projects to the current electricity rate scenario, in addition to favoring measures that help irrigators mitigate current energy costs, such as promoting energy production from renewable sources and self-consumption in irrigable areas.

2.4. Concluding Remarks

Irrigation plays a fundamental role in addressing challenges to food security, both globally and in the case of Spain. Increasing food production to meet the growing future demands of the world population will require not only the expansion of the irrigated area (through the conversion of rainfed land) and better water availability (both in space and time), but also greater efficiency in the use of resources. Both these drivers of growth represent serious challenges worldwide, thus requiring complex institutional and policy reforms, especially in those countries and regions where water scarcity is most acute. In the particular case of Spain, the growth of agricultural production as a guarantor of food security appears to lack the support of an increase in resources (land and water), but not that of an increase in agricultural productivity. The analysis carried out in this chapter has shown that the achievement of a higher production per unit of productive factor used (especially per unit of water used) has been decisive in guaranteeing the maintenance of production in the Spanish agricultural sector.

Globally, if food production needs are not met through the efficient and sustainable expansion of irrigation, then population growth and economic development will increase pressure on resources and accelerate the process of environmental degradation.[27] Likewise, the close links between food, water, and energy security require the suitable coordination of public policies that encourage increased productivity in the irrigation sector in a context of the circular economy, of saving water and energy resources, of developing crops with lower water needs, and of greater tolerance to pests, that is, by putting the focus on biotechnology.[28]

Finally, we should bear in mind that climate change will affect both future agricultural production and water availability, and will therefore exert an impact on food security on a global level. However, most studies indicate that climate change will have a neutral or relatively modest effect on agricultural production processes, at least until 2050. In contrast, future water availability for agriculture will suffer a greater impact from climate change, thereby rendering the reliability of water supply a serious challenge to food security.

References

1. Food and Agriculture Organization of the United Nations (2011). *El estado de la inseguridad alimentaria en el mundo*. Rome, Italy: Food and Agriculture Organization of the United Nations.
2. Food and Agriculture Organization of the United Nations (2012). *Statistical Yearbook 2012. World Food and Agriculture*. Rome, Italy: Food and Agriculture Organization of the United Nations.
3. Food and Agriculture Organization of the United Nations (2002). The State of Food Insecurity in the World. *When People Must Live with Hunger and Fear Starvation*. Rome, Italy: Food and Agriculture Organization of the United Nations.
4. Schlosser, C. A. et al. (2014). The Future of Global Water Stress: An Integrated Assessment, *Earth's Future*, 2(8): 341–361, 2014. DOI:10.1002/2014EF000238.
5. Alexandratos, N. and Bruinsma, J. (2012). World Agriculture Towards 2030/2050: The 2012 Revision (ESA Working paper, vol. 12–03). Rome, Italy: Food and Agriculture Organization of the United Nations.
6. Du, T., Kang, S., Zhang, J., and Davies, W. J. (2015). Deficit Irrigation and Sustainable Water-resource Strategies in Agriculture for China's Food Security, *Journal of Experimental Botany*, 66(8), 2,253–2,269. DOI:10.1093/jxb/erv034.
7. Dillon, A. (2007). Do Differences in the Scale of Irrigation Projects Generate Different Impacts on Poverty and Production?, Discussion Paper 01022. International Food Policy Research Institute, Washington, D.C.
8. Mangisoni, B. (2008). Impact of Treadle Pump Irrigation Technology on Small Holder Poverty and Food Security in Malawi. A Case Study of Blantyre and Mchinji Districts, *International Journal of Agricultural Sustainability*, 6, 248–266. DOI:10.3763/ijas.2008.0306.
9. Omilola, B. (2009). Estimating the Impact of Irrigation on Poverty Reduction in Rural Nigeria, IFPRI Discussion Paper 00902. International Food Policy Research Institute, Washington, D.C.

10. Anwar, M. R., Liu, D. L., Macadam, I., and Kelly, G. (2013). Adapting Agriculture to Climate Change: A Review. *Theoretical and Applied Climatology*, *113*, 225–245. DOI:10.1007/s00704-012-0780-1.
11. Darko, R. O., Yuan, S., Hong, L., Liu, J., and Yan, H. (2016). Irrigation, a Productive Tool for Food Security: A Review, *Acta Agriculturae Scandinavica, Section B — Soil & Plant Science*, *66*(3), 191–206. DOI:10.1 080/09064710.2015.1093654.
12. Freibauer, A., et al. (2011). *Sustainable Food Consumption and Production in a Resource-constrained World*. Brussels, Belgium: European Commission-Standing Committee on Agricultural Research (SCAR).
13. López-Gunn, E., Mayor, B., and Dumont, A. (2012). Implications of the Modernization of Irrigation Systems. In L. De Stefano and M. R. Llamas (Eds.). *Water, agriculture and the environment in Spain: Can we square the circle?*, CRC Press, ch. 19, pp. 241–256.
14. Brown, L. R. (2005). Outgrowing the Earth. The Food Security Challenge in an Age of Falling Water Tables and Rising Temperatures. New York, U.S.A.: Earth Policy Institute, Earthscan.
15. United Nations (2018). *World Population Prospects: The 2017 Revision, Key Findings and Advance Tables*. ESA/P/WP/248. New York, U.S.A.: UN Population Division, Department of Economic and Social Affairs.
16. Nachtergaele, F., Bruinsma, J., Valbo-Jorgensen, J., and Bartley, D. (2011). *Anticipated Trends in the Use of Global Land and Water Resources*. (SOLAW Background Thematic Report–TR01). Rome, Italy: Food and Agriculture Organization of the United Nations.
17. Penning de Vries, F. W. T. (2001). Food Security? We Are Losing Ground Fast. In J. Nosberger, H. H. Geiger and P. C. Struik (Eds.) *Crop Science: Progress and Prospects*, Wallingford, UKCABI Publishing, ch. 1, pp. 1–14.
18. Berbel, J., Expósito, A., and Gutiérrez-Martín, C. (2019). Effects of the Irrigation Modernization in Spain 2002–2015, *Water Resources Management*, *33*, 1,835–1,849. DOI:10.1007/s11269-019-02215-w.
19. Expósito, A., and Berbel, J. (2019). Drivers of Irrigation Water Productivity and Basin Closure Process: Analysis of the Guadalquivir River Basin (Spain), *Water Resources Management*, *33*(4), 1,439–1,450. DOI:10.1007/ s11269-018-2170-7.
20. Grindlay, A. L., Lizárraga, C., Rodríguez, M. I., and Molero, E. (2011). Irrigation and Territory in the Southeast of Spain: Evolution and Future Perspectives Within New Hydrological Planning. *WIT Transactions on Ecology and the Environment*, *150*, 623–638.
21. Economic Research Service of the US Department of Agriculture (2020). *Agricultural Productivity in the U.S.* Retrieved from: https://www.ers.usda. gov/data-products/agricultural-productivity-in-the-us/.

22. Virginia Tech, College of Agriculture and Life Sciences (2020). 2020 Global Agricultural Productivity Report (GAP Report Initiative). Retrieved from: https://globalagriculturalproductivity.org.
23. Food and Agriculture Organization of the United Nations (2017). *The Future of Food and Agriculture: Trends and Challenges*. Rome, Italy: Food and Agriculture Organization of the United Nations.
24. Fischer, R. A., Byerlee, D., and Edmeades, G. O. (2009). Can Technology Deliver on the Yield Challenge to 2050?, Retrieved from Food and Agriculture Organization of the United Nations website: https://www.fao. org/3/ak977e/ak977e.pdf.
25. Berbel, J., Gutierrez-Marín, C., and Expósito, A. (2018). Impacts of Irrigation Efficiency Improvement on Water Use, Water Consumption and Response to Water Price at Field Level, *Agricultural Water Management*, *203*, 423–429. DOI: 10.1016/j.agwat.2018.02.026.
26. Espinosa-Tasón, J., Berbel, J., and Gutiérrez-Martín, C. (2020). Energized Water: Evolution of Water-energy Nexus in the Spanish Irrigated Agriculture, 1950–2017, *Agricultural Water Management*, *233*, 106,073. DOI: 10.1016/j. agwat.2020.106073.
27. Berbel, J., Expósito, A., Gutiérrez-Martín, C., and Pérez Blanco C. D. (2020). Water, Where Do We Stand? In D. Vasel-Be-Hagh, and S. K. Ting (Eds.), *Environmental Management of Air, Water, Agriculture, and Energy*, CRC Press, Taylor & Francis Group, ch. 2, pp. 7–31.
28. Fereres, E., Orgaz, F., and Gonzalez-Dugo, V. (2011). Reflections on Food Security Under Water Scarcity, *Journal of Experimental Botany*, *62*(12), 4,079–4,086. DOI: 10.1093/jxb/err165.

Chapter 3

Measuring the Sustainability of Water, Energy and Food Resources in the Context of the WEF Nexus

Luxon Nhamo*,††, Henerica Tazvinga†, Sylvester Mpandeli*,§,
Stanley Liphadzi*,§, Joel Botai†,‡, and Tafadzwanashe Mabhaudhi‡,¶

*Water Research Commission of South Africa,
Lynnwood Manor, Pretoria 0081, South Africa
†South Africa Weather Services (SAWS), Ecoglades,
Centurion 0157, Pretoria, South Africa
‡Centre for Transformative Agricultural and Food Systems,
School of Agricultural, Earth and Environmental Sciences,
University of KwaZulu-Natal (UKZN), Scottsville, Pietermaritzburg
3209, South Africa
§Faculty of Science, Engineering, and Agriculture, University of
Venda, Thohoyandou, South Africa
¶International Water Management Institute (IWMI-GH), West Africa
Office, Accra, Ghana
††luxonn@wrc.org.za

Abstract

As the Sustainable Development Goals (SDGs) are a call to action by countries to promote prosperity while protecting the planet, the 17 goals aim to monitor 169 targets that collectively describe the progress towards achieving a sustainable future. This was necessitated by global challenges in resources degradation and depletion due to population growth, urbanization, migration, and improved living standards, demographic shifts, changing lifestyles, a burgeoning middle class, and the growing influence of climate change on the demand and supply chains of mainly water, energy, and food. As the SDGs were formulated in the context of challenges related to resource insecurity, climate change, and human wellbeing, and are designed to recognize the inter-linkages between human wellbeing, economic prosperity, and a healthy environment, this chapter establishes the linkages between resources. It provides pathways to assess progress towards sustainability. The principal question addressed in the chapter is whether the water–energy–food (WEF) nexus is an appropriate approach for linking and monitoring progress towards related SDGs, particularly Goals 2, 6, and 7, as reflected in the pursuit of water, energy, and food security.

Keywords: Global challenge, sustainability, drivers of change, nexus planning, resource management.

3.1. Introduction

Natural resources continue to deplete worldwide due to increasing demand from a growing population.[1,2] In 2017, more than 1.06 billion people, most of whom reside in rural areas, had no access to safe and affordable energy, and half of these people lived in Sub-Saharan Africa.[3] As of 2016, some 793 million people in the world were still undernourished, and 2.4 billion had no access to improved sanitation.[4] Also, ecosystems have significantly been degraded, as evidenced by a declining trend in the productivity of one-fifth of the Earth's land surface covered by vegetation between 1998 and 2013.[5] The main drivers of these changes include the increasing demand for food resources and dietary transitions, accelerated economic development, urbanization, lack of transboundary cooperation, and climate variability and change.[6] Projections indicate that by 2050, the world population will increase to about nine billion people, which will result in a rise of 80% in energy consumption, and a 60% increase in food demand.[4] These negative trends are happening when agriculture is already consuming most of the available freshwater resources.[7] The gloomy outlook for world resources is exacerbated by climate change, and socioeconomic inequities.[8,9] As a result, to mitigate these challenges, the United Nations General Assembly (UNGA) launched the 2030 Agenda on Sustainable Development Goals (SDGs) in an attempt to promote sustainability in resource management.[10] Pre-dating this event, the water–energy–food (WEF) nexus had already emerged as a polycentric and transformative approach promoting sustainability in resource management.[11]

As the post-2015 focus has shifted towards implementing the SDGs and evaluating progress towards achieving sustainability by 2030, the main challenge has been to develop models capable of monitoring and assessing the implementation progress by individual countries.[12] Various models are being developed and tested on how effectively they can monitor and evaluate progress towards achieving sustainability.[12–14] The SDGs were developed so that each target is assessed through one or more indicators that keep track of progress towards set targets.[15] These indicators are the backbone of monitoring progress in the SDGs, which depends on data availability and model development.[13,16] However, a major global challenge has been data unavailability and heterogeneity.[16,17] Although, to some extent, remote sensing has provided solutions to many challenges associated with data unavailability.[16,18] Several indicators related to land use/cover and the environment can be monitored through remote sensing, including the SDG indicators related to water, energy, and food (WEF) resources.[19] Remote sensing technology is assisting many end users to access data at different scales, particularly for areas where data has been scanty.[20]

The congruence between the WEF nexus and the SDGs has many advantages, and the WEF nexus is here proposed as a "fitting approach" for integrating and assessing SDG implementation. The essence of the WEF nexus is to monitor water, energy, and food security, without compromising ecosystem services.[11] The WEF nexus is envisaged to simplify the intricate and systematic interactions between the natural environment and human activities in the three sectors of water, energy, and food.[9,13] Recently developed WEF nexus analytical models provide pathways to enhance the coordination and sustainable planning and management of natural resources across sectors, at all levels and spatial scales.[12–14,21]

Therefore, nexus planning is relevant in implementing and assessing the progress of the essence of the SDGs over time.[12,22] Sectoral approaches in resource management will only result in inequality, exacerbate poverty, widen the gap between the rich and the poor, and increase vulnerability.[23] Yet, a cross-sectoral approach embedded in the WEF nexus's conceptual approach could be seen as a precondition for achieving the SDGs.[9,14]

Because of its inherent suitability, the WEF nexus has become a significant policy and management analytical framework for sustainable development.[24,25] There is currently a surge in global recognition of the importance of the WEF nexus in leveraging the SDGs' implementation process with its monitoring and evaluation, particularly towards making informed decisions on goals, targets, and indicators.[13,26] As a cross-sectoral and interlinking approach, the WEF nexus supports the integration of indicators across sectors and clarifies how resources can be best allocated between competing needs, thus making SDG more efficient and cost-effective.[12] The approach provides data to integrate the three intricately related resources' management, clarifying the complex and dynamic interlinkages between the securities of water, energy, and food.[13,16] Thus, linking directly to SDGs 2, 6, and 7. The WEF nexus has evolved into a decision support tool to guide interactions between the natural environment and the biosphere.[27–30] The approach has evolved to be a critical systems approach in sustainability circles and a multi-purpose and polycentric approach for sustainable resource management.[16,31]

As the data on SDGs starts to be available, an unprecedented range of data describing water, energy and food security, gender issues, climate change, and the state of natural resources, etc., are also becoming available. The challenge then will be how to interpret these data, and adjust policies and practices so that each country contributes towards the global goal of achieving a sustainable future. The sheer span of the anticipated data and the disparate format of its reporting will challenge interpretation. A likely approach using themes is already being noted in the literature as an appropriate way for reporting on the SDGs.[16,32–35] The WEF nexus analytical model establishes the numerical relationships between indicators and addresses one of the pivotal themes to the planet, i.e., water, energy, and food security.[12,13] This nexus is the coalescence of the main factors that bring together natural resources in the quest for integrated resource security.

The WEF nexus is preferred for this kind of analysis because of its transformative and polycentric nature, capable of integrating different but interlinked and interdependent components. Unlike monocentric and linear approaches like the Integrated Water Resources Management (IWRM), which is water-centric, the WEF nexus considers all sectors on equal terms without putting more value on one sector.[13]

3.2. The Concept of Sustainable Development

Before considering the WEF nexus concerning the SDGs, it is necessary first to define the dominant concepts of sustainable development, sustainability indicators, sustainable agriculture, food security, sustainable water resources, and energy accessibility.

The term 'sustainable development' has been defined many times, with essential descriptions presented here. Leopold (1949), one of the founding fathers of ecological science, stated that sustainable development is the organizing principle for sustaining finite resources necessary to provide for future generations of life on the planet. It is a process that envisions a desirable future state for human societies in which living conditions and resource use continue to meet human needs without undermining the "integrity, stability, and beauty" of natural biotic systems.[36] This definition emphasizes an overriding principle of sustainable development — the need to balance the use and protection of resources. The World Commission on Environment and Development (1987), also called the Brundtland Commission, defined sustainable development as "development that meets the needs of the present without compromising the ability of future generations to meet their own needs."[37] This definition served as a basis during the United Nations Earth's Summit meeting in 1992, the World Summit (Johannesburg, South Africa) 2002, and the UN Conference on Sustainable Development (Rio+20) Rio de Janeiro) in 2012. The Brundtland definition of sustainable development comprises two key elements: (i) needs — aiming at meeting the needs of all humans in the present and future, and (ii) limits — limitations to environmental resources and the biosphere's ability to meet requirements of present and future generations.[37]

In providing context, the Brundtland Report describes that the intervention in natural systems was previously on a small scale with limited impact during our development. Today, these interventions threaten our very life-support system. Sustainable development is all about development without endangering the systems that support life on Earth.[37]

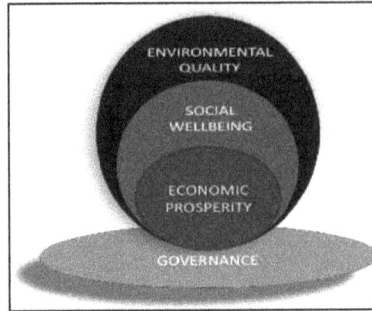

Figure 3.1.   Three dimensions of sustainable development within a governance framework.

The concept "sustainable development" was divided during the World Summit in Johannesburg into the three pillars of environment, economy, and social.[38] The idea is that sustainability can be attained by balancing these three dimensions so that sustainable development should be considered from a "holistic and integrated approach" as illustrated in Figure 3.1.[39]

The UN report titled *The Future We Want* from which the SDGs program evolved, defined sustainable development as "promoting sustained, inclusive and equitable economic growth, creating greater opportunities for all, reducing inequalities, raising basic standards of living; fostering equitable social development and inclusion; and promoting integrated and sustainable management of natural resources and ecosystems that support inter alia economic, social and human development while facilitating ecosystem conservation, regeneration and restoration and resilience in the face of new and emerging challenges."[40] This concept is well illustrated in Figure 3.1.

The seminal work by Rockstrom et al. addressed the environmental issues of sustainable development by introducing the concept of planetary boundaries, which are biophysical boundaries used for estimating a safe operating space for humanity concerning the functioning of the Earth system.[41] These boundaries were unashamedly biophysical, acknowledging that it would be up to society's socio-political systems to work out how to live within these boundaries. These boundaries thus describe the conditions for sustainable development from a biophysical perspective.

3.3. Measuring the Sustainability of Food Resources

Various definitions of food security have been developed, such as the World Food Summit,[42] which stated that "food security exists when all people, at all times, have physical, social and economic access to sufficient, safe and nutritious food that meets their dietary needs and food preferences for an active and healthy life." Thus, food security clearly cannot be divorced from agriculture and the whole food value chain. With a growing human population and the decline in natural resources used to support agriculture, agriculture's very sustainability is now being questioned, with solutions involving either its extensification or intensification. Over the ages, extensification has been the approach, merely clearing more land when more crops were required. However, with the room to move on the planet now reduced due to land degradation, climate change, and population increase, among other drivers, attention has been turned to the intensification of agriculture, producing more food with the same or less land (and other resources).

The Brundtland Report noted that agriculture often takes center stage in the debate about sustainable development because of the scale of its intervention in the natural ecosystems throughout the world.[37] Yet they are definite in their recommendation that development should not stop, but rather that it should be in harmony with the ecosystem. This remains the status quo for today. For example, agricultural land pressures from crop and livestock production can be partly relieved by increasing productivity. Still, the authors recognized that short-sighted, short-term improvements in productivity could create different forms of ecological stress, such as the loss of genetic diversity in standing crops, salinization and alkalization of irrigated lands, nitrate pollution of groundwater, and pesticide residues in food.[37] The report stressed that more benign alternatives are available ecologically and need to be sourced. In both developed and developing countries, future productivity increases should thus be based on best agricultural practices and extensive use of organic manures and non-chemical means of pest control. An agricultural policy based on ecological realities can promote these alternatives only.

Throughout the Brundtland strategy for sustainable development, a common theme is the need to integrate economic and ecological considerations in decision-making, which requires a change in attitude and objectives at every level of institutional arrangement. Nevertheless, the compatibility of environmental and economic goals is often lost in the pursuit of individual or group gains, with little regard for the impacts on others, blind faith in science's ability to find solutions, and ignorance of the distant consequences of today's decisions.

Rockstrom et al. provide a similar perspective. The authors reviewed the importance of agriculture in the world, including its importance to society and its impact on the environment, which is key to the success of agriculture and, thus, society.[41] In particular, they note that agriculture is key to attaining the UN SDGs of eradicating hunger and securing food for a growing world population, which may require an increase in global food production of between 60% and 110% in a world of rising global environmental risks. Rockstrom et al. cautioned about the prevailing thinking of sustainable intensification of agriculture-based mostly on enhancing agricultural productivity while reducing its environmental impacts, i.e., how to produce more food with fewer resources. The authors recognize that while sustainable agriculture is a useful and essential feature, it remains focused on improving resource use efficiency.[41] There is therefore, a need to shift from this paradigm of "focusing on productivity first and sustainability as a question of reducing environmental impacts, to a paradigm where sustainability constitutes the core strategy for agricultural development."[41] Thus, the recommendation to that, "sustainable principles as the entry point for generating productivity enhancements, which fundamentally requires real progress in increasing agricultural output by capitalizing on ecological processes in agro-ecosystems." The authors note that such a shift would "recognize that the biophysical boundaries of Planet Earth impose a hierarchy of criteria on the definition of sustainability: that sustainability is not a relative concept or an act of balancing competing claims; it sets absolute biophysical limits. It is only within such biophysically defined boundaries that we stand a high probability of avoiding irreversible shifts in environmental conditions." While this perspective has been unopposed in some quarters, it provides a critical view that may guide further considerations of what constitutes sustainability.

Thus, Rockstrom et al. (2016) proposed a new paradigm for the sustainable intensification of agriculture, which "aims at hunger-reduction through biodiversity conservation that secures ecological functions in agricultural landscapes." They note a strong case for adopting sustainable agricultural intensification as the strategy to meet the twin objectives for people and the planet. The "human goal," adopted by the UN Sustainable

Development Goals (SDGs) in 2015, is to eradicate hunger and poverty by 2030. They note that the SDG Agenda's objectives can only be understood as the world community sets out to feed humanity within a safe operating space of a stable and resilient Earth system.

The whole essence is to achieve sustainable food systems. As food systems are complex socio-ecological systems that involve various interactions between human (economic and political trends, food price volatility, population dynamics, changes in diets and nutrition, and advances in science and technology), and natural components (landcover changes, land and soil degradation, climate change, biodiversity loss, sea-level rise, and air pollution),[43–45] it is paramount to understand these relationships holistically to transition towards sustainable food systems (Figure 3.2). In between the socio-ecological systems are external drivers, including exposure and sensitivity, driving food systems. Knowledge of these drivers and how they relate to influencing activities and outcomes of food systems is vital for informing policy decisions.[46] Food and nutrition security are the primary outcomes of any food system. Thus, a food system is considered vulnerable or resilient when it fails to deliver food security or ensures food security, respectively.[47] Nexus planning connects these interactions by defining, measuring, and modeling progress towards sustainability, through a set of indicators formulated around resource utilization,

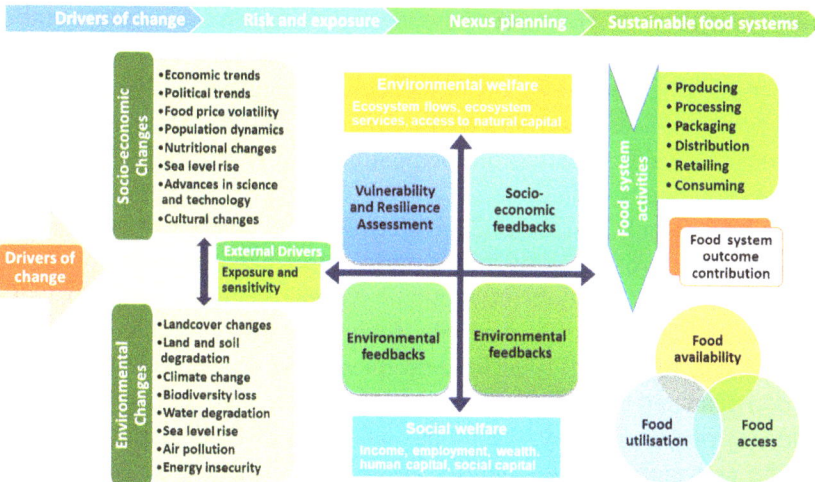

Figure 3.2.   A sustainable food system conceptual framework illustrating the processes and interactions involved in achieving a sustainable food system through nexus planning.

accessibility, and availability.[13,23] These developments facilitate modeling, monitoring, and simulating some aspects of sustainability.

Various studies have developed indicators to measure agricultural systems' overall sustainability (economic, social, and environmental).[13,48] A previous study defined five domains, all of which need to be considered to increase productivity and be sustainable[49,50] (Figure 3.2).

- Productivity (e.g., yield, input efficiency, water efficiency, animal health, etc.).
- Economic sustainability (e.g., agricultural income, crop, and livestock value).
- Environmental sustainability (e.g., biodiversity, carbon sequestration, erosion, nutrient dynamics, soil biological activity, soil quality).
- Social sustainability (e.g., information access, gender and other equity, social justice).
- Human wellbeing (e.g., food and nutrition security, risk, quality of life).

Implementation of each of these domains will require monitoring, reporting, verification, and evaluation. Indicators of each can be defined, which will have multiple metrics that can be directly measured, thus providing data that can be used to manage sustainable agriculture's sustainable intensification. However, the challenge is an almost infinite number of management options, and a combination of options open to farmers. Innovation is needed to support farmers in adopting and applying the most appropriate choices according to their biophysical, social, and economic contexts. However, guiding, implementing, and monitoring such implementation remains a challenge.

In particular, Goal 2 (End hunger, achieve food security, improve nutrition, and promote sustainable agriculture) has included monitoring the sustainability of food production.[10] Goal 2 includes indicators that consider the nourishment of people, food insecurity, production of food, income, and sustainability of food producers and production, the security of food plant and animal genomes, the orientation of governments towards food security, and finally, the proper operation of food markets.[10] These are the indicators that the authors of the SDGs suggest are sufficient to monitor the overall status of food security in terms of sustainability.

Target 2.3 of the SDGs is described as "by 2030 double an indicator of productivity measures the agricultural productivity and the incomes of small-scale food producers, particularly women, indigenous peoples, family farmers, pastoralists and fishers, including through secure and equal access to land, other productive resources and inputs, knowledge, financial services, markets and opportunities for value addition and non-farm employment."[51] SDG Target 2.4 is a direct measure of sustainable agriculture and is described as "by 2030 ensure sustainable food production systems and implement resilient agricultural practices that increase productivity and production, that help maintain ecosystems, that strengthen capacity for adaptation to climate change, extreme weather, drought, flooding, and other disasters, and that progressively improve land and soil quality". Note how the wording of this indicator emphasizes progressive improvement of land and soil quality. This will be a challenge for the world's farmers.[51]

3.4. Measuring the Sustainability of Water Resources

The first component of the WEF nexus considers the use of 'water resources' and how it can be used to determine if a country is on a path to a sustainable future in water resources. Shilling et al. developed a framework to measure sustainable water resources use that covers the three dimensions of economy, social, and environmental.[52] Inspired by the Brundtland definition of 'sustainable development', Shilling et al. defined 'water resources sustainability' as "the dynamic state of water use and supply that meet today's needs without compromising the long-term capacity of the natural and human aspect of the water system to meet the needs of future generations."[52]

The report selected indicators that fall under five categories— (i) water supply reliability, (ii) water quality, (iii) ecosystem health, (iv) adaptive and sustainable management, and (v) social benefits and equity—are given in Table 3.1. A definition of each category is shown in Table 3.1. It can be noticed that these categories in Table 3.1 mirror, to some extent, the targets to be found in SDG 6.

As represented in Goal 6, the SDG for water is intended to provide a sufficient indication of the overall status of water in terms of sustainability. The targets and indicators within Goal 6 have a good cross-section of resource indicators as well as indicators designed to decrease human dependence on the resource, including 6.3.1 on waste discharge, 6.3.2 on

Table 3.1.   Categories of water sustainability indicators.

Category	Definition
Water supply reliability	The provision of water of sufficient quantity and quality to meet water needs for health and economic well-being and functioning.
Water quality	The chemical and physical quality of water to meet the ecosystem and drinking water standards and requirements.
Ecosystem health	The condition of the natural system, including terrestrial systems interacting with aquatic systems through runoff pathways.
Adaptive and sustainable management	A management system that can nimbly and appropriately respond to changing conditions and that is equitable and representative of the various needs for water in the country in question.
Social benefits and equity	The health, economic, and equity benefits realized from a well-managed water system, including management of water withdrawal and water renewal.

ambient water quality, 6.4.1 on water use efficiency, 6.4.2 on water stress, 6.5 on water resource governance, and 6.6.1 on water-related ecosystems.

3.5. Measuring the Sustainability of Energy Resources

The second component of the WEF nexus considers the use of 'energy resources' on determining if a country is on a path towards a sustainable future regarding sustainable energy resources.[13] Sustainable energy, which incorporates energy access, efficiency, and renewable energy, is a critical enabler for all SDGs.[53] Globally, there has been tremendous progress made towards the achievement of SDG 7 and a shift towards sustainable energy supply options. There has been continued progress in electricity access in developing countries, with a global increase from 83% in 2010 to 87% in 2017.[53] Energy efficiency and renewable energy use increased globally from 8.6% in 2010 to 10.2% in 2016. However, more effort is still required to improve access to clean and safe cooking fuels and technologies for three billion people globally, as this has increased at a meagre rate of 0.5% per annual from 2010 to 2017.[53] There is a need to increase electrification in sub-Saharan Africa, as only 44% of the population has access to electricity.[54] Tracking the progress of SDGs is a big challenge as this requires collecting, processing, analyzing, and disseminating vast amounts of data and statistics at subnational, national, regional, and global levels.[54]

Providing secure, clean, affordable, and reliable energy for all is an essential element of sustainable development, which improves the quality of life by providing socio-ecological and economic benefits such as improved air quality and preservation of vegetation.[55] Figure 3.3 shows the linkages among the four sustainability dimensions of the energy system. The environmental status of the energy system is shown to be influenced by economic and social dimensions factors. In turn, the social situation is affected by economic dimension factors, while the institutional dimension can influence social, economic, and environmental dimensions through policies that affect the energy system sustainability.[56,57] To measure the sustainability of the social, economic, environmental, and institutional state, the International Atomic Energy Agency (IAEA), in cooperation with various organizations, developed a set of globally recommended energy indicators known as the Indicators for Sustainable Energy Development (ISED/EISD).[58] This is an analytical tool for assessing energy systems and measuring progress towards a more sustainable future energy system that serves as the reference point for energy indicators.[56]

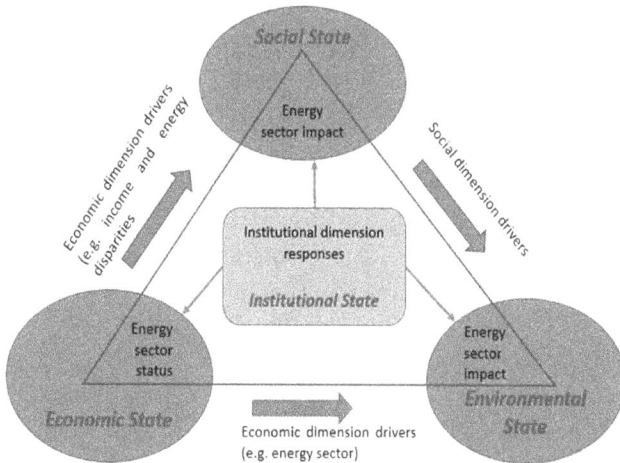

Figure 3.3.   Linkages among sustainability dimensions of the energy sector.[56]

Mainali and Silveira[59] assessed energy technologies for rural electrification using an energy technology sustainability index and established those mature technologies have better sustainability performance. Razmjoo et al.[60] presented suitable indicators to measure energy sustainability in line with the SDGs (17 UN Goals), UN-Habitat III (14 goals), and SEDI for developing countries. The study by Razmjoo et al.[60] analyzed the relationship between energy sustainability and selected groups of indicators for developing countries. The indicators' groups included environmental impact, policy, renewable energy, use of energy, transport, resource access, and resilience. This work recommends that should policymakers implement the selected group of indicators, an improvement in energy sustainability could be achieved, leading to changes in energy supply, intensity, and consumption. In the development of the framework for measuring the sustainability of energy resources,[60] similar to the IAEA/ IEA,[56] also assessed linkages among the social, economic, environmental, institutional sustainability and they also considered the technical sustainability, stated as follow:

- *Social sustainability* can be measured by assessing energy's accessibility and distributional effect on society. This is achieved by considering the residential sector's per capita consumption of clean energies and income inequality (gross national income [GNI] coefficient).
- *Economic sustainability* considers the gross domestic product (GDP) for the country as one of the key indicators as it reflects the people's growth and welfare. This sustainability dimension involves calculating per capita consumption of commercial energy, which requires final energy intensity (i.e., amount of energy per unit of output) and share of productive use of energy data.
- *Environmental sustainability* is measured by considering the share of dirty fuels in residential energy consumption and carbon intensity.
- *Technical sustainability* performance data requirements include the share of non-renewable (i.e., depletable) energies, depletion coefficient of local energy resource, and overall system conversion efficiency. This improves energy efficiency by ensuring that the energy supply system can provide society's current and future needs in a practical, reliable, and sustainable way, and from clean sources. Resources include coal, natural gas, crude oil, hydropower, nuclear, and renewables.

- *Institutional sustainability* is a critical sustainability dimension that can indicate the level of local participation in the management and control of the energy system, is measured through the calculation of the overall self-efficiency, and takes into account sector factors such as local skill base, local regulation, public participation and protection of consumers and investors.[60]

In line with the sustainability measures, the United Nations Economic Commission for Europe (UNECE) specifies the SDG 7 indicators for energy access as the proportion of the population with physical access to electricity and the population's ratio with access to clean cooking fuels and technology.[53] The challenge observed with this measure is that even in countries with high proportions of access to electricity, many people face affordability, quality of access, and quality of service problems necessitating the addition of indicators that would address these challenges, such as affordability and reliability of electricity access.[53] To address these challenges, for both cooking and electrification, the World Bank has developed a multi-tier framework methodology (i.e., a measure of energy access) covering seven quality dimensions (including affordability and reliability) and places households in one of five tiers of access through the use of a methodology based on surveys, that picks up deficiencies in service.[61] In this respect, complementary indicators to indicate progress would include heat demand, affordability of heating, and heating services quality.

The SDG 7 indicator for energy efficiency is the rate of growth in energy intensity measured as the ratio of total primary energy supply (TPES), to GDP, with the latter measured in a way designed to avoid distortions caused by exchange rate fluctuations.[10] For renewable energy, Indicator 7.2.1 talks about the renewable energy share in the total final energy consumption.[54] Measuring renewable energy as a proportion of energy is an indicator of progress in reducing global greenhouse gas (GHG) emissions and local pollution sources; a country's progress in developing and utilizing available resources sustainably; and improving sustainability over the entire energy value chain. Renewable energy can also be expressed as a proportion of TPES, but this indicator measurement does not consider the losses incurred in fossil fuel combustion. Another indicator that can be used accounts for exergy, or entropy, whereby renewable energy's contribution is evaluated concerning total primary energy requirements (TPER), based on the primary energy required to provide its equivalent input to the energy system.[60] This indicator reflects the real contribution of renewable energy to reducing GHG emissions and the displacement of non-renewable energy sources.

Other indicators for renewable energy progress would be renewable energy generating capacity additions, and investments in energy capacity with the tracking of investments in renewable energy capacity providing insight into developing renewable energy.[53] The United Nations Economic Commission for Europe (UNECE) recommends that the new indicators should address the water, food, energy, and climate nexus areas, that is, linking these sectors, and hence the need to develop a WEF nexus framework to assess progress towards the SDGs about the sustainability dimensions.[53] Therefore, it is important to determine resource sustainability by understanding the impacts of diversification and linkages of resources, economics.[62]

This could be measured by tracking WEF resources in an integrated manner. Based on the literature, the framework shown in Figure 3.4 could be adopted and applied across all nexus areas.[53,56,59,60,63]

The proposed framework (Figure 3.4) for integrated sustainability assessment involves the selection of environmental, techno-economic, and social indicators, selection, and specification of WEF technologies;

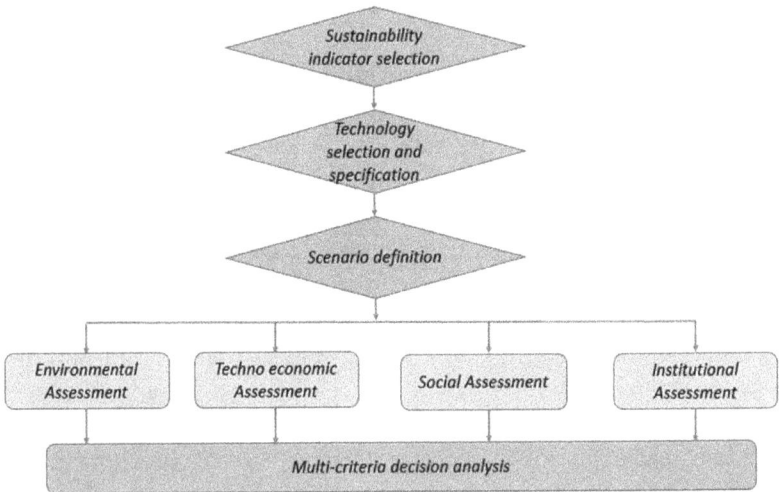

Figure 3.4.   Decision-support framework for integrated sustainability assessment of WEF nexus.

definition of scenarios, environmental, social, and techno-economic evaluation; and integration of sustainability or SDG indicators through a multi-criteria decision analysis to determine the most sustainable options for the future. The selection of the indicators is driven by existing research findings from various literature on the sustainability of the systems and the institutional and/or global policies. The different sustainability dimensions will be considered (Figure 3.4). The outputs of the assessments will then provide the data for the multi-criteria decision analysis. Many multi-criteria decision analysis approaches exist in the literature that can be used within the decision-support framework and are useful when dealing with multiple and sometimes conflicting factors.

3.6. The WEF Nexus and Sustainable Development

The WEF nexus is all about integration, implying that all the WEF sectors should be managed in a holistic and integrated manner, and not in isolation, as the latter only transfers challenges from one sector to the other and creates optimal efficiencies in one sector at the expense of the others.[23] Thus, as a key aspect of the SDGs, food security should not be considered in isolation from water and energy resources. Without considering each of these, sustainable management of food resources and achieving security is patently impossible.[64] The WEF nexus basis ensures that developments in one sector should not impact other sectors.[18] This improves resource use efficiency and is a catalyst for sustainable development. However, the challenge is integrating the diverse and disparate data and information into a form that is both intelligible and intuitive. Such a challenge forms the basis of the nexus approach as exemplified by the water, energy food (WEF) nexus that has recently become so topical in sustainability issues.[12,13,30]

The WEF nexus is useful (i) for understanding the interactions between water, energy, and food systems and the linkages with human resources in any given context, and (ii) to evaluate the performance of technical or policy interventions in this context (Figure 3.5). According to the FAO, the concept of a nexus "has emerged as a useful concept to describe and address the complex and interrelated nature of our global resource systems, on which we depend to achieve different social, economic, and environmental goals. In practical terms, it presents a conceptual approach to better understand and systematically analyze the

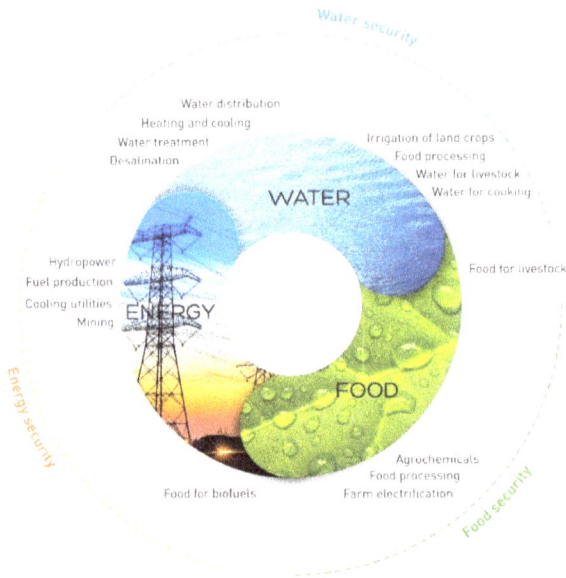

Figure 3.5.   Linkages between WEF sectors and importance of nexus planning in the security of resources.

interactions between the natural environment and human activities and work towards a more coordinated management and use of natural resources across sectors and scales. This can help us to identify and manage trade-offs and to build synergies through our responses, allowing for more integrated and cost-effective planning, decision-making, implementation, monitoring and evaluation."[30]

The essence of the WEF nexus is to ensure the security of the component resources. It is a three-dimensional framework used either as an analytical tool, a conceptual framework, or a discourse.[31] As an analytical tool, the nexus systematically applies quantitative and qualitative methods to understand the interactions among WEF resources; as a conceptual framework, it simplifies an understanding of WEF linkages to promote coherence in policy-making and enhances sustainable development; and as a discourse, it is a tool for problem framing and enabling cross-sectoral collaboration.[24] Thus, the WEF nexus approach is a pathway for understanding complex and dynamic interlinkages between water, energy, and food security issues. In this regard, it can also be used to monitor the performance of the WEF nexus components in the 2030 Agenda of the

Sustainable Development Goals (SDGs), particularly SDGs 2 (zero hunger), 6 (clean water and sanitation), and 7 (affordable and clean energy).[65] The WEF nexus is an innovative integrated approach that derives cross-sectoral sustainability indicators. Sectoral policies to resources management risk significant and unintended consequences as they always fail to manage cross-sectoral synergies and trade-offs.[66,67]

3.7. Indicators of Sustainability

WEF nexus components come from distinct subject areas, with different units of measurements, which makes their integration very complicated. This complexity has delayed the development of analytical tools capable of quantifying resource management in an integrated manner and not in isolation. However, recent developments have seen indicators as an option to assess WEF resource use and management as seen together through the multi-criteria decision making (MCDM).[68] The advantage of using sustainability indicators is that they can be expressed as composite indices and related to indices from different subject areas.[69,70] Sustainability indicators convey information on resources' performance and the current status at a given spatial scale[71-73] and quantify the state or trend of resource utilization.[74]

Sustainability indicators are increasingly being described to measure and assess society's sustainability concerning many issues (Singh et al., 2012). An indicator quantifies the state or trend of a phenomenon.[75,76] Indicators can be used alone or together in indices, where all individual indicator scores are combined into one single score.[77]

The Environmental Sustainability Index (ESI) established by Esty et al.[78] (Figure 3.6) is an example of a complex sustainability index, based on the compilation of 21 indicators derived from 76 underlying data sets. Examples of indicators are, among others, water quality, water quantity, air quality, biodiversity, and greenhouse gas emissions. The 21 indicators are also spread among five components—environmental systems, reducing environmental stresses, reducing human vulnerability to environmental stresses, societal and institutional capacity to respond to environmental challenges, and global stewardship. The final index score is the average value of all 21 unweighted indicator values (Figure 3.6). While acknowledging the limitations of an index that combines dissimilar data and information, the authors feel that by facilitating comparative analysis across

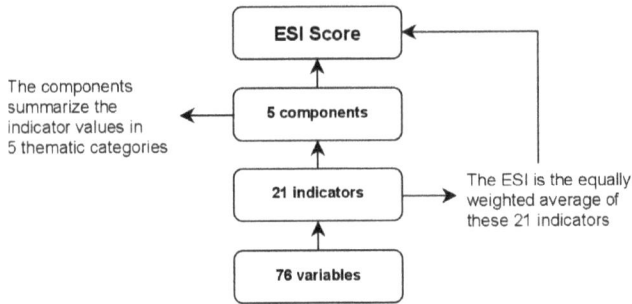

Figure 3.6.   Aggregation of the final Environmental Sustainability Indicators (ESI) index score.

national jurisdictions, these metrics provide a mechanism for making environmental management more quantitative, empirically grounded, and systematic.

WEF nexus sustainability indicators provide decision-making with a critical analytical framework that indicates the state of WEF resources, both in short-term and long-term perspectives.[68] As essential components of the WEF nexus, sustainability indicators provide the needed parameters to balance resource planning, governance, and technology development to enhance human wellbeing, now and in the future.[79] They are measurable parameters that indicate the performance of ecological, social, or economic systems,[52] hence their relationship with SDGs progress assessment. They connect statements of intent (objectives) and measurable aspects of natural and human systems.[80]

3.8. Concluding Remarks

The water–energy–food (WEF) nexus has gained international recognition as an intersectoral framework for resource management and sustainable development. It has evolved into a useful decision support tool that provides evidence on integrated solutions on resource security and management, enhancing resilience and adaptive capacities to climate change, and promoting awareness of the need for proper resource governance. It has developed into an indispensable analytical model for quantifying WEF sector interactions, identifying possible trade-offs and synergies, and as a lens to identifying areas for intervention. Such analyses are useful for monitoring progress towards the SDGs and informing the much-needed corrective measures. Resource management remains on the lower end of unsustainability, as evidenced by increasing poverty and hunger at the household level, water scarcity, and energy insecurity. Apart from being a decision support tool for integrated resources management, the WEF nexus is an important pathway to addressing the challenges of poverty, unemployment, and inequality. The approach promotes integrated planning, decision-making, governance, and management of resources. The approach enhances cross-sectoral cooperation and mitigates conflicts, increasing resource-use efficiencies in the process. The following common principles guide both the WEF nexus and the SDGs: (i) promotion of sustainable and efficient resource use, (ii) access to resources for vulnerable population groups, (iii) maintenance and support of underlying ecosystem services, and (d) improving human wellbeing. Importantly, the WEF nexus approach addresses the five key elements of the SDGs, i.e. People, Planet, Prosperity, Peace, and Partnership.

References

1. Pimentel, D. (1991). Global Warming, Population Growth, and Natural Resources for Food Production. *Society & Natural Resources, 4*(4), 347–363.
2. Lampert, A. (2019). Over-exploitation of natural resources is followed by inevitable declines in economic growth and discount rate. *Nature Communications, 10*(1), 1–10. DOI: 10.1038/s41467-019-092.
3. International Energy Agency. (2017). *Energy Access Outlook 2017: From Poverty to Prosperity*. Paris, France: International Energy Agency.
4. FAO-IFAD-UNICEF-WFP-WHO (2018). *The State of Food Security and Nutrition in the World 2018. Building climate resilience for food security and nutrition*. Rome, Italy: Food and Agriculture Organisation of the United Nations.
5. MacDicken, K. G. (2015). Global forest resources assessment 2015: What, why and how? *Forest Ecology and Management, 352*, 3–8.
6. Avtar, R., Tripathi, S., Aggarwal, A. K., and Kumar, P. (2019). Population–Urbanization–Energy Nexus: A Review. *Resources, 8*(3), 136.
7. Cosgrove, W. J. and Loucks, D. P. (2015). Water management: Current and future challenges and research directions. *Water Resources Research, 51*(6), pp. 4, 823–4,839. DOI: 10.1002/2014WR016869.
8. Liphadzi, S., Mpandeli, S., Mabhaudhi, T., Naidoo, D., and Nhamo, L. (2021). The Evolution of the Water–Energy–Food Nexus as a Transformative Approach for Sustainable Development in South Africa, in *The Water–Energy–Food Nexus: Concept and Assessments*, S. Muthu Ed. Kowloon, Hong Kong: Springer, ch. 2, pp. 35–67.

9. Mpandeli, S. et al. (2018). Climate Change Adaptation Through the Water-Energy-Food Nexus in Southern Africa. *International Journal of Environmental Research and Public Health, 15*(10), 2,306. DOI: 10.3390/ijerph15102306.

10. United Nations (2015). Transforming Our World: The 2030 Agenda for Sustainable Development. Resolution adopted by the General Assembly (UNGA). Retrieved from: http://www.un.org/ga/search/view_doc.asp?symbol=A/RES/70/1&Lang=E.

11. Hoff, H. (2021). Understanding the Nexus: Background Paper for the Bonn 2011 Conference, Stockholm Environment Institute (SEI), Stockholm, Sweden. Retrieved from: https://mediamanager.sei.org/documents/Publications/SEI-Paper-Hoff-UnderstandingTheNexus-2011.pdf.

12. Mabhaudhi, T. et al. (2021). Assessing Progress towards Sustainable Development Goals through Nexus Planning, *Water, 13*(9), 1,321. DOI: 10.3390/w13091321.

13. Nhamo, L. et al. (2020). An Integrative Analytical Model for the Water-Energy-Food Nexus: South Africa Case Study. *Environmental Science and Policy, 109*(109), 15–24. DOI: 10.20944/preprints201905.0359.v1.

14. Mabhaudhi, T. et al. (2019). The Water–Energy–Food Nexus as a Tool to Transform Rural Livelihoods and Well-Being in Southern Africa. *International Journal of Environmental Research and Public Health, 16*(16), p. 2,970.

15. Miola, A. and Schiltz, F. (2019). Measuring Sustainable Development Goals Performance: How to Monitor Policy Action in the 2030 Agenda Implementation? *Ecological Economics, 164*(106), 373. DOI: 10.1016/j.ecolecon.2019.106373.

16. Naidoo, D. et al. (2021). Operationalising the water-Energy-Food Nexus Through the Theory of Change," *Renewable and Sustainable Energy Reviews, 149*(111416), 10. DOI: 10.1016/j.rser.2021.111416.

17. Nhamo, L. et al. (2021). Urban Nexus and Transformative Pathways Towards a Resilient Gauteng City-Region, South Africa, *Cities, 116*(103, 266). DOI: 10.1016/j.cities.2021.103266.

18. Nhamo, L., Ndlela, B., Nhemachena, C., Mabhaudhi, T., Mpandeli, S., and Matchaya, G. (2018). The Water-Energy-Food Nexus: Climate Risks and Opportunities in Southern Africa, *Water, 10*(5), 567. DOI: 10.3390/w10050567.

19. Mariathasan V., Bezuidenhoudt, E., and Olympio, K. R. (2019). Evaluation of Earth Observation Solutions for Namibia's SDG Monitoring System. *Remote Sensing, 11*(13), 1,612. DOI: 10.3390/rs11131612.

20. Sheffield, J. et al. (2018). Satellite Remote Sensing for Water Resources Management: Potential for Supporting Sustainable Development in Data-Poor Regions. *Water Resources Research, 54*(12), 9,724–9,758. DOI: 10.1029/2017WR022437.

21. Nhamo, L., Ndlela, B., Mpandeli, S., and Mabhaudhi, T. (2020). The Water-Energy-Food Nexus as an Adaptation Strategy for Achieving Sustainable Livelihoods at a Local Level. *Sustainability, 12*(20), p. 8,582.
22. Nhamo, L. et al. (2021). Transitioning Toward Sustainable Development Through the Water–Energy–Food Nexus, in *Sustaining Tomorrow via Innovative Engineering*, D. Ting and R. Carriveau, Eds. Singapore: World Scientific, ch. 9, pp. 311–332.
23. Nhamo, L. and Ndlela, B. (2021). Nexus Planning as a Pathway Towards Sustainable Environmental and Human Health Post Covid-19. *Environment Research*, 110376, 7. DOI: 10.1016/j.envres.2020.110376.
24. Albrecht, T. R., Crootof, A., and Scott, C. A. (2018). The Water-Energy-Food Nexus: A systematic review of methods for nexus assessment. *Environmental Research Letters, 13*(4), 043002.
25. Terrapon-Pfaff, J., Ortiz, W., Dienst, C., and Gröne, M.-C. (2018). Energising the WEF Nexus to Enhance Sustainable Development at Local Level. *Journal of Environmental Management, 223*, 409–416.
26. Yumkella, K. and Yillia, P. (2015). Framing the Water-Energy Nexus for the Post-2015 Development Agenda. *Aquatic Procedia, 5*, 8–12.
27. Biggs, E. M. et al. (2015). Sustainable Development and the Water–Energy–Food Nexus: A Perspective on Livelihoods. *Environmental Science & Policy, 54*, 389–397.
28. Nhamo, L., Ndlela, B., Nhemachena, C., Mabhaudhi, T., Mpandeli, S., and Matchaya, G. (2018). The Water-Energy-Food Nexus: Climate Risks and Opportunities in Southern Africa. *Water, 10*(567), 18. DOI: 10.3390/w10050567.
29. Mpandeli, S. et al. (2018). Climate Change Adaptation through the Water-Energy-Food Nexus in Southern Africa. *International Journal of Environmental Research and Public Health, 15*(2,306).
30. Food and Agriculture Organization of the United Nations. (2014). *The Water-Energy-Food Nexus: A New Approach in Support of Food Security and Sustainable Agriculture*. Rome, Italy: Food and Agriculture Organization of the United Nations.
31. Keskinen, M., Guillaume, J. H., Kattelus, M., Porkka, M., Räsänen, T. A., and Varis, O. (2016). The Water-Energy-Food Nexus and the Transboundary Context: Insights from Large Asian Rivers. *Water, 8*(5), p. 193.
32. Cutter, A., Osborn, D., Romano, J., and Ullah, F. (2015). Sustainable Development Goals and Integration: Achieving a Better Balance between the Economic, Social and Environmental Dimensions. In *Stakeholder Forum and German Council for Sustainable Development*, Bonn, Germany, 2015. Retrieved from: https://nbsapforum.net/sites/default/files/Stakeholder%20Forum.%202015.%20Sustainable%20Development%20Goals%20and%20Integration_Achieving%20a%20better%20balance%20between%20the%20economic.pdf.

33. Elder, M., Bengtsson, M., and Akenji, L. (2016). An Optimistic Analysis of the Means of Implementation for Sustainable Development Goals: Thinking about Goals as Means. *Sustainability, 8*(9), p. 962.
34. Le Blanc, D. (2015). Towards Integration at last? The Sustainable Development Goals as a Network of Targets. *Sustainable Development, 23*(3), 176–187.
35. International Council for Science (2015). Review of the Sustainable Development Goals: The Science Perspective. Paris: International Council for Science.
36. Leopold, A. (1989). *A Sand County Almanac, and Sketches Here and There* (Outdoor Essays & Reflections). New York, USA: Oxford University Press.
37. World Commission on Environment and Development. (1987). *Our Common Future; Report of the World Commission on Environment and Development (WCED).* Oxford University Press, Oxford and New York, 10.
38. Warbrick, C., McGoldrick, D., and Gray, K. R. (2003). Accomplishments and New Directions? *International & Comparative Law Quarterly, 52*(1), 256–268.
39. Department of Environmental Affairs. (2017). About Green Economy. (Ed.) Pretoria, South Africa: Department of Environmental Affairs.
40. United Nations General Assembly. (2012). The future we want. 66th Session of the UN General Assembly. New York, U.S.A.: United Nations General Assembly. Retrieved from: www.un.org/ga/search/view_doc.asp?symbol= A/RES/66/288&Lang=E.
41. Rockström, J. et al. (2016). The World's Biggest Gamble. *Reviews of Geophysics, 4*(10), 465–470.
42. Food and Agriculture Organization of the United Nations. (1996). Rome Declaration on World Food Security and World Food Summit Plan of Action: World Food Summit. Rome, Italy: Food and Agriculture Organization of the United Nations, 9251039399.
43. Marshall, G. (2015). A Social-Ecological Systems Framework for Food Systems Research: Accommodating Transformation Systems and their Products. *International Journal of the Commons, 9*(2).
44. Mabhaudhi, T. et al. (2019). Prospects of Orphan Crops in Climate Change. *Planta*, pp. 1–14.
45. Nhamo, L., Mabhaudhi, T., and Magombeyi, M. (2016). Improving Water Sustainability and Food Security through Increased Crop Water Productivity in Malawi. *Water, 8*(9), p. 411.
46. Béné, C. et al. (2019). When Food Systems Meet Sustainability — Current Narratives and Implications for Actions. *World Development, 113*, 116–130. DOI: 10.1016/j.worlddev.2018.08.011.

47. Ericksen, P. J. (2008). Conceptualizing Food Systems for Global Environmental Change Research. *Global Environmental Change, 18*(1), 234–245.

48. Bizikova, L., Roy, D., Swanson, D., Venema, H. D., and McCandless, M. (2013). *The water-energy-food security nexus: Towards a practical planning and decision-support framework for landscape investment and risk management.* Winnipeg, Manitoba, Canada: International Development Research Centre (IDRC), p. 28.

49. Smith, A., Snapp, S., Chikowo, R., Thorne, P., Bekunda, M., and Glover, J. (2017). Measuring Sustainable Intensification in Smallholder Agroecosystems: A Review. *Global Food Security, 12*, 127–138.

50. Willett, W. et al. (2019). Food in the Anthropocene: The EAT–Lancet Commission on Healthy Diets from Sustainable Food Systems. *The Lancet, 393*(10170), pp. 447–492.

51. United Nations General Assembly (2015). Transforming Our World: The 2030 Agenda for Sustainable Development. In *Resolution adopted by the General Assembly (UNGA).* New York, U.S.A.: United Nations General Assembly.

52. Shilling, F., Khan, A., Juricich, R., and Fong, V. (2013). Using indicators to Measure Water Resources Sustainability in California. In *World Environmental and Water Resources Congress 2013: Showcasing the Future,* pp. 2,708–2,715.

53. United Nations Economic Commission for Europe (2017). Global Tracking Framework: UNECE Progress in Sustainable Energy. New York, U.S.A. and Geneva, Switzerland: United Nations Economic Commission for Europe. Retrieved from: https://unece.org/DAM/energy/images/CSE/ publications/Global_Tracking_Framework_-_UNECE_Progress_in_ Sustainable_Energy.pdf.

54. United Nations (2019). *The Sustainable Development Goals Report 2019.* New York, U.S.A.: United Nations.

55. Amigun, B., Musango, J. K., and Stafford, W. (2011). Biofuels and Sustainability in Africa. *Renewable and sustainable energy reviews, 15*(2), 1,360–1,372.

56. Vera, I., Langlois, L., and Rogner, H. (2007). Indicators for Sustainable Energy Development. *Energy Indicators for Sustainable Development: Country Studies on Brazil, Cuba, Lithuania Mexico, Russian Federation, Slovakia and Thailand,* pp. 5–16.

57. Spalding-Fecher, R., Winkler, H., and Mwakasonda, S. (2005). Energy and the World Summit on Sustainable Development: What next? *Energy Policy, 33*(1), 99–112.

58. Abdallah, K. B., Belloumi, M., and De Wolf, D. (2013). Indicators for Sustainable Energy Development: A Multivariate Cointegration and Causality

Analysis from Tunisian Road Transport Sector. *Renewable and Sustainable Energy Reviews, 25*, 34–43.

59. Mainali, B. and Silveira, S. (2015). Using a Sustainability Index to Assess Energy Technologies for Rural Electrification. *Renewable and Sustainable Energy Reviews, 41*, 1,351–1,365.

60. Razmjoo, A. A., Sumper, A., and Davarpanah, A. (2019). Development of Sustainable Energy Indexes by the Utilization of New Indicators: A Comparative Study. *Energy Reports, 5*, 375–383.

61. Bhatia, M. and Angelou, N. (2015). *Beyond Connections: Energy Access Redefined.* Washington D.C, USA: The World Bank.

62. Santoyo-Castelazo, E. and Azapagic, A. (2014). Sustainability Assessment of Energy Systems: Integrating Environmental, Economic and Social Aspects. *Journal of Cleaner Production, 80*, 119–138.

63. Bizikova, L. et al. (2014). *Water-Energy-Food Nexus and Agricultural Investment: A Sustainable Development Guidebook.* Winnipeg, Canada: International Institute for Sustainable Development (IISD).

64. Pérez-Escamilla, R. (2017). Food Security and the 2015–2030 Sustainable Development Goals: From Human to Planetary Health: Perspectives and Opinions. *Current Developments in Nutrition, 1*(7), p. e000513.

65. Stephan, R. M. et al. (2018). Water–Energy–Food Nexus: a Platform for Implementing the Sustainable Development Goals. *Water International, 43*(3), 472–479.

66. Leck, H., Conway, D., Bradshaw, M., and Rees, J. (2015). Tracing the Water–Energy–Food Nexus: Description, Theory and Practice. *Geography Compass, 9*(8), 445–460.

67. Mohtar, R. H. and Daher, B. (2016). Water-Energy-Food Nexus Framework for Facilitating Multi-stakeholder Dialogue. *Water International, 41*(5), 655–661.

68. Nhamo, L. et al. (2019). Sustainability Indicators and Indices for the Water-Energy-Food Nexus for Performance Assessment: WEF Nexus in Practice — South Africa Case Study. *Preprint,* vol. 2019050359, p. 17. DOI: 10.20944/preprints201905.0359.v1.

69. Dizdaroglu, D. (2017). The Role of Indicator-based Sustainability Assessment in Policy and the Decision-Making Process: A Review and Outlook. *Sustainability, 9*(6), p. 1,018.

70. Farinha, F., Oliveira, M. J., Silva, E. M., Lança, R., Pinheiro, M. D., and Miguel, C. (2019). Selection Process of Sustainable Indicators for the Algarve Region — OBSERVE Project. *Sustainability, 11*(2), p. 444.

71. Singh, R. K., Murty, H. R., Gupta, S. K., and Dikshit, A. K. (2012). An Overview of Sustainability Assessment Methodologies. *Ecological Indicators, 15*(1), 281–299.

72. Bell, S. and Morse, S. (2018). Sustainability Indicators Past and Present: What Next? *Sustainability, 10*(5), p. 1,688.
73. Warhurst, A. (2002). Sustainability Indicators and Sustainability Performance Management, in "Mining, Minerals and Sustainable Development [MMSD] Project Report," University of Warwick, UK, 43.
74. Garnåsjordet, P. A., Aslaksen, I., Giampietro, M., Funtowicz, S., and Ericson, T. (2012). Sustainable Development Indicators: From Statistics to Policy. *Environmental Policy and Governance, 22*(5), 322–336.
75. Bell, S. and Morse, S. (2012). *Sustainability Indicators: Measuring the Immeasurable?* London, UK: Routledge.
76. Zhen, L. and Routray, J. K. (2003). Operational Indicators for Measuring Agricultural Sustainability in Developing Countries. *Environmental Management, 32*(1), 34–46.
77. Nardo, M., Saisana, M., Saltelli, A., and Tarantola, S. (2008). *Handbook on Constructing Composite Indicators: Methodology and User Guide.* Paris, France: OECD publishing, p. 24.
78. Esty, D. C., Levy, M., Srebotnjak, T., and De Sherbinin, A. (2005). *Environmental Sustainability Index: Benchmarking National Environmental Stewardship.* New Haven, U.S.A.: Yale Center for Environmental Law & Policy.
79. Bizikova, L., Roy, D., Swanson, D., Venema, H. D., and McCandless, M. (2013). *The Water-energy-food Security Nexus: Towards a Practical Planning and Decision-support Framework for Landscape Investment and Risk Management.* Winnipeg, Manitoba, Canada: International Institute for Sustainable Development.
80. Fiksel, J. R., Eason, T., and Frederickson, H. (2012). *A Framework for Sustainability Indicators at EPA.* Washington D.C., U.S.A.: Citeseer.

Chapter 4

Green Compressed Air Energy Storage Technology

Mehdi Ebrahimi, David S.-K. Ting, and Rupp Carriveau

Turbulence and Energy Lab., Ed Lumley Centre for Engineering Innovation, University of Windsor, Windsor, Ontario, Canada

Abstract

Green Compressed Air Energy Storage (GCAES) is a new concept that combines thermal energy storage with traditional compressed air energy storage. The goal is to recover the heat of compression and reuse it during the expansion phase, thus eliminating the need for external heat. This chapter compares the overall performance of GCAES with its traditional Compressed Air Energy Storage (CAES) counterpart and estimates the amount of greenhouse gas reduction. Generally, a small change in one of the parameters of CAES systems can propagate to other factors, significantly altering the performance of the plant. A change in the system status can also alter the process parameters of the thermal sub-systems. The key process parameters are air and thermal fluid mass flow rates, temperature, and pressure of the system at each design point. These parameters can significantly influence the performance of a CAES plant. This chapter compares the effect of variations in these parameters on the performance of GCAES and traditional CAES plants with three stages of expansion. Thermodynamic model simulations were carried out over a pressure from 40 bar to 80 bar and hot water mass flow between 176 kg/s and 216 kg/s. The obtained results show that the net generated power for the GCAES and the traditional CAES systems are about 110 megawatts and 65 megawatts, with maximum efficiencies of 78.6% and 70.5%, respectively. This study also reveals that a GCAES not only makes accessible greater energy generation but can also reduce up to 80 tons of carbon dioxide (CO_2) per discharge cycle. This is equivalent to more than 270,000 tons of CO_2 emission per operating year for the modeled CAES plant.

Keywords: Green air energy storage, CAES, CO_2 emission, thermodynamic analysis

4.1. Introduction

Increasing demand for electricity with economic development and population growth is a significant challenge. According to the Statistical Review of World Energy 2021, globally more than 80% of the generated electricity comes from fossil fuels.[1] This report reveals that the COVID-19 pandemic had a dramatic impact on energy markets, with both greenhouse gas emissions and primary energy falling at their fastest rates since the Second World War. Nevertheless, renewable energy generation has conspicuously increased with the expansion of industrial divisions in recent decades.[2,3]

In spite of the numerous advantages of renewable energy resources, their unstable and intermittent nature and dependency on location and weather conditions are deliberated as their major drawbacks. This uncertainty can be removed through the deployment of energy storage systems; however, storage technology has thus far lagged generation. Among various energy storage technologies, compressed air energy storage (CAES) is one of the most competent large-scale concepts.[4,5] CAES has a low capital cost, relatively long lifetime, high reliability, and reasonable efficiency, making this technology a feasible solution for grid-scale energy storage applications. However, the waste heat of the compressors and turbines is a significant environmental issue in CAES plants. Also, to heat up compressed air during the discharging cycles external heat is required, which causes heat pollution. This energy can be recovered (in GCAES) in charge cycles from compression phases to reduce both thermal pollution and greenhouse gas emissions.[6,7]

The CAES technology has been in use for more than four decades, however, there is only one GCAES plant currently under operation in the world in Goderich, Canada[8] with zero-emission. The world's first CAES was built in 1978 and is still under operation in Huntorf, Germany, with the capacity of 320 megawatts and 750 megawatts per hour energy. In traditional CAES plants such as Huntorf, air is injected into natural gas burners to heat up before entering the air expanders, thus resulting in greenhouse gas emissions. In 1991, the second large CAES plant was constructed with a capacity of 110 megawatts in McIntosh, United States of America (US). The cavern volume in this plant is about 19M cubic meters, and the nominal discharge cycle can be up to 26 hours. This project is also equipped with gas burners. However, a portion of waste heat in the compression phase is recovered by a recapture unit, resulting in a fuel consumption rate about 25% less than that of the Huntorf CAES Plant.[9] This modification also improved the round-trip efficiency (RTE) from 42% to 54%.

Different approaches have been proposed to recover wasted heat of CAES plants to improve RTE and exergetic RTE.[10] In Ref. 11, a high-temperature Kalina cycle system 6 (KCS6) was used to recover the waste heat of the CAES gas turbine. This improvement increased RTE and exergetic RTE by 8.77% and 8.76% compared to a standalone CAES system. Wang et al.[12] presented a new technology with solar collectors and reduced the outlet temperature of the turbine from about 370 K to 280 K by using an ORC cycle with R245fa as the working fluid. They reported energy efficiency and exergy efficiency of about 98% and 68% for the integrated system, respectively. Mohammadi et al.[13] used an organic Rankine cycle (ORC) to increase power production and cooling capacity by recovering the dissipated heat. They used toluene as the working fluid and an absorption refrigeration cycle and reached an RTE of 53.94%. Meng et al.[14] proposed using a recuperator-equipped ORC cycle to recover the heat dissipated in the expansion and compression phases. They examined various ORC working fluids and reported that R123 improves the RTE by 6.7%. Razmi et al.[15] used a multi-effect desalination cycle to recover the waste heat of the turbine and compressors. Their method produces 38 kg/s and 62.5 kg/s freshwaters during the off-peak and peak times, as well as generating about 80 megawatts of power for peak shaving. Zeynalian et al.[16] proposed a hybrid system based on a combination of CAES, ORC, and CO_2 capture units to improve the environmental feasibility of the CAES. By using this method, they could remove about 88% of the produced CO_2. Javidmehr et al.[17] proposed a hybrid system composed of a micro gas turbine, a compressed air energy storage, a solar dish collector, an organic Rankine cycle, and a multi-effect distillation system. They applied energy and exergy analyses to investigate the thermodynamic performance of the system. They achieved an additional power of about 17 kW by coupling the exhaust heat from the turbine with ORC units. Raju and Khaitan[18] simulated the thermodynamic behavior of air streams by using mass and energy conservation equations. They developed a flexible method for the calculation of the heat transfer coefficient in CAES systems. They calibrated the proposed model with the real data from the Huntorf CAES plant. Liu et al.[19] thermodynamically simulated a CAES system and analyzed the performance of the system. They showed that the overall efficiency of the system can be improved by about 10% if a combined cycle is being used. They reported that the recovered heat from compressors' intercoolers can be utilized for improving system performance by keeping the steam part of the plant on the hot standby mode. In Ref. 20, an adiabatic CAES system with thermal energy storage was proposed. In this model, no external heat was added to the CAES system. With this modification, the RTE of the system was improved to around 69%.

Despite the numerous studies on CAES from different perspectives, the early-stage development of CAES technology has been slow with only a few industrial plants realized over the past few decades. Studying the effect of waste heat recovery on the performance of CAES systems is also lacking. This is a key field of study that can expedite the process of moving from traditional CAES plants to plants with no greenhouse gas emissions. Globally, there is only one relatively small under operation GCAES plant with the capacity of 0.7 megawatt in Goderich, Canada.[8] To fill this gap, this work aims to propose the most efficient GCAES for various operating conditions. To improve our knowledge about the thermal subsystem of GCAES systems, the thermal performance and the thermal fluid mass flow rate at various system statuses are also studied in this work, which can assist this technology's development and penetration into the energy industry. This work also calculates how much CO_2 emission can be reduced by modifying traditional CAES to GCAES systems at various operating conditions.

4.2. CAES Technology Description

The overall operation of the CAES systems can be considered as a variation of gas power plants where the compression and expansion cycles operate independently. In CAES plants, the compression phase takes place in electricity off-peak periods and the expansion phase comes about in peak periods.[21] The schematic flow diagram of a CAES system is shown in Figure 4.1. Generally, in CAES systems, air is compressed via compressors and is transferred to the air accumulator. Since the storage volume of air in industrial plants is enormous, usually an underground cavern is used for the air storage.[22]

At each stage of the compression phase, the air temperature rises. In GCAES systems, this energy is recovered. This process has three advantages. First, the required power is reduced. Secondly, a smaller CAES vessel can be used. Most importantly, it eliminates gas burners from the system and thus also greenhouse gas production. This modification also can significantly improve the performance of CAES systems.

The discharge phase starts at the electricity demand peak times when the pressurized air from the air storage returns to the system. Air is then heated up before entering the expansion stages (in GCAES systems, the return air is heated by the stored hot thermal fluid). The hot, pressurized air then enters the turbines and generates electricity. It is worth mentioning that in GCAES systems, the size of the machine cooling equipment could also be reduced dramatically if the plant's waste heat is used to heat up the air in the expansion phase.

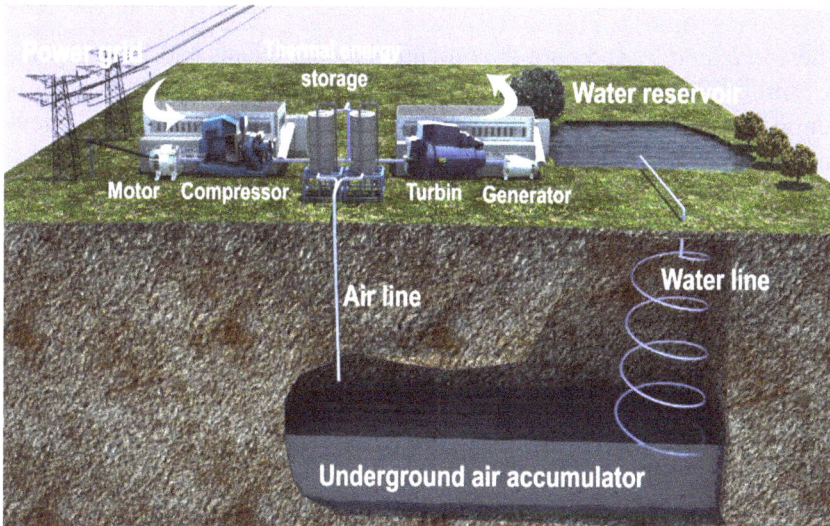

Figure 4.1.   Process flow diagram for GCAES concept.

4.3. Mathematical Analysis

In this section, the governing equations for the thermodynamic analysis of the system and its subsystems are presented. To have a fair comparison between generated models, it was assumed that the mass flow rate and the temperature of the hot water stream for all cases are the same. Upper and lower limits for process parameters in the thermodynamic model were also selected based on our experience with the operation of the Toronto Island Underwater CAES plant and the Goderich advanced CAES plant, such that the simulated system is one that has an acceptable RTE and could be built at a reasonable cost.[5,23]

In this work, air is treated as an ideal gas, and it is assumed that the kinetic and potential energies of fluid streams are negligible. The specific heat of air is also considered as a function of temperature and the ratio of specific heats. For a fair comparison between generated models, it is assumed that the isentropic efficiency of the turbine for both traditional CAES and GCAES systems is 89% at any stage.[24] The outlet pressure of the system is fixed at 100 kPa and the system discharges until it runs out of hot water in a hot tank with the capacity of 3,000 cubic meters. A logarithmic polynomial function[25] was used for the calculation of the specific heats of fluid streams as:

$$c_p(T) = a + bT + cT^2 + dT^3 + eT^4 + fT^5 \tag{4.1}$$

where a, b, c, d, e, and f are the polynomial coefficients and T is the temperature in Kelvin (K). Considering that the outlet air from the cavern is humid, the Enthalpy of humid air at any temperature then can be calculated as:

$$h = c_{p\text{-}air}(T) + (2500 + c_p - h_2o(T)) \tag{4.2}$$

Thermodynamic evaluation of the models has been performed by simulating the different components of the system based on energy and mass conservation equations.

4.4. System Efficiency and Heat Efficiency

Generally, GCAES systems have two different energy inputs — (i) air stream energy and (ii) hot water stream energy. They have two energy outputs — (i) the mechanical energy of expanders and (ii) the heat energy of water leaving heat exchangers. To compare CAES and GCAES systems and to be able to study the effect of air property changes on the system performance, two efficiencies including discharge cycle efficiency and heat recovery efficiency are defined in this work as:

$$Cycle\ Efficiency = \frac{\dot{E}_{Turb1} + \dot{E}_{Turb2} + \dot{E}_{Turb3}}{\dot{E}_{CavernOutlet} + \dot{E}_{HotW\_in}} \tag{4.3}$$

$$Heat\ Efficiency = \frac{\dot{E}_{HotW\_in} - \dot{E}_{HotW\_out}}{\dot{E}_{HotW\_in}} \tag{4.4}$$

where $\dot{E}_{Turb\#}$ is the k-th expander's energy, $\dot{E}_{CavernOutlet}$ air stream energy at the outlet of the cavern and $\dot{E}_{HotW\_in}$ the energy of hot water at the inlet of the heat exchangers.

For calculating other thermodynamic parameters of the system such as specific entropy, enthalpy, and exergy additional assumptions are required. These assumptions are summarized in Table 4.1.

Table 4.1.   The assumption of parameters considered for calculation of system status.

Component	Parameters
First stage expander	$\eta_{Isentropic}$ = 89%
First heat exchanger	*Air pressure loss = Equ (4); up to 11kPa* *Water pressure loss $\cong$ 3.5kPa* *Water inlet temperature = 220°C* *Hot water mass flow rate portion = 0.333*
Second stage expander	$\eta_{Isentropic}$ = 89%
Second heat exchanger	*Air pressure loss = Equ (4); up to 10kPa* *Water pressure loss $\cong$ 3.5kPa* *Water inlet temperature = 220°C* *Hot water mass flow rate portion = 0.327*
Third stage expander	$\eta_{Isentropic}$ = 89%
Third heat exchanger	*Air pressure loss = Equ (4); up to 7kPa* *Water pressure loss $\cong$ 3.5kPa* *Water inlet temperature = 220°C* *Hot water mass flow rate portion = 0.340*
Air storage pipeline	*Air pressure loss $\cong$ 6.5% of the cavern pressure*

4.5. Results and Discussion

In this work, the effect of air temperature and pressure, expander pressure ratios, hot water mass flow rate, and air mass flow rate on the performance of traditional CAES and GCAES systems are investigated for a wide range of operating conditions. The thermodynamic simulations for both traditional CAES and GCAES systems were performed for systems with air mass flow rate of 270 kg/s and hot water mass flow rate between 176 kg/s and 216 kg/s, air storage pressures between 40 bar and 80 bar, cavern air temperatures between 5°C and 25°C. According to the generated thermodynamic model and the considered assumptions, the generated power for GCAES and traditional CAES were between 87 megawatts to 109 megawatts, and 44 megawatts to 67 megawatts, respectively. The water in the hot tank was fixed at 220°C. In Table 4.2, the thermodynamic state of the system (Figure 4.2) is presented at the inlets and outlets of every major component of the system, and how they respond to different air storage conditions for two sample state points.

In Figure 4.3, the total exergy of the system and in Figure 4.4, the exergy of the air stream are compared at various air temperatures and pressures for air and water mass flow rate of 270 kg/s and 216 kg/s, respectively. From these figures, it can be observed that the effect of temperature compared to the pressure change is insignificant for the considered operation range. Moreover, for all the cases, the exergy of the GCAES system is greater than the traditional CAES system since wasted heat in the GCAES system is recovered and reused in the system.

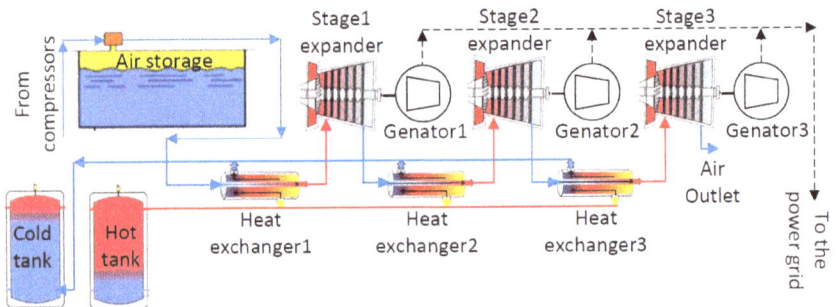

Figure 4.2.   The schematic diagram of the applied GCAES for the thermodynamic analysis.

Table 4.2. Air thermodynamic properties for two different states of the system.

State point	Fluid	State point 1 air storage pressure 60 bar and T = 15°C			State point 2 air storage pressure 80 bar and T = 25°C		
		T (°C)	P (kPa)	Mass (kg/s)	T (°C)	P (kPa)	Mass (kg/s)
Air storage outlet	Air	15.0	60.0	270.10	25	80.0	270.10
Air storage pipeline outlet	Air	7.6	50.3	270.10	17.8	68.5	270.10
Inlet of Stage 1 expander	Air	194.4	50.1	270.10	192.4	68.2	270.10
Outlet of Stage 1 expander	Air	57.28	12.5	270.10	55.4	17.1	270.10
Inlet of Stage 2 expander	Air	201.0	12.2	270.10	200.3	16.8	270.10
Outlet of Stage 2 expander	Air	63.2	3.1	270.10	62.5	4.2	270.10
Inlet of Stage 3 expander	Air	203.0	2.7	270.10	202.9	3.8	270.10
Outlet of Stage 3 expander	Air	69.5	0.7	270.10	69.5	1.0	270.10
Hot tank outlet	Water	220.0	50.0	189.0	220	50.0	189.0
High-pressure line	Water	22.6	49.8	62.9	32.7	49.8	62.9
Intermediate pressure line	Water	72.2	49.9	61.8	70.39	49.9	61.8
Low-pressure line	Water	83.2	49.8	64.3	82.5	49.8	64.3
Cold tank inlet	Water	59.5	49.8	189.0	62.0	49.8	189.0

In Figure 4.5, the net generated power by GCAES system is plotted at various system pressures and temperatures. Figure 4.6 shows the corresponding power when the generated heat in the compression phase is not recovered (traditional CAES). It can be observed that the net generated power in the traditional CAES is significantly less than GCAES system. For example, at $T = 10°C$, $P = 40$ bar, and $T = 25°C$, $P = 80$ bar, the net generated power by GCAES are 9.1% and 5.6% larger than the traditional CAES system. It also can be seen that the net generated power is more sensitive to pressure change at lower air temperatures. For example, when at 15°C, air pressure changes from 40 bar to 50 bar, the net generated

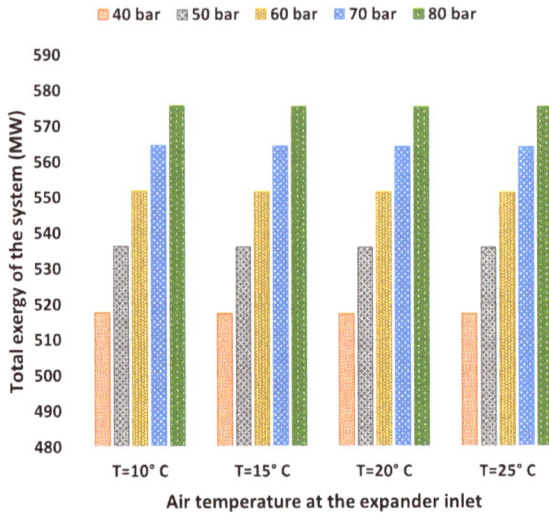

Figure 4.3.   Total exergy of the system at various operating pressure and temperature.

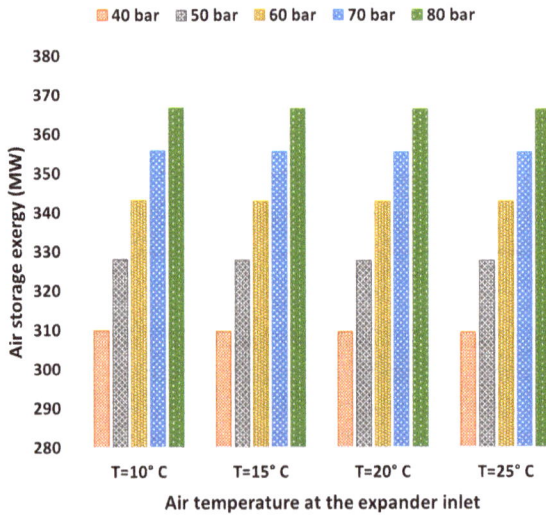

Figure 4.4.   Exergy of the airstream at the inlet of the system at various operating pressure and temperature.

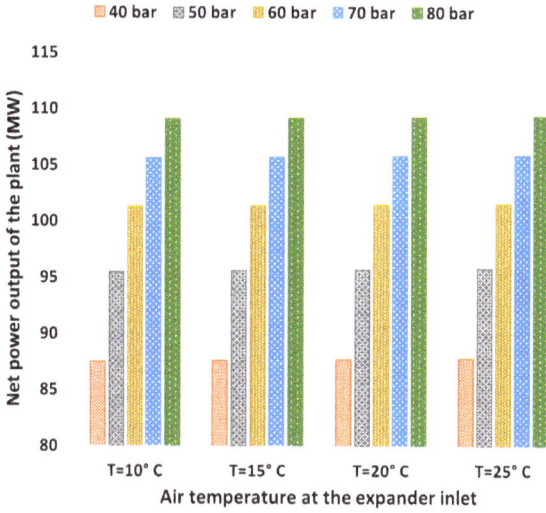

Figure 4.5. The net output power for a GCAES system at various operating pressure and temperature.

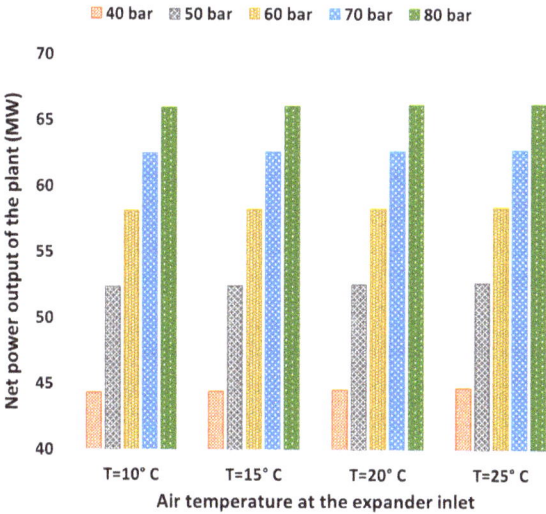

Figure 4.6. The net output power for a traditional CAES system at various operating pressure and temperature.

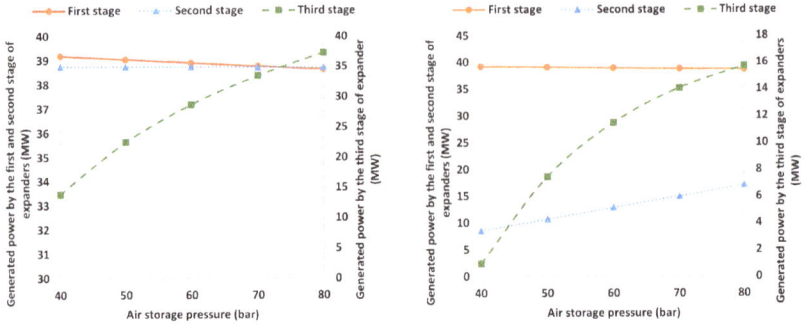

Figure 4.7.   The generated power by each stage of the expander for the (left) GCAES and (right) traditional CAES systems for instance at the air storage temperature of $T = 10°C$.

power increases about 9.5%, while when pressure changes from 70 bar to 80 bar, the net output power change is about 3.5%.

In Figure 4.7, the generated power by each of the three expansion stages is shown for GCAES and traditional CAES systems for an air storage temperature of $T = 10°C$. From this figure, it can be seen that the performance of Stages 1 and 2 are pretty consistent for the GCAES system. It should be noted that since the outlet of Stage 3 is the local environment, the generated power by this stage has a dependency on the ambient weather condition. On the other hand, the power generation graph for the traditional CAES system indicates that for all the expansion stages, the generated power is less than the same expansion stage of GCAES system, specifically for Stages 2 and 3. This is due to the fact that a portion of the generated power is being used for reheating air before entering the expanders.

In Figure 4.8, the efficiency of the discharge cycle for GCAES and traditional CAES system under various operating conditions are compared. This figure shows that at the considered operation range, GCAES can have an efficiency of up to 21% more than traditional CAES systems. It also reveals that every 5°C increase in cavern air temperature, the efficiency of GCAES and CAES systems are improved by about 0.17% and 0.35%. As expected, the efficiency of both systems improved with increasing air pressure; however, it should be noted that higher operating pressure implies higher capital costs. Therefore, a cost-benefit analysis is required for each project to determine the optimum pressure range.

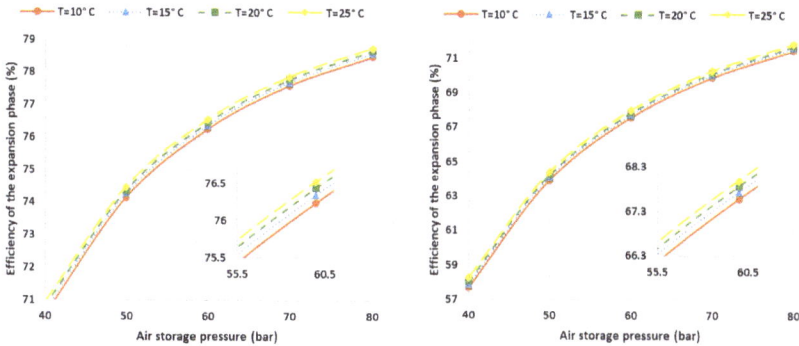

Figure 4.8.   Efficiency of the discharge cycle at various system status.

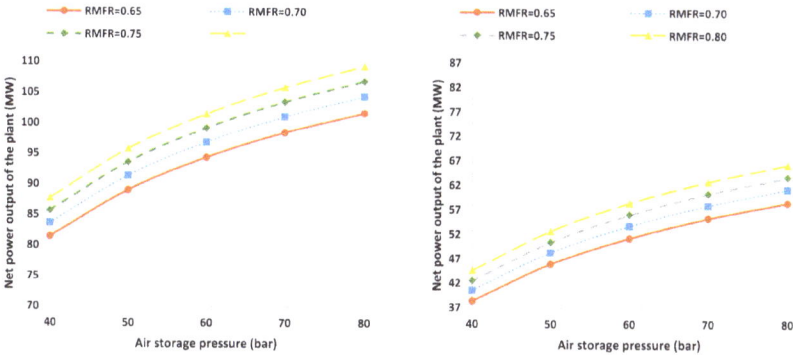

Figure 4.9.   Net power output from GCAES and traditional CAES systems for various air storage pressure at the air mass flow rate of 270 kg/s.

In Figure 4.9, the net generated power for the GCAES and the traditional CAES systems are compared for the mass flow rate of 176 kg/s to 216 kg/s at various system pressures. To have a fair comparison between the two models, it is assumed that in the traditional CAES plant, the air before entering the expanders is being heated up with the same thermal fluid stream as GCAES. In this case, the required energy to heat up the hot stream is provided by the system internally. From this figure, it can be observed that for both systems, at higher mass flow rates, greater power can be generated. However, it should be considered that at this condition,

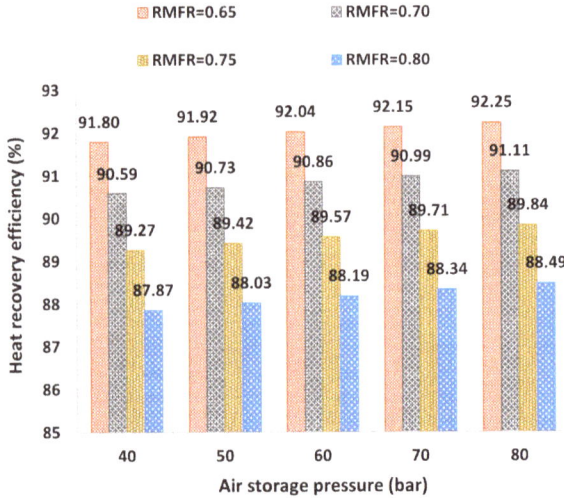

Figure 4.10.   Heat recovery efficiency of the discharge cycle at various hot water mass flow rates and operating conditions.

either hot tank or air storage tank pressure reaches their minimum level faster, and consequently, the discharge duration will be shorter. Therefore, for each plant, the optimum air and thermal fluid mass flow rate should be determined according to the air storage size (usually a cavern) and hot tank size and temperature.

As mentioned earlier, it is assumed that the heat transfer efficiency for both systems is the same. According to Equation 4.3, and the thermodynamic status of the system at various operating conditions the heat recovery efficiency can be determined (Figure 4.10). From this figure, it can be seen that at lower hot water mass flow rates, higher heat recovery efficiencies are obtainable. It can also be observed that the heat transfer efficiency slightly improves as air pressure increases. For example, the heat transfer efficiency is about 0.5% more at the system pressure of $P = 80$ bar compared to the $P = 40$ bar.

The total efficiency of the GCAES and traditional CAES systems is shown in Figure 4.11 at various system pressures and hot water mass flow rates. This figure shows that the performance of the system improves by about 0.28% for every one bar system pressure increase. This figure also indicates that at higher system pressure, the mass flow rate has a greater impact on the total efficiency of the system. By comparing the efficiency

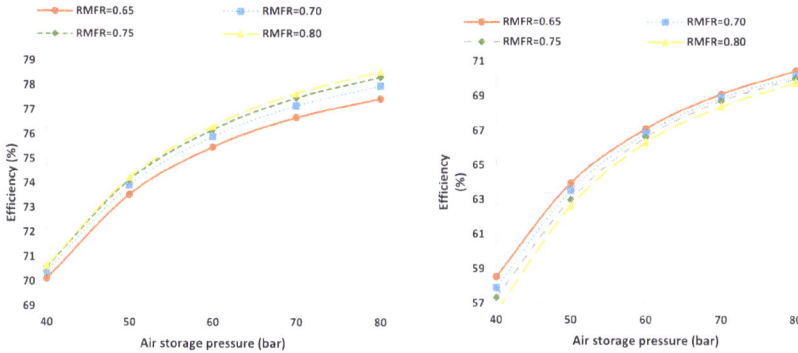

Figure 4.11.   Total efficiency of the discharge cycle at various hot water mass flow rates and operating conditions.

Table 4.3.   The discharge duration for various hot water mass flow rates.

	Hot water mass flow rate			
	$\dot{m} = 176$ kg/s	$\dot{m} = 189$ kg/s	$\dot{m} = 203$ kg/s	$\dot{m} = 216$ kg/s
Discharge duration (hr)	4.67	4.33	4.05	3.79

of GCAES and traditional CAES systems, it can be concluded that at the considered operating range, the efficiency of GCAES is around 24.9% and 12.6% more than traditional CAES systems at the air storage pressure of 40 bar and 80 bar, respectively.

As discussed, after each stage of the expansion phase, air needs to be heated up before entering the next stage. In CAES plants, the required energy for such a process supplies from an external energy resource which is usually a fossil fuel burner. In this part, it will be analyzed how much CO_2 emission can be reduced by the construction of GCAES instead of a CAES system.

For the first step, according to the hot water mass flow rate and thermodynamic specification of each model, the discharge duration was calculated (Table 4.3). Then by estimating the equivalent CO_2 emission for heating up that volume of water in the cold tank to a certain temperature, the reduction in CO_2 dispersion can be calculated for the generation of each megawatt-hour of energy (CO_2-e/MWhr) for each operating condition.

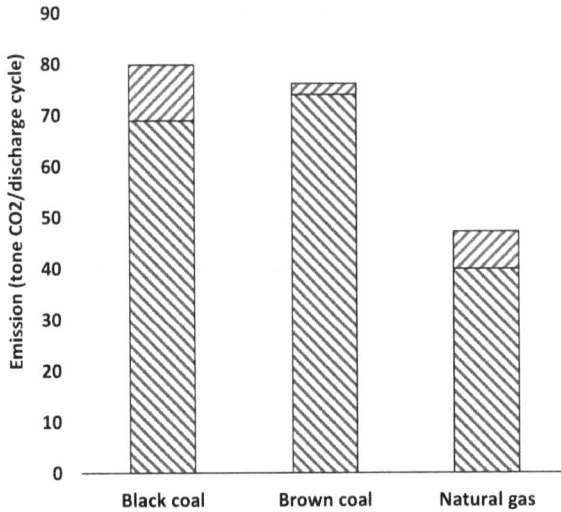

Figure 4.12.   Total CO_2 emissions from the combustion of brown coal, black coal, and natural gas, for the heating up of the water in the cold tank for a discharge cycle of a traditional CAES system with the duration of 4.7 hours and 465 MWhr energy generation.[26]

It should be noted that the CO_2 emissions in burners vary depending on the burning process and the feedstock. In this part, CO_2 production rate is calculated as an example for the system pressure of 60 bar and the discharge duration of 4.7 hours with 465 MWhr energy generation, using black and brown coal and natural gas as the most common fossil fuel globally. In all these fuels, hydrogen is separated from the carbon, producing large amounts of CO_2 emissions, roughly 342 kg to 396 kg CO_2-e/MWhr for black coal, 367 kg to 378 kg CO_2-e/MWhr for brown coal and 198 kg to 234 kg CO_2-e/MWhr for natural gas.[26] In Figure 4.12, the CO_2 emissions resulting from burning these fuels are compared. The values in this figure have been calculated according to the required energy in each case. The red error bars in this figure reflect the variation in emissions that happens to the natural variation in the carbon content of different fossil fuels. This value is estimated from the IPCC emission factors.[27] From this figure, it can be observed that black coal can produce up to 80 tons of CO_2 per discharge cycle. The lowest value is for natural gas with about 40 tons of CO_2 per discharge cycle.

Table 4.4. Generated energy by expanders at various operating conditions.

Fossil fuel type	Tone CO_2 emission reduction			
	$\dot{m} = 176\,\text{kg/s}$	$\dot{m} = 189\,\text{kg/s}$	$\dot{m} = 203\,\text{kg/s}$	$\dot{m} = 216\,\text{kg/s}$
Black coal	27,123	25,185	23,507	22,039
Brown coal	27,388	25,431	23,736	22,254
Natural gas	15,877	14,743	13,760	12,901

In Table 4.4, the yearly CO_2 emission for instance for the system pressure of 60 bar is given. This table indicates that by recovering heat and going with a green CAES system, at least 12,901 and up to 27,123 tons of CO_2 emissions can be prevented.

4.6. Conclusion

Renewable energy and green energy storage technologies are key elements for mitigating climate change. Compressed air energy storage systems are considered as one of the clean energy storage technologies. However, fossil fuels are the main energy source for heating air in existing CAES plants.[9] Burning fossil fuels produce a significant amount of greenhouse gas emissions. The main objective of this work was to compare the performance of the traditional and green compressed air energy storage systems and to determine how much CO_2 emission occurs when running a traditional CAES system. Modeling was performed for systems with a pressure range of 40 bar to 80 bar and hot water mass flow between 176 kg/s and 216 kg/s with a fixed air mass flow rate of 270 kg/s. The obtained result showed that for the defined models the net generated power and efficiency of GCAES systems can be up to 35% and 8%, respectively, more than that of the traditional CAES systems. This study also revealed that for the studied conditions, the traditional CAES can produce up to 80 tons of CO_2 per discharge cycle while the GCAES are emission-free with higher efficiencies and greater energy generation. This work also indicated that clean engineering of GCAES comes with a higher initial cost. Therefore, some incentives from the government are required to realize such energy storage technology with zero-carbon emissions.

Nomenclature

		Subscripts and superscripts	
C_p	specific heat at constant pressure	cold	cold medium
e	specific exergy (J/kg)	f	final state
E	energy (J)	hot	hot medium
E_D	exergy destruction (J)	i	initial state
E_x	exergy (J)	in	inlet
h	enthalpy (J)		
m	mass flow rate (kg/s)	*Abbreviations*	
P	pressure (kPa)	CAES	compressed air energy storage
Q	heat (J)	GCAES	green compressed air energy storage
R	gas constant (J/(kg K))	EX	turbo expander
s	entropy (J/K)	GN	generator
T	temperature (K)		
W	work (J)	*Greek letters*	
W_{exp}	expander's power (kW)	γ	polytropic exponent
		η	pressure loss constant

References

1. B. p.l.c, "Statistical Review of World Energy," Whitehouse Associates, London, 2021.
2. Razmi, A., Soltani, M., Tayefeh, M., Torab, M., and Dusseault, M. B. (2019). Thermodynamic Analysis of Compressed Air Energy Storage (CAES) Hybridized with a Multi-Effect Desalination (MED) System. *Energy Conversion and Management, 199*, 1–11.
3. Ebrahimi, M., Ting, D. S.-K., Carriveau, R., and McGillis, A. (2021). Efficiency and Sensitivity Analysis of Cavern-Based CAES Systems During off-Design Operating Conditions, in *Sustaining Tomorrow; Proceedings of Sustaining Tomorrow 2020 Symposium and Industry Summit*, Springer.
4. Ebrahimi, M., Carriveau, R., Ting, D. S.-K., McGillas, A., and Young, D. (2019). Transient Thermodynamic Assessment of the World's First Grid Connected UWCAES Facility by Exergy Analysis, in *IEEE*, BREST, France.

5. Ebrahimi, M., Ting, D. S.-K., Carriveau, R., McGillis, A., and Young, D. (2020). Optimization of a Cavern-Based CAES Facility with an Efficient Adaptive Genetic Algorithm. *Energy Storage*, 2(6), 1–12.

6. Hartmann, N., Vöhringer, O., Kruck, C., and Eltrop, L. (2012). Simulation and Analysis of Different Adiabatic Compressed Air Energy Storage Plant Configurations. *Applied Energy*, 93, 541–548.

7. Liu, W., Li, Q., Liang, F., Liu, L., Xu, G., and Yang, Y. (2014). Performance Analysis of a Coal-Fired External Combustion Compressed Air Energy Storage System. *Entropy*, 16, 5,935–5,953.

8. Hydrostor (2021). Retrieved from: https://hydrostor.ca/.

9. Khan. N., Dilshad, S., Khalid, R., Kalair, A. R., and Abas, N. (2019). Review of Energy Storage and Transportation of Energy. *Energy Storage*, 1(3), 1–49.

10. Alsagri, A. S., Arabkoohsar, A., RezaRahbari, H., and Alrobaian, A. A. (2019). Partial Load Operation Analysis of Trigeneration Subcooled Compressed Air Energy Storage System. *Journal of Cleaner Production*, 238, 1–12.

11. Zhao, P., Wang, J., and Dai, Y. (2015). Thermodynamic analysis of an Integrated Energy System Based on Compressed Air Energy Storage (CAES) System and Kalina Cycle. *Energy Conversion and Management*, 98, 161–172.

12. Wang, X., Yang, C., Huang, M., and Ma, X. (2018). Off-design Performances of Gas Turbine-Based CCHP Combined with Solar and Compressed Air Energy Storage with Organic Rankine Cycle. *Energy Conversion and Management*, 156, 626–638.

13. Mohammadi, A., Ahmadi, M. H., Bidi, M., Joda, F., Valero, A., and Uson, S. (2017). Exergy Analysis of a Combined Cooling, Heating and Power System Integrated with Wind Turbine and Compressed Air Energy Storage System. *Energy Conversion and Management*, 131, 69–78.

14. Meng, H., Wang, M., Aneke, M., Luo, X., Olumayegun, O., and Liu, X. (2018). Technical Performance Analysis and Economic Evaluation of a Compressed Air Energy Storage System Integrated with an Organic Rankine Cycle. *Fuel*, 211, 318–330.

15. Razmi, A., Soltani, M., Tayefeh, M., Torabi, M., and Dusseault, M. (2019). Thermodynamic Analysis of Compressed Air Energy Storage (CAES) Hybridized with a Multi-Effect Desalination (MED) System. *Energy Conversion and Management*, 199, 1–11.

16. Zeynalian, M., Hajialirezaei, A. H., Reza Razmi, A., and Torabi, M. (2020). Carbon Dioxide Capture from Compressed Air Energy Storage System. *Applied Thermal Engineering*, 178, 1–10.

17. Javidmehr, M., Joda, F., and Mohammadi, A. (2018). Thermodynamic and Economic Analyses and Optimization of a Multi-Generation System Composed by a Compressed Air Storage, Solar Dish Collector, Micro Gas

Turbine, Organic Rankine Cycle, and Desalination System. *Energy Conversion and Management, 168*, 467–481.

18. Raju, M. and Khaitan, S. K. (2012). Modeling and Simulation of Compressed Air Storage in Caverns: A Case Study of the Huntorf Plant. *Applied Energy, 89*(1), 474–481.

19. Liu, W., Liu, L., Zhou, L., Huang, J., Zhang, Y., Xu, G., and Yang, Y. (2014). Analysis and Optimization of a Compressed Air Energy Storage-Combined Cycle System. *Entropy, 16*(6), 3,103–3,120.

20. Sciacovelli, A., Li, Y., Chen, H., Wu, Y., Wang, J., Garvey, S., and Ding, Y. (2017). Dynamic Simulation of Adiabatic Compressed Air Energy Storage (A-CAES) Plant with Integrated Thermal Storage — Link between Components Performance and Plant Performance. *Applied Energy, 185*, 16–28.

21. Luo, X., Wang, J., Dooner, M., Clarke, J., and Krupke, C. (2016). Modelling Study, Efficiency Analysis and Optimisation of Large-Scale Adiabatic Compressed Air Energy Storage Systems with Low-Temperature Thermal Storage. *Applied Energy, 162*, 589–600.

22. Ebrahimi M., Ting, D. S.-K., Carriveau, R. and McGillis, A. (2020). Hydrostatically Compensated Energy Storage Technology. In *Green Energy and Infrastructure*, Taylor and Francis, pp. 1–25.

23. Ebrahimi, M., Carriveau, R., Ting, D. S.-K., and McGillis, A. (2019). Conventional and Advanced Exergy Analysis of a Grid Connected Underwater Compressed Air Energy Storage Facility. *Applied Energy, 242*, 1,198–1,208.

24. Carriveau, R., Ebrahimi, M., Ting, D. S.-K., and McGillis, A. (2019). Transient Thermodynamic Modeling of an Underwater Compressed air Energy Storage Plant: Conventional Versus Advanced Exergy Analysis. *Sustainable Energy Technologies and Assessments, 31*, 146–154.

25. Lanzafame, R. and Messina, M. (2000). A New Method for the Calculation of Gases Enthalpy. In *35th Intersociety Energy Conversion Engineering Conference and Exhibit*.

26. Longden, T., Beck, F. J., Jotzo, F., Andrews, R., and Prasad, M. (2022). 'Clean' hydrogen? — Comparing the Emissions and Costs of Fossil Fuel Versus Renewable Electricity Based Hydroge. *Applied Energy, 306*, 1–14.

27. Muradov, N. (2017). Low to Near-Zero CO_2 Production of Hydrogen from Fossil Fuels: Status and Perspectives. *International Journal of Hydrogen Energy, 42*(20), 14,058–14,088.

Chapter 5

Marinas can Co-function as Biodiversity Sanctuaries

Loke Ming Chou*, Chin Soon Lionel Ng*,
Kok Ben Toh*, Karenne Tun†, Pei Rong Cheo†,
and Juat Ying Ng†
*Department of Biological Sciences,
National University of Singapore, Singapore
†National Parks Board, Singapore

Abstract

Marinas serve the need of the boating community for a sheltered environment. Their development involves major shoreline modification that often includes land reclamation and the construction of seawalls and breakwaters. These have an impact on the natural shore ecosystem. The semi-enclosed waters become a modified environment with different hydrodynamic and physico-chemical conditions to the adjacent open sea. Various structures such as pontoons, piles, seawalls and breakwaters abound in the marina, increasing structural complexity of its water body. They provide opportunities for the colonization of biological communities that are not found in open water. Surveys of three Singapore marinas were conducted to examine if this highly modified environment in the tropics could promote the development of marine biodiversity. The soft bottom sediment of these marinas supported 73 macrobenthic taxa from eight phyla dominated by polychaetes. Submerged sides of the berthing pontoons were occupied by 94 epibiotic taxa, including macroalgae, soft corals and bivalves. The seawalls of one marina were naturally colonized by 21 genera of reef-building corals. In addition, 49 fish species comprising estuarine or reef-associated ones, inhabited the marinas. With adequate planning during the design, construction, and operational phases, marinas can co-function effectively as biodiversity sanctuaries. These include ensuring sufficient water exchange throughout the marina; providing hard substrates to facilitate colonization of reef-building species; encouraging "green" and non-polluting practices among marina users; as well as information-sharing between stakeholders via a systematic and sustained biodiversity monitoring programme.

Keywords: Coastal biodiversity, marinas, macrobenthos, epibenthos, fishes.

5.1. Introduction

The increasing demand for marine recreation and boat-mooring facilities has led to the construction of marinas throughout the world.[1-3] Sydney Harbour alone has 40 marinas servicing 35,000 recreational vessels,[4] while the small island state of Singapore has seven. Two of Singapore's marinas were developed just less than 15 years ago, highlighting the growing attraction of marine recreation.

Marinas in the past were developed with the sole purpose of providing a sheltered environment for small craft. Less attention was given to maintaining adequate water flow within these created semi-enclosed bodies of water, usually flanked by breakwaters and seawalls. Whether the altered environment could support marine life remained inconsequential. The construction of marinas transformed natural shores to human-engineered structures suitable for the berthing of yachts but, at the same time, interfered with natural coastal physical processes, resulting in impacts to water quality, adjacent coastal habitats and marine biodiversity of the original shore.[5-8] Activities such as land reclamation, dredging, shoreline realignment and the installation of concrete walls and piles during marina development decimated original intertidal habitats and nearby coral reefs.[8,9] Specifically, bottom dredging directly removes benthic organisms from the seabed and increases the availability of trace metals in the sediments through re-suspension, causing them to accumulate in the food chain.[9,10] The resultant increased sediment load and turbidity of the waters then smother sediment-sensitive fauna such as bivalves and corals.[9,11]

Boat operation and maintenance also introduce contaminants into the water, thus altering other physico-chemical parameters of the marina environment. The contaminants include petroleum hydrocarbons from engine exhaust and fuel leakage,[12-14] heavy metals leached from antifouling paints and boat cleaning detergents,[15-18] and untreated sewage from toilet flushing.[19,20] These pollutants negatively impact marine life and are accumulated due to the semi-enclosed nature of marinas that limits flushing by currents[21,22] but nothing in the literature suggests that they have initiated microalgal blooms in marinas. Environmental deterioration is not conducive to the continued operation of marinas as the sight and odor of deoxygenated contaminated waters with floating dead organisms are repulsive.

5.2. Do Marinas Promote Biodiversity?

Modern marinas are now constructed to provide an aquatic environment that is less prone to pollution and more able to sustain marine life. Sheltered conditions are still maintained but water flow is improved through gaps in the breakwaters. Enhanced water circulation and increased water changeover will help to prevent pollution and promote marine life. The modified environment need not necessarily be equated with the decimation of marine biodiversity because for as long as there is adequate water circulation and exchange, species favored by the changed conditions can develop and form new biological communities.[8,23] This has been observed in some of Singapore's marinas and the boat canal system of a residential cove development.[24-28] The newly established marine life is surprisingly diverse and includes scleractinian corals and seahorses, indicating that marinas can perform a useful role in supporting rather than diminishing marine biodiversity.

Apart from seawalls and breakwaters, a marina's aquatic environment is usually filled with artificial structures such as piles and floating pontoons, which traverse the water column and offer a larger variety of niche habitats (Figure 5.1).[29-32] This structural complexity helps to support and sustain biological communities.[33] Vertical structures novel to the environment can potentially function as habitat for rich epibiotic assemblages e.g., sponges, corals, ascidians, bivalves, and algae.[31-36]

Figure 5.1.   Anthropogenic structures in marinas required for berthing of yachts can enhance biodiversity.

These in turn support higher trophic levels of the marine food chain and ultimately fish communities by functioning like artificial reefs, as has been observed from various marinas in subtropical areas.[37–43] However, information on the capacity of tropical (especially equatorial) marinas to support marine life is poor.

An assessment of the biodiversity present in three of Singapore's marinas was therefore initiated and compared with the biodiversity within a one-kilometer radius of the open sea. All three marinas were built on mature urban waterfronts that involved land reclamation and extensive seaward shift of the shoreline. Their influence on marine biodiversity is based on the overall modified environment of the new seafront. Also included was the correlation of biodiversity within the marinas with water quality and their physical design. The main research question driving the investigation was how much and what kind of marine biodiversity can modified coastal habitats such as tropical marinas support? Macrobenthos in the soft bottom sediment, epibiota on the berthing pontoons and sea-walls, and fishes throughout the water column of the three marinas were surveyed. The assessment was meant to provide an indication of biodiversity distribution within, and differences between marinas and the adjacent sea to determine the extent that marinas could serve as marine biodiversity sanctuaries.

The second question was how marina habitats can be made to enhance marine biodiversity and serve the additional purpose of a marine sanctuary. This was to help identify physical and chemical water quality parameters, as well as the design and layout of physical structures that promote marine biodiversity within a marina. Such information will be useful for the planning of future marinas and/or modifications of existing marinas to enhance marine biodiversity.

The investigation will establish a better understanding of the role of marinas as biodiversity sanctuaries. Enhancing this role will widen the utility of marinas so that they not only provide a safe environment for boats, but also a sanctuary for marine biodiversity. In this way, marinas and other similar development need not be viewed as coastal development that destroys or diminishes marine biodiversity but as development that eventually promotes marine biodiversity. Marinas could take lead roles in marine conservation and be a responsible partner in protecting and sustaining marine life.[44] To achieve this, the proper management of marinas should be complemented with relevant scientific inputs and proper monitoring to promote marine biodiversity.

5.3. Study Sites and Environmental Conditions

Raffles Marina (RM) situated at Tuas (Figure 5.2) in the west of mainland Singapore, was officially opened in 1994. Its 3.7-hectare area provided 152 berths for boats up to 20-meters in length, with subsequent modifications to house up to six super-yachts and four mega-yachts.[45] Four vertical concrete walls surround the marina, with a single opening for vessels facing the western Straits of Johor.[46] A 5-centimeters horizontal gap along the vertical wall facilitates water exchange during high tides. The sides of the mainly concrete pontoons submerge approximately 50-centimeters below the water surface.[47] The bottom depth of the marina increases gently towards the open waters, and is between 1.5 meters and four meters below chart datum.

The 7.3-hectare Marina at Keppel Bay (MKB) is situated on Keppel Island, south of mainland Singapore. It has 180 berths, some of which can accommodate superyachts up to 590 feet.[48] The berths were initially

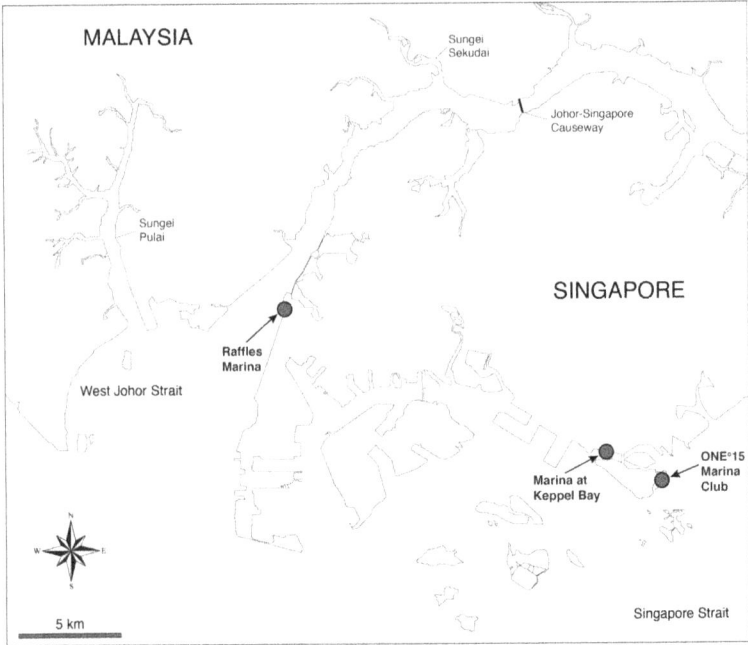

Figure 5.2. Location of the three Singapore marinas assessed for their marine biodiversity status.

constructed in two phases — 2007 and 2010. Seawalls bound the north and south of the marina, with water exchange facilitated by openings at the west and east ends.[46] Berthed vessels are protected from strong waves by breakwaters at the western end as well as two-meters deep wave-attenuator pontoons on the eastern end. The pontoons are similar in design and material to those in RM, and have a draft of approximately 50 centimeters. The bottom depth of the MKB ranges between six meters and 10 meters.

ONE°15 Marina Club (O15), the largest of the three marinas extending 11.3 hectares is located in Sentosa Cove, at the south-east of Sentosa Island.[49] Its perimeter is lined by vertical concrete walls alternating with sloping granite walls. The main entrance faces the Straits of Singapore, while a smaller gate inner to the entrance connects to canals serving residential housing. The marina supports 270 berths, including those for mega- and super-yachts. Construction was completed in 1999, but the berths were only completed in three phases — 2005, 2007, and 2011.

All pontoons are made of fiberglass with a shallow draft of approximately 15 centimeters.[46] Depth is generally between four and six meters below chart datum.

Salinity was significantly different among marinas with RM having the lowest salinity (29.25 ppt ± 0.29 ppt), compared to MKB (31.58 ppt ± 0.24 ppt) and O15 (31.03 ppt ± 0.05 ppt). Dissolved oxygen was also significantly higher in RM (3.62 mg/l ± 0.09 mg/l) than in O15 (3.32 mg/l ± 0.05 mg/l) and MKB (3.41 mg/l ± 0.04 mg/l). Water motion, as observed by clod card dissolution rates, was weaker in RM compared to MKB and O15.[46] In addition, the east and west entrances of MKB allow through-flow of water across it, resulting in higher tidal velocities than at O15. Analyses of the seabed showed that MKB and RM had similar sediment composition, with 65% to 70% of the sediment grains larger than one millimeter, including material such as shell fragments and leaf litter. In contrast, 45% of sediment in O15 were between 0.2 millimeters to 0.6 millimeters, consisting of a large proportion of fine sand.

5.4. Epibiota on Pontoons and Seawalls

Artificial structures such as pontoons, piles, and seawalls offer novel habitats for epibiotic diversity (Figure 5.3). They provide a wide range of habitat complexity and ecological niches, and can support a higher diversity of biota than even surrounding natural reefs.[50] Floating pontoons that demarcate berths for vessels in marinas are flanked by sides that stay

Figure 5.3.   Diverse assemblage of epibiotic species developing on submerged zones of vertical seawalls.

partially submerged all the time. Being close to the water surface, they provide a unique habitat that is rarely found in nature. The closest analogues are transient floating materials such as logs, leaf litter, and algae mats. Studies have indicated that pontoon epibiotic assemblages differed greatly from fixed pilings and nearby rocky reefs, as pontoons have a constant light source, altered hydrodynamic conditions, and encounter larvae that recruit near the water surface.[51–54]

Submerged pontoon surfaces are however, relatively understudied compared to other artificial substrates such as pier pilings and intertidal seawalls that extend vertically through the entire water column. With the variety of microhabitats that exist on pontoons, e.g., vertical sides and horizontal under-surfaces,[55–57] as well as depth zonation on the vertical sides,[52,58] pontoons can play important roles in supporting epibiotic assemblages.[35] Epibiotic organisms such as algae and sessile invertebrates (sponges, corals, ascidians, bivalves, barnacles, and tube worms) provide food and shelter for mobile crustaceans, gastropods, and echinoderms, which then attract fish and visiting shorebirds and turtles.[44,59] Information on the diversity and distribution of epibiota on marina pontoons is therefore necessary so that their role as unique habitats can be better understood. For example, an earlier study on pontoons in MKB documented 49 taxa including seven scleractinian coral genera, and attributed distribution differences of epibiotic assemblages to hydrodynamic and light intensity variation.[60]

Other researchers have also suggested that coastal structures such as seawalls and breakwaters provide more of a hard-bottom habitat compared to existing artificial reefs.[61,62] In addition, it was observed that these structures supported a variety of epibiotic organisms[23,31,62–64] and provided an alternative habitat for hard corals.[65–68] In Singapore, 17 genera of hard coral on the exposed seawalls of Singapore's southern offshore islands were recorded.[69] Compared to fringing reefs, mean coral species richness on seawalls was lower and comprised more species bearing massive and thick-plating growth forms.[70] Sloped seawalls supported more hard coral species (21 to 64 species) than vertical ones (11 to 35 species).[71,72] Interestingly, the seawalls within O15 and the SAF Yacht Club (another marina on the east coast of Singapore) supported 25 and 33 genera of hard corals respectively.[26,28,73] Observations at the latter marina are noteworthy because coral communities were not known to occur along the east coast of Singapore, which had been a sandy beach environment with no historical evidence of reefs in the vicinity. With the construction of the granite seawall within the SAF Yacht Club, larvae could settle and scleractinian communities were thus able to establish on a stable, solid substrate.

5.4.1. *Pontoons*

In our study, 94 taxa (including hexacorallians, hydrozoans, bryozoans, annelids, gastropods, arthropods, and echinoderms) were recorded from the pontoon surfaces in all marinas[47] (Figure 5.4). MKB supported the highest number of taxa (65), followed by RM (58), and O15 (43). Mean taxonomic richness was not different between marinas, while RM and MKB both registered Shannon indices that were higher than O15. The epibiotic assemblages differed significantly among the three marinas, with those at RM comprising more sediment-coated surfaces and shell fragments, and those at O15 and MKB dominated by macroalgae.

These trends can be explained by factors such as geographical location as well as marina design (e.g., pontoon material, submerged pontoon area). Sited in different parts of Singapore's coast, each marina is exposed to different hydrodynamic regimes due to shoreline contours, tidal

Figure 5.4.   Local conditions influence epibiotic community structure on the submerged sides of pontoons.

currents, bathymetry, and monsoon winds.[74] These affect current velocity and transport of sediment and larvae to the marina,[75,76] which in turn influences the epibiotic community within each. The slower tidal velocities in the Straits of Johor (0.5 ms^{-1})[77] compared to the Straits of Singapore (1–2 ms^{-1})[78] meant that flow within RM was much slower than the other two marinas. Additionally, MKB is connected to the Singapore Straits via openings at its eastern and western ends, accounting for the fastest flow rates among the marinas. These result in significantly higher sedimentation rates at RM compared to MKB and O15, which affected the composition of epibiotic colonizers. Runoff from construction sites, road surfaces, and reclamation works around RM at the time of the study were likely contributors to the high sediment load, as did sediment discharge from Malaysian rivers, most noticeably Sungei Sekudai and Sungei Pulai.[79] The impact of sedimentation on epibiotic communities was evident as 33.5% of RM's pontoon submerged area were coated with sediment, in contrast to 9.2% at MKB and 0.5% at O15. As chronically high suspended sediment concentrations reduce the light available for photosynthesis, zooxanthellate hard corals were almost absent inside RM. The thick sediment also coats hard substrate and inhibits settlement and survival of hard coral larvae.[80,81] The inner Johor Straits also receives waste discharge from human activities in South Johor and northern Singapore,[82] while heavy metals such as zinc, copper, chromium, and lead have been found in surface sediments of the Johor Straits.[83,84] Such problems are compounded in the semi-enclosed conditions of RM, where pollutants are more likely to accumulate and may account for the dominance of widespread, eurytopic species like *Caulerpa racemosa*, *Perna viridis*, *Crassostrea* sp., and cyanobacteria. In contrast, lower sedimentation rates and relatively clearer and faster-flowing waters promoted the abundance of zooxanthellate corals on the pontoons of MKB and seawalls of O15.

Pontoon material also influenced larval settlement of different species. The fiberglass pontoons of O15 had much higher crustose coralline algae (CCA) cover (8.7–10.8%) than on Singapore's reefs (0.3–8.6%).[85] High CCA cover was also observed on fibreglass reef enhancement units indicating that the material favored its growth. As CCA is useful for encouraging the settlement and metamorphosis of coral larvae,[86] their presence on artificial structures could encourage colonization and establishment of scleractinian corals in marinas. Pontoons in MKB and RM were made of concrete, and its rough texture facilitated the attachment of mytilid bivalves.[87] Additionally, enhanced recruitment of oysters on concrete surfaces had been reported.[88] These findings possibly account for the high abundance of bivalves at RM, which consisted mostly of the mytilid, *Perna viridis* and oyster, *Crassostrea sp.* Tropical soft coral planulae were also shown to favor settlement on rough surfaces and avoided smooth surfaces,[89] which may explain the lack of soft corals at O15.

Interestingly, despite favorable abiotic conditions that supported hard coral establishment on the seawalls of O15 and high CCA cover on pontoons that should promote coral settlement,[90] no coral recruitment on pontoons was observed. Thus, the low epibiotic species diversity and evenness in O15 could be attributed to the shallow draft (10 centimeters) of its pontoons, unlike RM and MKB where the draft is 50 centimeters. Species diversity is known to increase with habitat size,[91, 92] and a larger surface area also increases the availability of microhabitats across wider depth and light gradients.[31] The effect of depth on epibiotic colonization of pier pilings resulted in vertical zonation differences.[93] Such an influence would apply to pontoons as well. A small submerged surface area also disadvantages colonizing organisms which require a certain space to successfully establish and compete for resources.[94]

MKB was noteworthy in terms of coral establishment, as its pontoons supported seven genera of scleractinians, compared to one genus each at the other two marinas.[47] This represents approximately 10% of the total known genera in Singapore. Most genera were common to the reefs of the southern islands (*Pocillopora, Montipora, Favites, Porites*),[95] indicating the relative success of pontoons in supporting settlement and establishment of corals from surrounding reefs (Figure 5.5). *Pocillopora* was the dominant genus (>50%) and was most widely distributed throughout MKB, likely due to the genus being common on the southern islands' reefs,[85] and having the ability to recruit easily.[96] This was followed by *Tubastraea* (>30%), an azooxanthellate coral known to dominate benthic habitats.[97] It was thus unusual to observe such a large number of *Tubastraea* colonies on the surfaces of these floating pontoons, which are constantly less than 50 centimeters below the water surface. However, studies have shown that *Tubastraea* larvae have a preference for settling on concrete, making the concrete pontoons of MKB a suitable habitat, despite their proximity to the water surface.[97] All other genera occurred less frequently.

The central section of the marina supported higher scleractinian diversity than the periphery. The periphery was dominated by *Pocillopora* and

Figure 5.5.   With the right conditions, reef-associated species including scleractinian corals can dominate submerged pontoon zones.

to a lesser extent, *Tubastraea*, while the opposite was observed for the centre, indicating some variation of environmental conditions across the marina. However, no significant differences in light intensity between the periphery and central zone suggesting that diversity differences could be due to hydrodynamic variation, as the periphery had more exposure to wave action and stronger flow velocity.[60] Apart from transporting more food and nutrients, strong water flow also reduces sedimentation rate thus promoting coral growth and survivorship. *Pocillopora* is also known to have a higher net photosynthesis rate with faster water flow.[98]

Pontoons thus provide a habitat very different from other fixed structures. As they float, coral colonies on them remain at a constant depth close to the water surface and are unaffected by tide level exposure. These colonies do not experience fluctuating light levels, which could influence the diversity present.[52] These floating platforms also have limited connection to the seafloor and are less accessible by benthic predators.[99] Floating pontoons in marinas could therefore potentially serve as stepping stones for the colonization of hard corals in urbanized reef systems.

5.4.2. *Seawalls*

The seawall of O15 offers a unique habitat for epibiota and favors the colonization of scleractinian coral.[26] Built within the sheltered marina, it is not exposed to strong wave action and high water flow. A population of 346 coral colonies was counted, comprising 21 genera from 10 families based on updated scleractinian taxonomy.[73] This represents 40% of the total number of coral genera present in Singapore.[95] The data showed an increase in the number of colonies but a decrease in number of genera since 2006/2007 (25 genera from 13 families).[26,73]

Turbinaria dominated with a colony count of 110 (31.8%), followed by *Pectinia* with 50 (14.5%), and *Porites* with 40 (11.6%), while in 2006/2007, *Pectinia* was the most dominant genus with a total count of 121 (35.0%), followed by *Turbinaria* with 54 (15.6%). Since 2006/2007, the abundance of *Turbinaria* had doubled while that of *Pectinia* was halved. Although a significant increase in the overall colony size of *Turbinaria* and *Pectinia* was observed, there was an increase in the number of small *Turbinaria* colonies but not of *Pectinia*. This indicated that although both *Turbinaria* and *Pectinia* showed overall size growth of existing colonies, only *Turbinaria* showed continued recruitment of new colonies. The three most abundant genera — *Turbinaria*, *Pectinia*, and *Porites* — are well adapted to turbid waters. *Turbinaria* and *Pectinia* have good sediment rejection abilities, and make use of ciliary activity and mucus to remove small particles off the tissue surfaces.[100] *Porites*, despite being poor at sediment rejection, is able to rapidly regenerate sediment-damaged tissue.[101] Genera such as *Leptastrea*, *Astreopora*, *Pocillopora*, *Oxypora*, and *Trachyphyllia*, which were present in 2006/2007, were absent in the present study, while *Ctenactis*, *Herpolitha*, and *Pachyseris* were newly recorded. There was also a considerable increase in the occurrence of *Favites* and *Fungia*. The results were similar to the 2006/2007 study, but with slight improvements to diversity and evenness. This indicates success in sustaining overall diversity and abundance since 2006/2007. The presence of large coral colonies (with diameters of more than 50 centimeters) despite the large-scale thermal bleaching of South East Asia in 2010,[102] reflects the marina's suitable environment for hard coral growth and resiliency. With adequate management, marinas could potentially serve as a recipient site as well as a sanctuary for hard corals, while also contributing key functions such coral carbonate production and habitat provision for other organisms.[70] These findings indicate the potential of using artificial structures such as pontoons and seawalls for hard coral conservation to ultimately enhance the management of Singapore's largely modified coastal environment.

5.5. Soft-bottom Macrobenthos

Comprehensive studies of soft-bottom macrobenthic communities in Singapore's natural habitats were conducted between 1985 and 2015.[103,104] These included spatial and temporal analyses on the effects of coastal development and total petroleum hydrocarbons on soft-bottom macrobenthos.[105–108,141] Research on modified habitats, such as that of marinas, remains limited. The first and only study on macrobenthos in Singapore's marinas conducted almost 20 years ago documented 51 polychaete species in Raffles Marina.[109] It showed the potential of modified habitats for supporting biodiversity. The study however did not provide sufficient quantitative information for comparison.

In the present study, 73 macrobenthic taxa from eight phyla were documented from the three marinas,[110] demonstrating how these modified marine environments can support a wide range of soft-bottom macrobenthic organisms. The highest number of taxa (54) was found from sampling stations within MBK, although its surrounding waters supported the lowest (43 taxa). O15 and RM supported 50 taxa each, while the waters surrounding RM registered 48 taxa. Polychaetes comprised up to 80% of the total macrofaunal abundance and dominated the macrobenthic assemblages of all marinas and their surrounding waters, followed by arthropods, molluscs, and echinoderms. Other faunal groups included nematodes, sea anemones, and fishes. While mean taxonomic richness did not differ between marinas, mean macrofaunal abundance was highest in RM, followed by O15 and MKB. Shannon diversity index was also highest at MKB.

There was also a distinct separation of icrobenthic assemblages among marinas. Polychaetes from the families Cirratulidae, Spionidae, Nephtyidae, and Pilargidae were more associated with RM, while Lumbrinereidae and Capitellidae were associated with MKB. At O15, amphipods, along with polychaetes from the families Poecilochaetidae, Capitellidae, and Paraonidae, were more abundant. Across sampling locations, the abundance of polychaetes was highest within RM, while the waters outside RM registered the highest percentage of echinoderms (mainly ophiuroids).

The dominance of polychaetes is typical of soft-bottom habitats as these provide organic matter for deposit feeding and suitable substrate for burrowing.[111,112] Of interest was the high abundance of capitellid polychaetes in all marinas. In particular, *Capitella capitata* has often been identified as an indicator of environments stressed by anoxia and organic enrichment.[113,114] The ubiquity of this deposit-feeder in the three marinas was likely due to its preference for fine sediments and ability to tolerate high levels of organic matter, heavy metals, and hydrocarbons.[16,115–118] Therefore, this opportunistic species is particularly dominant in marinas, while pollutant-sensitive species are less so.[22,119] The second most abundant phylum was Arthropoda, mainly consisting of amphipods which are common in soft-bottom communities, where they can burrow into sediment to seek food or escape from predators, or use the sediment to form membranous tubes from which they can feed. They are known to be sensitive to pollutants and thus are considered "negative bioindicators" of polluted water.[17,120–122] The presence of this group suggests that the environment within the marinas could still accommodate pollutant-sensitive fauna.

Macrobenthic faunal composition was not uniform among marinas, indicating differences in benthic environment across the study sites. RM had the highest proportion of cirratulid and spionid polychaetes that are tolerant of high copper concentrations.[17,123] Meanwhile, O15 registered the lowest taxonomic richness and diversity per sampling station, amphipods — sensitive to organic enrichment, oil pollution, and toxicity[124] — were abundant. The coarser substrate at O15 possibly reduced adsorption to metal contaminants[125] and encouraged the dominance of poecilochaetid polychaetes, which are common in sandy areas.[126] At MKB, terebellid and lumberinereid polychaetes — indicators of unpolluted to slightly polluted environments[116,127] — were abundant. The high taxonomic richness and Shannon index at MKB demonstrated its greater suitability for supporting soft-bottom macrofauna compared to other marinas.

The differing rates of water movement and hence renewal within the three marinas could have also accounted for these observations.[47] Areas with limited water exchange are usually characterized by increased sediment contamination, along with reduced macrobenthic diversity and density.[15,119,128,129] As RM is sited at the western Strait of Johor, a water body of low tidal velocity and metal-contaminated seabed,[83,130] it therefore experiences further reduction in water flow and quality,[131] culminating in the domination of opportunistic polychaete species within it. On the other hand, O15 and MKB — both located in the Singapore Strait where currents can be faster than two meters s^{-1} [78] — supported faunal groups such as amphipods and terebellid polychaetes that are associated with good water quality.[124,127] Sufficient water renewal is essential in structuring macrobenthic assemblages,[132,133] and this was evident for MKB, where its two openings facilitated through-flow of water between mainland Singapore and Keppel Island.

There were also significant interactions in biodiversity indices between location (i.e., within versus outside) and marinas, highlighting how tropical marinas may not support less biodiversity than the surrounding waters outside. Compared to the surrounding waters outside, RM's enclosed environment supported a less diverse community with more opportunistic polychaetes. In contrast, Shannon diversity, taxonomic richness and abundance were higher within MKB than outside. Additionally, there was stark disparity in the abundance of opportunistic cirratulids and spionids within and outside RM, unlike at MKB, suggesting that water quality within and around the latter marina did not differ substantially. This was likely due to the openness of MKB which enabled good flushing and water quality renewal, leading to the development of a thriving macrobenthic community.[119] These observations reflect that of Ceuta Harbour in Spain, which has opposing entrances that encourage water movement and therefore support higher polychaete richness and diversity than other regional harbors.[134]

5.6. Fish Diversity

Marinas in subtropical areas have been known to function as habitats for diverse fish assemblages, due to the presence of artificial structures within them.[135] However, fish communities in such modified environments often differ from that of original or adjacent natural habitats. Coastal fish assemblages in human-modified habitats comprised more generalist species compared to reef-specialist species, which characterized nearby reefs.[136] Comparisons of fish communities around the breakwaters of marinas also revealed that distinct assemblage differences existed between open water-facing and inward-facing sides of seawalls,[137] as well as between seawalls and adjacent sandy habitats.[138] Information and literature sources for fish assemblages, however, are scarce for marinas in the equatorial zone. An earlier investigation of fish diversity in RM revealed 63 species spanning different trophic levels and niches, but no quantitative data were given.[109] The dearth of information on other marinas necessitates a thorough investigation into their role as a habitat for coastal fishes.

Using un-baited traps in a catch-and-release program, a total of 1,160 fishes from 49 species (31 families) were recorded.[46] Common across all study sites were: *Chelmon rostratus, Parachaetodon ocellatus, Gerres oyena, Diagramma pictum, Choerodon oligacanthus, Monacanthus chinensis,* and *Siganus javus*. RM yielded the highest average catch of all marinas. Assemblages between marinas were highly dissimilar.[46] Species that contributed most to dissimilarity between the marinas include: *Monacanthus chinensis* (fan-bellied leatherjacket), *Parachaetodon ocellatus* (six spine butterflyfish), *Etroplus suratensis* (pearlspot), *Arius oetik, Chelmon rostratus* (copperband butterflyfish), *Siganus javus* (streaked spinefoot), and *Choerodon oligacanthus* (white-patch tuskfish). While *M. chinensis* was the dominant species in all three marinas, their proportions were highest in RM and O15. *Etroplus suratensis* and *A. oetik* were found exclusively in RM. MKB had higher proportions of *Parachaetodon ocellatus, Siganus javus,* and *Choerodon oligocanthus. Chelmon rostratus* was mostly found in O15 but almost non-existent in RM.

The presence of various structures in marinas creates "human-generated heterogeneity,"[52] which is different from what was originally present. Although unintended, the floating pontoons, pilings and sloping seawalls functioned similarly to artificial reefs or fish aggregating devices in helping to host and shelter fish of various niche specialisations.[38,41,42] Overhangs created by the pontoons functioned as habitat for shade-preferring fish such as *Pempheris oualensis*,[139] a species not previously recorded from earlier surveys of Singapore's modified coastal areas. Pilings and seawalls submerged in the water column provide surfaces for epibiotic growth, which in turn serves as food for foraging fish such as *Monacanthus chinensis*, a species commonly associated with artificial structures,[39,99] and which was most abundant in all three marinas. By extending throughout the entire water column, these structures are able to accommodate species at all depths.[38]

While fish communities in modified environments can differ from that of the original or adjacent marine habitats,[136] the surrounding environment appears to have a role in influencing the community of fish in the new habitats. Fishes recorded in O15 and MKB, such as *Lethrinus lentjan,*

Choerodon schoenleinii, *Choerodon anchorago*, *Choerodon oligacan-thus*, and various species of pomacentrids, were mostly reef-associated species, while RM supported mainly species with a preference for brack-ish or estuarine conditions such as *Arius oetik*, *E. suratensis* and *Johnius belangerii*. This was likely due to the location of the marinas as well as the biotic components that they support. Both MKB and O15 are located at the south of mainland Singapore and are nearer the coral reefs fringing the surrounding offshore islands. This facilitated the natural colonization of up to 26 scleractinian genera on the pontoons and seawalls, creating an environment conducive toward the development of reef biota at both marinas.[26,60] The family Chaetodontidae consists of many obligate coral-livores such as *Chaetodon octofasciatus*, which favors corals of the gen-era *Acropora* and *Fungia*,[140] and as such, was only found in MKB and O15. This was supported by the presence of other reef-associated chaeto-dontids, *Chelmon rostratus* and *Parachaetodon ocellatus*, which were found in higher numbers in MKB and O15 than RM.

The average catch within RM was more than two times that in MKB and O15. While the age of RM (at 20 years) could have accounted for such high counts, it has been reported that there was no correlation between fish abundance with age and size of a marina.[135] Located along the western Straits of Johor and adjacent to mangrove and seagrass habi-tats, RM is generally more turbid and estuarine compared to the other two marinas. The absence of hard corals from previous surveys[109] makes the presence of corallivorous fish unlikely. The marina however supports numerous species of polychaetes and mollusks, which serve as food for bottom-dwelling fish scouring the soft bottom sediment.[109,110] The envi-ronment favors species such as *Ostracion nasus*, which is found in waters with heavy siltation and muddy substrate,[141,142] as well as *E. suratensis*, a euryhaline species thought to have entered Singapore via the narrow Straits of Johor. The latter is an extreme opportunist with a diet ranging from detritus and algal matter to sand and fish scales.[143] Other fish known to inhabit the Straits of Johor and which have moved into the marina include *A. oetik*.[144]

5.7. Mass Fish Kill and Recovery

In February 2014, a bloom of the harmful algae *Karlodinium sp.* and *Takayama sp.* triggered a mass fish kill in the Straits of Johor[145,146] and affected the fish assemblages in and around RM. It presented an opportunity to study the recovery of fish communities and whether recovery patterns differed between inside and outside of the marina. This will help in understanding the role that marinas perform in the aftermath of environmental disturbances.

Fish community composition, species richness, abundance, and diversity over time differed between the inside and outside of RM.[147] Outside the marina, there was no clear temporal pattern or clustering of fish assemblages. Although the harmful algal bloom (HAB) event was reported to have affected a wide stretch of the West Johore Straits,[146] it lasted just a few days during the neap tide period when the water was relatively stagnant. Its effect on the West Johore Straits, which is a large water body, may have been limited and fish assemblages outside the marina were less affected by the HAB event.

Fish communities within the marina were more severely affected. Algal cells tend to concentrate in areas with low water motion such as harbors.[148] Higher concentrations of *Karlodinium australe* (more than two million cells l^{-1}) were also observed in a marina (Puteri Harbour), which was in the inner reach of the West Johore Straits, compared to sampling stations located at the outermost parts of the Strait (approximately 0.3 million cells l^{-1}).[146] It is therefore postulated that the semi-enclosed nature of RM elevated the algal cell concentrations and exacerbated the impacts of the HAB event. Nevertheless, fish communities took about seven months to return to pre-impact levels. Richness and diversity index increased gradually throughout the year from December 2013 to November 2014, strongly suggesting that the steep drop in richness, abundance, and diversity was a consequence of the HAB event that occurred in February 2014, instead of a natural seasonal change in fish populations. This was well within the time taken for fish communities to recover from localized HAB disruptions, as reported by other researchers.[149–151]

While the impacts of a HAB event within a marina may be higher than the adjacent open waters, fish abundances within RM two to three months after were still higher than outside RM, a trend observed well before the HAB event. These results indicate that post-HAB, the marina was still able to support certain fish species, such as *Monacanthus chinensis*.[147] The marina's sheltered conditions could mitigate disturbances by providing shelter to biotic communities. While abundance and species richness reached pre-HAB levels in both inside and outside the marina, diversity and community composition were changed.[147]

The dominance of *Monacanthus chinensis* within the marina was replaced by a three-fold increase of *Parachaetodon ocellatus* and *Ostracion nasus*. Outside the marina, the abundance of *Plotosus lineatus* and *Scatophagus argus* increased at the expense of *M. chinensis*. Fish community structure was affected by the HAB event, although temporal recovery of abundance and diversity occurred both inside and outside the marina. Further monitoring of the fish communities is needed to confirm if such shifts in community structure are part of a natural succession or a permanent alteration of community composition.

5.8. Design and Management Considerations

Three key factors which had the greatest influence on marine biodiversity in marinas were identified — (i) geographical location, (ii) the design of artificial structures such as pontoons, and (iii) the distance from the marina entrance and water motion. The following recommendations may be of use to the planning of new marinas that can effectively function as biodiversity sanctuaries:

1. Hydrodynamic patterns and ecology of the surrounding environment should be thoroughly investigated. Both water motion as well as adjacent natural habitats play important roles in shaping the assemblage of flora and fauna that eventually inhabit a marina.
2. Marinas should be designed to allow sufficient water exchange throughout the whole marina without compromising the safety of berthed vessels. For example, an elongated marina with only an opening at one end may result in reduced water flow in the innermost section leading to accumulation of pollutants. A through-flow design with more openings will be useful for facilitating exchange with the adjacent open waters.
3. An external coating of concrete on the submerged parts of pontoons provides suitable surfaces that encourage the recruitment of biodiversity. In addition, larger submerged surface areas facilitate increased establishment of biodiversity.
4. The presence of hard substrates such as granite rock seawalls can also encourage the recruitment of organisms such as hard corals. For example, up to 25 hard coral genera (half of Singapore's total generic diversity) were naturally recruited onto the rock bunds along the

seawall of O15. This would not have been possible if the substrate was mainly soft sediment. The coral communities in turn would help to attract and recruit other reef biota. In addition, interventions such as habitat rehabilitation or ecological engineering, including the seeding of pontoons or seawalls with native organisms such as mussels or corals, have shown promise in augmenting biodiversity in other urbanized marine environments, and could similarly be used to enhance the provision of ecosystem functions in the marinas.

5. "Green" practices can help to improve water quality and overall biodiversity. Discharge of waste and pollutants generated by human and boating activities within the marina should be discouraged through regulation and education. While the study did not specifically explore the relationship between pollution and biodiversity, it showed that macrobenthic communities within the marina comprised more opportunistic polychaete species than outside the marina. As high numbers of opportunistic species are often associated with polluted environments and consequently a less diverse biological community, the results suggested that the environment within the marinas was affected to some extent. These efforts can also be supplemented with regular public outreach programs to raise awareness of the wide diversity of marine life that marinas can support.

6. A systematic, long-term biodiversity monitoring program and prompt information-sharing are critical to further understand the ecological and environmental processes within a marina. This will also enable prompt detection of issues, such as the deliberate or accidental introduction of non-native species from visiting recreational vessels, which could upset the ecological balance in the marinas.

5.9. Conclusion

In this study, 73 soft bottom macrobenthic taxa were recorded from the three marinas.[147] RM and O15 supported more taxa than their adjacent open waters. The pontoons in the marinas hosted at least 94 floral and faunal taxa, with seven and 21 genera of hard corals also recorded on the pontoons and seawalls, respectively.[47] The richness, abundance, and diversity of hard corals on the seawalls of O15 were similar to that of a previous survey conducted in 2006/2007, indicating the long-term prospect of the marina as an alternative coral reef.

A total of 49 fish species were documented within the three marinas with only the use of traps.[46] Species richness was higher within RM, with double the catch rate compared to its adjacent open waters. Following a harmful algal bloom (HAB) event in February 2014, the semi-enclosed nature of RM compounded the impacts to the resident fish population. However, the recovery of the fish communities was relatively fast, with richness and abundance recovering to pre-algal bloom levels in about seven months.[147]

As evidenced by the diverse communities of soft-bottom macrobenthos and fish within the marinas compared to those of the adjacent open waters, as well as the unique pontoon habitats which are absent outside the marinas, there is an overall indication that marinas are not entirely harmful to marine biodiversity. Rather, these modified marine environments can host reasonable levels of biodiversity and thus can serve as sanctuaries for marine life. To achieve this, the proper management of marinas, complemented with sound and regular scientific inputs, is key to supporting and enhancing marine biodiversity.

5.10. Acknowledgments

This study was funded by a grant (number R-154-000-557-490) from the Technical Committee for the Coastal and Marine Environment under a Research Collaboration Agreement between the National Parks Board and the National University of Singapore. The management of Raffles Marina, ONE°15 Marina Club, and Marina at Keppel Bay facilitated the research by granting full access and providing logistical support to the research team.

References

1. Airoldi, L. and Beck, M. W. (2007). Loss, Status and Trends for Coastal Marine Habitats of Europe. *Oceanography and Marine Biology — An Annual Review*, *45*, 345–405.
2. Hollin, D. (1992). Marinas are Big Business in Texas. *Parks Recreation*, *27*, 42–45.
3. Smith, C. and Jenner, P. (1995). Marinas in Europe. *Travel Tourism Analyst*, *6*, 56–72.
4. Widmer, W., Underwood, A., and Chapman, G. (2002). Recreational Boating on Sydney Harbour. *National Resource Management*, *5*, 22–27.

5. Biselli, S., Bester, K., Huhnerfuss, H., and Fent, K. (2000). Concentrations of the Antifouling Compound Irgarol 1051 and of Organotins in Water and Sediments of German North and Baltic Sea Marinas. *Marine Pollution Bulletin, 40*, 233–243.

6. Franco, A. D., Graziano, M., Franzitta, G., Felline, S., Chemello, R., and Milazzo, M. (2011). Do Small Marinas Drive Habitat Specific Impacts? A Case Study from Mediterranean Sea. *Marine Pollution Bulletin, 62*, 926–933.

7. Schiff, K., Brown, J., Diehl, D., and Greenstein, D. (2007). Extent and Magnitude of Copper Contamination in Marinas of the San Diego Region, California, USA. *Marine Pollution Bulletin, 54*, 322–328.

8. Turner, S. J. et al. (1997). Changes in the Epifaunal Assemblages in Response to Marina Operations and Boating Activities. *Marine Environmental Research, 43*, 181–199.

9. Iannuzzi, T. J., Weinstein, M. P., Sellner, K. G., and Barrett, J. C. (1996). Habitat Disturbance and Marina Development: An Assessment of Ecological Effects. 1. Changes in Primary Production due to Dredging and Marina Construction. *Estuaries, 19*, 257–271.

10. Lewis, M. A., Weber, D. E., Stanley, R. S., and Moore, J. C. (2001). Dredging Impact on an Urbanized Florida Bayou: Effects on Benthos and Algal-Periphyton. *Environmental Pollution, 115*, 161–171.

11. Ellis, J., Cummings, V., Hewitt, J., Thrush, S., and Norkko, A. (2002). Determining Effects of Suspended Sediment on Condition of a Suspension Feeding Bivalve (*Atrina zelandica*): Results of a Survey, a Laboratory Experiment and a Field Transplant Experiment. *Journal of Experimental Marine Biology and Ecology, 267*, 147–174.

12. Bianchi, A. P., Bianchi, C. A., and Varney, M. S. (1989). Marina Developments as Sources of Hydrocarbon Inputs to Estuaries. *Oil and Chemical Pollution, 5*, 477–488.

13. Goh, B., Nayar, S., and Chou, L. M. (2000). The Waters in Our Marinas: Does Green Really Mean Clean? *Marina 6*. Raffles Marina, 1–10.

14. Schaanning, M. T., and Harman, C. (2011). Release of Dissolved Trace Metals and Organic Contaminants During Deep Water Disposal of Contaminated Sediments from Oslo Harbour, Norway. *Journal of Soils and Sediments, 11*, 1,477–1,489.

15. Chen, K., Tian, S., and Jiao, J. J. (2010). Macrobenthic Community in Tolo Harbour, Hong Kong and its Relation with Heavy Metals. *Estuaries and Coasts, 33*, 600–608.

16. Inglis, G. J., and Kross, J. E. (2000). Evidence for Systematic Changes in the Benthic Fauna of Tropical Estuaries as a Result of Urbanization. *Marine Pollution Bulletin, 41*, 367–376.

17. Neira, C., Levin, L. A., Mendoza, G., and Zirino, A. (2014). Alteration of Benthic Communities Associated with Copper Contamination Linked to Boat Moorings. *Marine Ecology*, *35*, 46–66.

18. Ryu, J., Khim, J. S., Kang, S. G., Kang, D., Lee, C. H., and Koh, C. H. (2011). The Impact of Heavy Metal Pollution Gradients in Sediments on Benthic Macrofauna at Population and Community Levels. *Environmental Pollution*, *159*, 2,622–2,629.

19. McAllister, T. L., Overton, M. F., and Brill Jr, E. D. (1996). Cumulative Impact of Marinas on Estuarine Water Quality. *Environmental Management*, *20*, 385–396.

20. McMahon, P. J. T. (1989). The Impact of Marinas on Water Quality. *Water Science and Technology*, *21*, 39–43.

21. Hinwood, J. (1998). Marina Water Quality and Sedimentation. *Marinas 5*. Raffles Marina, Singapore, p. 4.

22. McGee, B. L., Schlekat, C. E., Boward, D. M., and Wade, T. L. (1995). Sediment Contamination and Biological Effects in a Chesapeake Bay Marina. *Ecotoxicology*, *4*, 39–59.

23. Bulleri, F. and Chapman, M. G. (2004). Intertidal Assemblages on Artificial and Natural Habitats in Marinas on the North-West Coast of Italy. *Marine Biology*, *145*, 381–391.

24. Chen, D. (2008). *The Effects of Pontoon Construction on the Fish Community in Sentosa Cove*. Undergraduate Research Opportunities in Science Project Report. National University of Singapore, Singapore.

25. Chou, L. M., Jaafar, Z., and Yatiman, Y. (2004). Marine Ecology of Raffles Marina and a Pilot Study on Bio-remediators to Improve Water Quality. In Phang, S. M., Chong, V. C., Ho, S. C., Noreieni, H. M. and Ooi, J. L. S. (Eds.). *Marine Science into the New Millennium*, Kuala Lumpur, Malaysia: University of Malaya Maritime Research Centre, pp. 649–660.

26. Chou, L. M., Ng, C. S. L., Chan, S. M. J., and Seow, L. A. (2010). Natural Coral Colonisation of a Marina Seawall in Singapore. *Journal of Coast Development*, *14*, 11–17.

27. Chou, L. M., Ng, C. S. L., Toh, K. B., Cheo, P. R., Ng, J. Y., and Tun, K. (2020). *Hidden Havens: Exploring Marine Life in Singapore's Marinas*. Singapore, National Parks Board.

28. Tan, Y. Z., Ng, C. S. L., and Chou, L. M. (2012). Natural Colonisation of a Marina Seawall by Scleractinian Corals along Singapore's East Coast. *Nature in Singapore*, *5*, 177–183.

29. Burdick, D. M., and Short, F. T. (1999). The Effects of Boat Docks on Eelgrass Beds in Coastal Waters of Massachusetts. *Environmental Management*, *23*, 231–240.

30. Chapman, M. G. (2003). Paucity of Mobile Species on Constructed Seawalls: Effects of Urbanization on Biodiversity. *Marine Ecology Progress Series*, *264*, 21–29.

31. Connell, S. D., and Glasby, T. M. (1999). Do Urban Structures Influence Local Abundance and Diversity of Subtidal Epibiota? A Case Study from Sydney Harbour, Australia. *Marine Environmental Research, 47*, 373–387.

32. Thompson, R. C., Crowe, T. P., and Hawkins, S. J. (2002). Rocky Intertidal Communities: Past Environmental Changes, Present Status and Predictions for the Next 25 years. *Environmental Conservation, 29*, 168–191.

33. Chou, L. M. (2021). Enhancing Marine Biodiversity with Artificial Structures. In Ting, D. S. K., and Stagner, J. A. (Eds.). *Sustainable Engineering Technologies and Architectures*, New York, U.S.A.: AIP Publishing, pp. 5-1–5-22.

34. Bulleri, F., Chapman, M. G., and Underwood, A. J. (2005). Intertidal Assemblages on Seawalls and Vertical Rocky Shores in Sydney Harbour, Australia. *Austral Ecology, 30*(6), 655–667.

35. Chapman, M. G., Blockey, D., People, J., and Clynick, B. (2009). Effect of Urban Structures on Diversity of Marine Species. In McDonnell, M. J., Hahs, A. K., and Breuste, J. H. (Eds.). *Ecology of Cities and Towns: A Comparative Approach,* London: Cambridge University Press, pp. 156–176.

36. People, J. (2006). Mussel Beds on Different Types of Artificial Structures Support Different Macroinvertebrate Assemblages. *Austral Ecology, 31*, 271–281.

37. Clynick, B. G., Chapman, M. G., and Underwood, A. J. (2007). Effects of Epibiota on Assemblages of Fish Associated with Urban Structures. *Marine Ecology Progress Series, 332*, 201–210.

38. Fabi, G., Grati, F., Puletti, M., and Scarcella, G. (2004). Effects of Fish Community Induced by Installation of Two Gas Platforms in the Adriatic Sea. *Marine Ecology Progress Series, 273*, 187–197.

39. Hair, C. A. and Bell, J. D. (1992). Effects of Enhancing Pontoons on Abundance of Fish: Initial Experiments in Estuaries. *Bulletin of Marine Science, 5*, 30–36.

40. Hair, C. A., Bell, J. D., and Kingsford, M. J. (1994). Effects of Position in the Water Column, Vertical Movement and Shade on Settlement of Fish to Artificial Habitats. *Bulletin of Marine Science, 55*, 2–3.

41. Rilov, G. and Benayahu, Y. (1998). Vertical Artificial Structures as an Alternative Habitat for Coral Reef Fishes in Disturbed Environments. *Marine Environmental Research, 45*, 431–451.

42. Rilov, G. and Benayahu, Y. (2000). Fish Assemblage on Natural Versus Vertical Artificial Reefs: The Rehabilitation Perspective. *Marine Biology, 136*, 931–942.

43. Rilov, G. and Benayahu, Y. (2002). Rehabilitation of Coral Reef-Fish Communities: The Importance of Artificial-Reef Relief to Recruitment Rates. *Bulletin of Marine Science, 70*, 185–197.

44. Chou, L. M. (1998). Marinas and Marine Conservation. *Marinas 5*. Raffles Marina, Singapore, pp. 1–5.
45. Raffles Marina (2021). Berthing Facilities. Retrieved from: https://www.rafflesmarina.com.sg/marina/group-1/berthing-facilities.html.
46. Toh, K. B., Ng, C. S. L., Leong, W. K. G., Jaafar, Z., and Chou, L. M. (2016). Assemblages and Diversity of Fishes in Singapore's Marinas. *Raffles Bulletin of Zoology, S32*, 85–94.
47. Toh, K. B., et al. (2017). Spatial Variability of Epibiotic Assemblages on Marina Pontoons in Singapore. *Urban Ecosystems, 20*, 183–197.
48. Marina at Keppel Bay (2021). Marina facilities: Marina at Keppel Bay. Retrieved from: https://marinakeppelbay.com/berth/marina-facilities/.
49. One015 Marina (2021). Marina overview: Fact sheet. Retrieved from: https://one15marina.com/marina/overview/#fact.
50. Chou, L. M. (2006). Marine habitats in one of the world's busiest harbours. In Wolanksi, E. (Ed.). *The Environment in Asia Pacific Harbours*, Netherlands: Springer, pp. 377–391.
51. Connell, S. D. (2000). Floating Pontoons Create Novel Habitats for Subtidal Epibiota. *Journal of Experimental Marine Biology and Ecology, 247*, 183–194.
52. Holloway, M. G. and Connell, S. D. (2002). Why do Floating Structures Create Novel Habitats for Subtidal Epibiota. *Marine Ecology Progress Series, 235*, 43–52.
53. Perkol-Finkel, S., Zilman, G., Sella, I., Miloh, T., and Benayahu, Y. (2006). Floating and Fixed Artificial Habitats: Effects of Substratum Motion on Benthic Communities in a Coral Reef Environment. *Marine Ecology Progress Series, 317*, 9–20.
54. Perkol-Finkel, S., Zilman, G., Sella, I., Miloh, T., and Benayahu, Y. (2008). Floating and Fixed Artificial Habitats: Spatial and Temporal Patterns of Benthic Communities in a Coral Reef Environment. *Estuarine, Coastal and Shelf Science, 77*, 491–500.
55. Connell, S. D. (1999). Effects of Surface Orientation on the Cover of Epibiota. *Biofouling, 14*, 219–226.
56. Glasby, T. M. (2000). Surface Composition and Orientation Interact to Affect Subtidal Epibiota. *Journal of Experimental Marine Biology and Ecology, 248*, 177–190.
57. Glasby, T. M. and Connell, S. D. (2001). Orientation and Position of Substrate have Large Effects on Epibiotic Assemblages. *Marine Ecology Progress Series, 214*, 127–135.
58. Cole, V. J., Glasby, T. M., and Holloway, M. G. (2005). Extending the Generality of Ecological Models to Artificial Floating Habitats. *Marine Environmental Research, 60*, 195–210.

59. Qiu, J. W., Thiyagarajan, V., Leung, A. W. Y., and Qian, P. Y. (2003). Development of a Marine Subtidal Epibiotic Community in Hong Kong: Implications for Deployment of Artificial Reefs. *Biofouling, 19*, 37–46.

60. Lam, L. M. and Todd, P. A. (2013). Spatial Differences in Subtidal Epibiotic Community Structure in Marina at Keppel Bay, Singapore. *Nature in Singapore, 6*, 197–206.

61. Airoldi, L., et al. (2005). An Ecological Perspective on the Deployment and Design of Low-Crested and other Hard Coastal Defence Structures. *Coastal Engineering, 52*, 1,073–1,087.

62. Bacchiochi, F. and Airoldi, L. (2003). Distribution and Dynamics of Epibiota on Hard Structures for Coastal Protection. *Estuarine, Coastal and Shelf Science, 56*, 1,157–1,166.

63. Chapman, M. G. and Bulleri, F. (2003). Intertidal Seawalls — New Features of Landscape in Intertidal Environments. *Landscape and Urban Planning, 62*, 159–172.

64. Davis, J. L. D., Levin, L. A., and Walther, S. M. (2002). Artificial Armoured Shorelines: Site for Open-coast Species in a Southern California bay. *Marine Biology, 140*, 1,249–1,262.

65. Burt, J., Bartholomew, A., Bauman, A., Saif, A., and Sale, P. F. (2009). Coral Recruitment and Early Benthic Community Development on Several Materials used in the Construction of Artificial Reefs and Breakwaters. *Journal of Experimental Marine Biology and Ecology, 373*, 72–78.

66. Burt, J., Bartholomew, A., Usseglio, P., A., Bauman, and Sale, P. F. (2009). Are Artificial Reefs Surrogates of Natural Habitats for Corals and Fish in Dubai, United Arab Emirates? *Coral Reefs, 28*, 663–675.

67. Viyakarn, V., Chavanich, S., Raksasab, C., and Loyjiw, T. (2009). New Coral Community on a Breakwater in Thailand. *Coral Reefs, 28*, 427.

68. Wen, K. C., Hsu, C. M., Chen, K. S., Liao, M. H., Chen, C. P., and Chen, C. A. (2007). Unexpected Coral Diversity on the Breakwaters: Potential Refuges for Depleting Coral Reefs. *Coral Reefs, 26*, 127.

69. Ng, C. S. L., Chen, D., and Chou, L. M. (2012). Hard Coral Assemblages on Seawalls in Singapore. In K. S. Tan (Ed.). *Contributions to Marine Science*, Singapore: Tropical Marine Science Institute, National University of Singapore, pp. 75–79.

70. Ng, C. S. L., et al. (2021). Coral Community Composition and Carbonate Production in an Urbanized Seascape. *Marine Environmental Research*, 168: 105322. Retrieved from: https://doi.org/10.1016/j.marenvres.2021.105322.

71. Kikuzawa, Y. P., et al. (2020). Diversity of Subtidal Benthic and Hard Coral Communities on Sloping and Vertical Seawalls in Singapore. *Marine Biodiversity, 50*, 95. Retrieved from: https://doi.org/10.1007/s12526-020-01118-z.

72. Lee, Y.-L., Lam, S. Q. Y., Tay, T. S., Kikuzawa, Y. P., and Tan, K. S. (2021). Composition and Structure of Tropical Intertidal Hard Coral Communities on Natural and Man-made Habitats. *Coral Reefs*, *40*, 685–700.

73. WoRMS Editorial Board (2021). World Register of Marine Species. Available from http://www.marinespecies.org at VLIZ. Accessed 2021-07-11. DOI:10.14284/170.

74. Tkalich, P., Pang, W. C., and Sundarambal, P. (Year). Hydrodynamics and Eutrophication Modelling for Singapore Straits. *Proceedings of the 7th OMISAR Workshop on Ocean Models for the APEC Region.*

75. Pineda, J., Hare, J. A., and Sponaugle, S. (2007). Larval Transport and Dispersal in the Coastal Ocean and Consequences for Population Connectivity. *Oceanography*, *20*, 22–39.

76. van Marren, D. S., and Gerritsen, H. (2012). Residual Flow and Tidal Asymmetry in the Singapore Strait, with Implications for Resuspension and Residual Transport of Sediment. *Journal of Geophysical Research*, *117*, C04021.

77. Thomas, G. S. (1991). Geology and Geomorphology. In Chia, L. S. and Rahman, A. (Eds.). *The Biophysical Environment of Singapore*, Singapore: National University of Singapore Press, pp. 50–88.

78. Chen, M., Murali, K., Khoo, B. C., Lou, J., and Kumar, K. (2005). Circulation Modelling in the Strait of Singapore. *Journal of Coastal Research*, *21*, 960–972.

79. Chan, E. S., Tkalich, P., Gin, K. Y. H., and Obbard, J. P. (2006). The Physical Oceanography of Singapore Coastal Waters and Its Implications for Oil Spills. In Wolansky, E. (Ed.). *The Environment in Asia Pacific Harbours,* Netherlands: Springer, pp. 393–412.

80. Hodgson, G. (1990). Sediment and the Settlement of Larvae of the Reef Coral *Pocillopora damicornis*. *Coral Reefs*, *9*, 41–43.

81. Rogers, C. (1990). Responses of Coral Reefs and Reef Organisms to Sedimentation. *Marine Ecology Progress Series*, *62*, 185–202.

82. Koh, H. L., Lim, P. E., and Midun, Z. (1991). Management and Control of Pollution in Inner Johore Strait. *Environmental Monitoring and Assessment*, *19*, 349–359.

83. Cuong, D. T., Karuppiah, S., and Obbard, J. P. (2008). Distribution of Heavy Metals in the Dissolved and Suspended Phase of the Sea-Surface Microlayer, Seawater Column and in Sediments of Singapore's Coastal Environment. *Environmental Monitoring and Assessment*, *138*, 255–272.

84. Zulkifli, S. Z., Ismail, A., Mohamat-Yusuff, F., Arai, T., and Miyazaki, N. (2010). Johor Strait as a Hotspot for Trace Elements Contamination in Peninsular Malaysia. *Bulletin of Environmental Contamination and Toxicology*, *84*, 568–573.

85. Loh, T. L., Tanzil, J. T. I., and Chou, L. M. (2006). Preliminary Study of Community Development and Scleractinian Recruitment on Fibreglass Artificial Reef Units in the Sedimented Waters of Singapore. *Aquatic Conservation: Marine and Freshwater Ecosystems, 16*, 61–76.
86. Morse, A. N. C., Iwao, K., Baba, M., Shimoike, K., Hayashibara, T., and Omori, M. (1996). An Ancient Chemosensory Mechanism Brings New Life to Coral Reefs. *Biology Bulletin, 191*, 149–154.
87. Vekhova, E. E. (2006). Reattachment of Certain Species of Mytilid Bivalves to Various Substrates. *Russian Journal of Marine Biology, 32*, 308–311.
88. Anderson, M. J., and Underwood, A. J. (1994). Effects of Substratum Type on the Recruitment and Development of an Intertidal Estuarine Fouling Assemblage. *Journal of Experimental Marine Biology and Ecology, 184*, 217–236.
89. Benayahu, Y. and Loya, Y. (1984). Substratum Preferences and Planulae Settling of Two Red Sea Alcyonaceans: *Xenia macrospiculata* Gohar and *Parerythropodium fluvum fluvum* Forskal. *Journal of Experimental Marine Biology and Ecology, 3*, 249–261.
90. Harrington, L., Fabricius, K., De'ath, G., and Negri, A. (2004). Recognition and Selection of Settlement Substrata Determine Post-Settlement Survival in Corals. *Ecology, 85*, 3,428–3,437.
91. Butler, A. J. (1991). Effect of Patch Size on Communities of Sessile Invertebrates in Gulf St. Vincent, South Australia. *Journal of Experimental Marine Biology and Ecology, 153*, 255–280.
92. Keough, M. J. (1985). Effects of Patch Size on the Abundance of Sessile Marine Invertebrates. *Ecology, 65*, 423–437.
93. Ong, J. L. J. and Tan, K. S. (2012). Observations on the Subtidal fouling community on jetty pilings in the southern islands of Singapore. In Tan, K. S. (Ed.). *Contributions to Marine Science*, Singapore: Tropical Marine Science Institute, National University of Singapore, pp. 55–71.
94. Jackson, J. B. C. (1977). Competition on Marine Hard Substrata: The Adaptive Significance of Solitary and Colonial Strategies. *The American Naturalist, 111*, 743–767.
95. Huang, D., Tun, K. P. P., Chou, L. M., and Todd, P. A. (2009). An Inventory of Zooxanthellate Scleractinian Corals in Singapore, including 33 new records. *Raffles Bulletin of Zoology, 22*, 69–80.
96. Stoddart, J. and Black, R. (1985). Cycles of Gametogenesis and Planulation in the Coral *Pocillopora damicornis*. *Marine Ecology Progress Series, 23*, 153–164.
97. Creed, J. C. and Paula, A. F. D. (2007). Substratum Preference During Recruitment of Two Invasive Alien Corals onto Shallow-Subtidal Tropical Rocky Shores. *Marine Ecology Progress Series, 330*, 101–111.

98. Lesser, M. P., Weis, V. M., Patterson, M. R., and Jokiel, P. L. (1994). Effects of Morphology and Water Motion on Carbon Delivery and Productivity in the Reef Coral, *Pocillopora damicornis* (Linnaeus): Diffusion Barriers, Inorganic Carbon Limitation, and Biochemical Plasticity. *Journal of Experimental Marine Biology and Ecology, 178*, 153–179.

99. Connell, S. D. (2001). Urban Structures as Marine Habitats: An Experimental Comparison of the Composition and Abundance of Subtidal Epibiota among Pilings, Pontoons and Rocky Reefs. *Marine Environmental Research, 52*, 115–125.

100. Stafford-Smith, M. and Ormond, R. (1992). Sediment-Rejection Mechanisms of 42 Species of Australian Scleractinian Corals. *Marine and Freshwater Research, 43*, 683–705.

101. Dikou, A. and van Woesik, R. (2006). Survival Under Chronic Stress from Sediment Load: Spatial Patterns of Hard Coral Communities in the Southern Islands of Singapore. *Marine Pollution Bulletin, 52*, 1,340–1,354.

102. Guest, J. R. et al. (2012). Contrasting Patterns of Coral Bleaching Susceptibility in 2010 Suggest an Adaptive Response to Thermal Stress. *PLOS ONE, 7*(3): e33353.

103. Chou, L. M. and Loo, M. G. (1994). A Review of Marine Soft-bottom Benthic Community Studies from Singapore. In Wilkinson, C. R., Sudara, S., and Chou, L. M. (Eds.). *Proceedings of the Third ASEAN–Australia Symposium on Living Coastal Resources*. Bangkok, Thailand: Chulalongkorn University, 365–371.

104. National Parks Board (2020). *Comprehensive Marine Biodiversity Survey*. Retrieved from: https://www.nparks.gov.sg/biodiversity/community-in-nature-initiative/cmbs.

105. Chong, E. C., and Chou, L. M. (1992). Effects of Reclamation on Benthic Communities in an Estuary (Sungei Punggol) in Singapore. *Third ASEAN Science and Technology Week Conference Proceedings, 6, Marine Science: Living Coastal Resource*. Singapore: National University of Singapore and National Science and Technology Board, 205–211.

106. Lu, L. (2005). Seasonal Variation of Macrobenthic Infauna in the Johor Strait, Singapore. *Aquatic Ecology, 39*, 107–111.

107. Lu, L. (2005). The Relationship between Soft-Bottom Macrobenthic Communities and Environmental Variables in Singaporean Waters. *Marine Pollution Bulletin, 51*, 1,034–1,040.

108. Lu, L., Goh, B. P., and Chou, L. M. (2002). Effects of Coastal Reclamation on Riverine Macrobenthic Infauna (Sungei Punggol) in Singapore. *Journal of Aquatic Ecosystem Stress and Recovery, 9*, 127–135.

109. Jaafar, Z., Chou, L. M., and Yatiman, Y. (2004). Marine Ecology of Raffles Marina and a Pilot Study on Bio-remediators to Improve its Water Quality. In Phang, S. M., Chong, V. C., Ho, S. C., Noraieni, H. M., and J. Ooi, L. S.

(Eds.). *Marine Science into the New Millennium*. Kuala Lumpur, Malaysia: University of Malaya Maritime Research Centre, 649–660.

110. Ng, C. S. L., et al. (2019). Distribution of Soft Bottom Macrobenthic Communities in Tropical Marinas of Singapore. *Urban Ecosystems, 22*, 443–453.

111. Hutchings, P. (1998). Biodiversity and Functioning of Polychaetes in Benthic Sediments. *Biodiversity and Conservation, 7*, 1,133–1,145.

112. Rhoads, D. C., and Young, D. K. (1970). The Influence of Deposit-Feeding Organisms on Sediment Stability and Community Trophic Structure. *Journal of Marine Research, 28*, 150–178.

113. Tsutsumi, H. (1995). Impact of Fish Net Pen Culture on the Benthic Environment of a Cove in South Japan. *Estuaries, 18*, 108–115.

114. Yokoyama, H. (2002). Impact of Fish and Pearl Farming on the Benthic Environments in Gokasho Bay: Evaluation from Seasonal Fluctuations of the Macrobenthos. *Fisheries Science, 68*, 258–268.

115. Belan, T. A. (2003). Benthos Abundance Pattern and Species Composition in Conditions of Pollution in Amursky Bay (the Peter the Great Bay, the Sea of Japan). *Marine Pollution Bulletin, 9*, 1,111–1,119.

116. Dean, H. K. (2008). The Use of Polychaete (Annelida) as Indicator Species of Marine Pollution: A Review. *Revista de Biologica Tropical, 56*, 11–38.

117. Forbes, V. E., Forbes, T. L., and Holmer, M. (1996). Inducible Metabolism of Fluoranthene by the Opportunistic Polychaete *Capitella* sp. *Marine Ecology Progress Series, 132*, 63–70.

118. Sanders, H. L. (1958). Benthic studies in Buzzards Bay. 1. Animal Sediment Relationships. *Limnology and Oceanography, 3*, 245–258.

119. Grimes, S., Ruellet, T., Dauvin, J. C., and Boutiba, Z. (2010). Ecological Quality Status of the Soft-bottom Communities on the Algerian Coast: General Patterns and Diagnosis. *Marine Pollution Bulletin, 60*, 1969–1977.

120. Jewett, S. C., Dean, T. A., Smith, R. O., and Blanchard, A. (1999). 'Exxon Valdez' Oil Spill: Impacts and Recovery in the Soft-Bottom Benthic Community in and Adjacent to Eelgrass Beds. *Marine Ecology Progress Series, 185*, 59–83.

121. Rygg, B. (1985). Distribution of Species along Pollution-Induced Diversity Gradients in Benthic Communities in Norwegian Fjords. *Marine Pollution Bulletin, 16*, 469–474.

122. Rygg, B. (1985). Effects of Sediment Copper on Benthic Dauna. *Marine Ecology Progress Series, 25*, 83–89.

123. Ganesh, T., Rakhesh, M., Raman, A. V., Nanduri, S., Moore, S., and Rajanna, B. (2014). Macrobenthos Response to Sewage Pollution in a Tropical Inshore Area. *Environmental Monitoring and Assessment, 186*, 3,553–3,566.

124. Gómez Gesteira, J. L., and Dauvin, J. C. (2000). Amphipods are Good Bioindicators of the Impact of Oil Spills on Soft-bottom Microbenthic Communities. *Marine Pollution Bulletin, 40*, 1,017–1,027.

125. Hinkey, L. M., Zaidi, B. R., Volson, B., and Rodriguez, N. J. (2005). Identifying Sources and Distributions of Sediment Contaminants at Two US Virgin Islands Marinas. *Marine Pollution Bulletin, 50*, 1,244–1,250.

126. Fauchald, K. and Jumars, P. A. (1979). The diet of worms: A study of polychaete feeding guilds. In M. Barnes (Ed.). *Oceanography and Marine Biology: An Annual Review*, Aberdeen: Aberdeen University Press, pp. 193–284.

127. Borja, A., Franco, J., and Pérez, V. (2000). A Marine Biotic Index to Establish the Ecological Quality of Soft-bottom Habitats within European Estuarine and Coastal Environments. *Marine Pollution Bulletin, 40*, 1,100–1,114.

128. Je, J. G., Belan, T., Levings, C., and Koo, B. J. (2004). Changes in Benthic Communities along a Presumed Pollution Gradient in Vancouver Harbour. *Marine Environmental Research, 57*, 121–135.

129. Moreno, M., Albertelli, G., and Fabiano, M. (2009). Nematode Response to Metal, PAHs and Organic Enrichment in Tourist Marinas of the Mediterranean Sea. *Marine Pollution Bulletin, 58*, 1,192–1,201.

130. Wood, A. K. H. J., Ahmad, Z., Shazili, N. A. M., Yaakob, R., and Carpenter, R. (1997). Geochemistry of Sediments in Johor Strait between Malaysia and Singapore. *Continental Shelf Research, 17*, 1,207–1,228.

131. Hadibarata, T., Abdullah, F., Yusoff, A. R. M., Ismail, R., Azman, S., and Adnan, N. (2012). Correlation Study between Land Use, Water Quality, and Heavy Metals (Cd, Pb, and Zn) Content in Water and Green Lipped Mussels *Perna viridis* (Linnaeus) at the Johor Strait. *Water, Air, & Soil Pollution, 223*, 3,125–3,136.

132. Guerra-García, J. M., and García-Gómez, J. C. (2005). Oxygen Levels Versus Chemical Pollutants: Do They Have Similar Influence on Macrofaunal Assemblages? A Case Study in a Harbour with Two Opposing Entrances. *Environmental Pollution, 135*, 281–291.

133. Lardicci, C., Abbiati, M., Crema, R., Morri, C., Bianchi, C. N., and Castelli, A. (1993). The Distribution of Polychaetes along Environmental Gradients: An Example from the Orbetello Lagoon, Italy. *Marine Ecology, 14*, 35–52.

134. Guerra-García, J. M. and García-Gómez, J. C. (2004). Soft Bottom Mollusc Assemblages and Pollution in a Harbour with Two Opposing Entrances. *Estuarine, Coastal and Shelf Science, 60*, 273–283.

135. Clynick, B. G. (2008). Characteristics of an Urban Fish Assemblage: Distribution of Fish Associated with Coastal Marinas. *Marine Environmental Research, 65*, 18–33.

136. Wen, C. C., Pratchett, M. S., Shao, K. T., Kan, K. P., and Chan, B. K. K. (2010). Effects of Habitat Modification on Coastal Fish Assemblages. *Journal of Fish Biology*, *77*, 1,674–1,687.

137. Clynick, B. G. (2006). Assemblages of Fish Associated with Coastal Marinas in North-Western Italy. *Journal of the Marine Biological Association of the United Kingdom*, *86*, 847–852.

138. Guidetti, P. (2004). Fish Assemblages Associated with Coastal Defence Structures in South-Western Italy (Mediterranean Sea). *Journal of the Marine Biological Association of the United Kingdom*, *84*, 669–670.

139. Fishelson, L., Popper, D., and Gunderman, N. (1971). Diurnal Cyclic Behavior of *Pempheris oualensis* Cuv. & Val. (Pempheridae, Teleostei). *Journal of National History*, *5*, 503–506.

140. Madduppa, H. H., Zamani, N. P., Subhan, B., Aktani, U., and Ferse, S. C. (2014). Feeding Behavior and Diet of the Eight-banded Butterflyfish *Chaetodon octofasciatus* in the Thousand Islands, Indonesia. *Environmental Biology of Fishes*, *97*, 1,353–1,365.

141. Chou, L. M., Yu, J. Y., and Loh, T. L. (2004). Impacts of Sedimentation on Soft-bottom Benthic Communities in the Southern Islands of Singapore. *Hydrobiologia*, *515*, 91–106.

142. Allen, G. R., and Erdmann, M. V. (2012). *Reef Fishes of the East Indies.* Honolulu: University of Hawaii Press.

143. Ng, T. H. and Tan, H. H. (2010). The Introduction, Origin and Life-History Attributes of the Non-native Cichlid *Etroplus suratensis* in the Coastal Waters of Singapore. *Journal of Fish Biology*, *76*, 2,238–2,260.

144. Ng, T. H. (2012). The Ariid Catfish of Singapore. *Nature in Singapore*, *5*, 211–222.

145. Leong, S. C. Y., Lim, L. P., Chew, S. M., Kok, J. W. K., and Teo, S. L. M. (2015). Three New Records of Dinoflagellates in Singapore's Coastal Waters, with Observations on Environmental Conditions Associated with Microbial Growth in the Johor Straits. *Raffles Bulletin of Zoology*, *S31*, 24–36.

146. Lim, H. C., et al. (2014). A bloom of *Karlodinium austral* (Gymnodiniales, Dinophyceae) Associated with Mass Mortality of Cage-cultured Fishes in West Johor Strait, Malaysia. *Harmful Algae*, *40*, 51-62.

147. Ng, C. S. L., et al. (2017). Influence of a Tropical Marina on Nearshore Fish Communities During a Harmful Algal Bloom Event. *Raffles Bulletin of Zoology*, *65*, 525–538.

148. Garcés, E., Fernandez, M., Penna, A., van Lenning, K., Gutierrez, A., Camp, J., and Zapata, M. (2006). Characterization of NW Mediterranean *Karlodinium sp.* (Dinophyceae) Strains using Morphological, Molecular, Chemical, and Physiological Methodologies. *Journal of Phycology*, *42*, 1,096–1,112.

149. Reis-Filho, J. A., da Silva, E. M., da Costa Nunes, J. deA. C., and Barros, F. (2012). Effects of a Red Tide on the Structure of Estuarine Fish Assemblages in Northeastern Brazil. *International Review of Hydrobiology*, *97*, 389–404.

150. Rhodes, K., and Hubbs, C. (1992). Recovery of Peco River Fishes from a Red Tide Fish Kill. *The Southwestern Naturalist*, *32*, 178–187.

151. Zamor, R. M., Franssen, R. N., Porter, C., Patton, T. M., and Hambright, K. D. (2014). Rapid Recovery of a Fish Assemblage following an Ecosystem Disruptive Algal Bloom. *Freshwater Science*, *33*, 390–401.

Chapter 6

Biomaterials for Water Purification: Integrating *Chlorella Vulgaris* and *Monoraphidium Conortum* in Architectural Systems for the Biodegradation of Sulfamethoxazole from Wastewater

Yomna K. Abdallah[*,†,‡] and Alberto T. Estevez[*]

*iBAG-UIC Barcelona, Institute for Biodigital Architecture & Genetics
Universitat Internacional de Catalunya, Spain
†Helwan University, Faculty of Applied Arts
Department of Interior Design, Egypt
‡*yomnaabdallah@uic.es*

Abstract

Pharmaceutical pollution is a universal hazardous issue that affects humans and ecosystems through the release of active pharmaceutical ingredients (APIs).[1,2] One of their major sources is antibiotics.[3] These compounds survive classical treatment processes, which trigger the development of antibiotic-resistant bacterial strains in aquatic systems.[4] Sulfamethoxazole is a widely used antibiotic that faces a change in the resistance patterns of the various bacterial strains that are sensitive to it. Since biodegradation is one of the main methods for pharmaceutical degradation, it requires the identification of the microbial strains that biodegrade various antibiotics to enable feasible upscale applications. In the work presented in this chapter, *Chlorella vulgaris* and *Monoraphidium conortum* were tested separately to degrade 3g/100 ml of sulfamethoxazole. A kinetic study was conducted to separately measure the degradation of the antibiotic by both strains. The results revealed that sulfamethoxazole increased the growth of *Chlorella vulgaris* and promoted its ability to produce both types of chlorophyll *a* and *b*, achieving maximum production of chlorophyll *a* of 0.135 mg/L and chlorophyll *b* of 0.095 mg/L at the 25th day of growth while reaching only 0.1 mg/L of chlorophyll *a* and 0.034 mg/L of chlorophyll *b* in the control group. This was unlike the growth of *Monoraphidium conortum* that was inhibited by sulfamethoxazole achieving 0.09 mg/L of chlorophyll *a* and 0.079 of chlorophyll *b* in comparison to 0.094 mg/L of chlorophyll *a* and 0.12 of chlorophyll *b* in the control group. On the other hand, *Chlorella vulgaris* had a higher capacity to degrade sulfamethoxazole, reaching 0.024 mg/L of sulfamethoxazole after 20 days of introducing it to the algal culture, while *Monoraphidium conortum* degraded the antibiotic to 0.036 mg/L. These results suggest the supremacy of *Chlorella vulgaris* in the degradation of sulfamethoxazole while maintaining boosted growth behavior. A system of cell-immobilization support was designed and 3D-printed using micro texture to employ these systems in wastewater treatment from pilot to urban scales.

Keywords: Biodegradation, *Chlorella vulgaris*, *Monoraphidium conortum*, drug pollution, antibiotics, sulfamethoxazole, ecologic architecture, sustainability, biodigital design, micro texture, cell immobilization, 3d printing.

6.1. Introduction

Water pollution is a broad environmental issue caused by the discharge of various chemical compounds from a vast number of human activities. This affects the aquatic environment.[5] "Drug pollution" or "pharmaceutical pollution" is one of the main sources of water pollution, significantly affecting the aquatic biota and the entire ecosystem. However, the estimation of these pharmaceutical pollutants, their types, and effects are not yet sorted due to the increased use of various pharmaceuticals triggered by the recent development of medicine.[1,2] Among the highest

Integrating *Chlorella Vulgaris* **AND** *Monoraphidium Conortum* **in Architectural Systems of Photocatalysis For The Degradation of Sulfamethoxazole From Wastewater.**

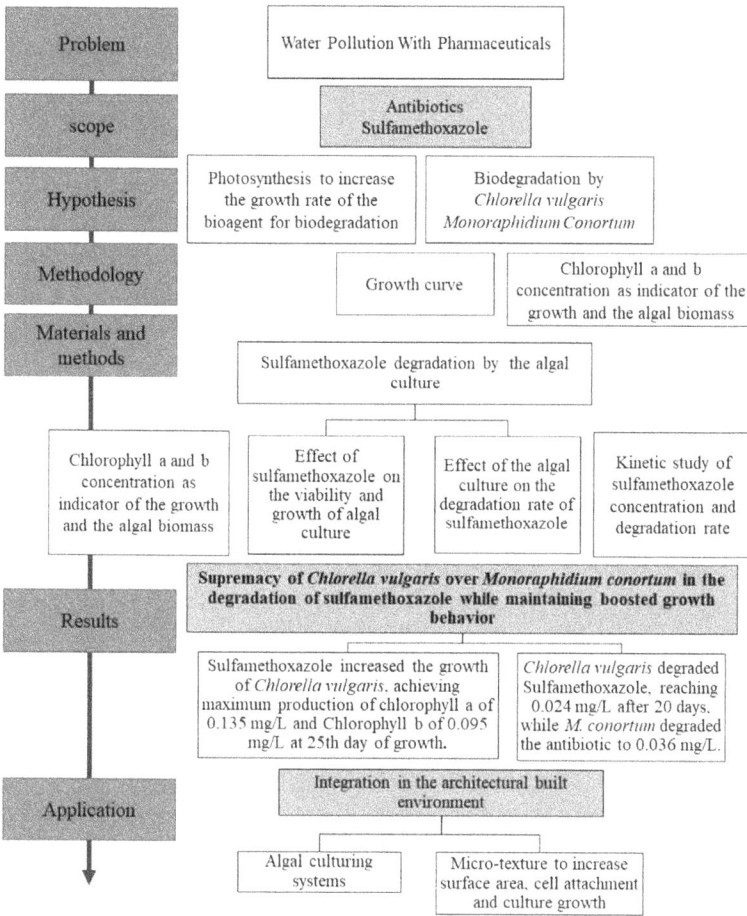

Figure 6.1. Diagram of the methodology and objective of integrating *Chlorella vulgaris* and *Monoraphidium contortum* in architectural systems for the biodegradation of sulfamethoxazole from wastewater (by the authors).

consumption levels are analgesics, antibiotics, antiseptics, hormone replacements, and antidepressants.[2] This enormous global use of pharmaceuticals with varied physicochemical properties has significantly triggered the release of active pharmaceutical ingredients (APIs) to the aquatic environment, including surface waters. It has been reported that APIs derived from antibiotics have reached a higher ppt (parts per trillion) and ppb (parts per billion) concentration range in the surface water.[3] The "priority substances" were recognized by the European Union (EU) as an issue equal to other micropollutants (EU 2013). This issue becomes more complicated because of the great number of the chemical compounds used to produce these drugs and the various transformational processes they undergo, challenging the ability to identify various APIs, and their level and effect in the environment.[2]

Pharmaceuticals are released in the environment by the effluents from manufacturing sites, hospital waste, excretion by livestock treated with antibiotics, domestic grey waters, and the flushing of unused prescriptions.[6] These sources release high concentrations of a large number of pharmaceuticals into the aquatic ecosystems.[7] Usually, these compounds survive classical treatment processes, which impose a persistence problem for these compounds that has not been completely studied yet. Consequently, this causes changes in the resistance patterns for various antibiotics, triggering antibiotic resistance genes in various bacterial strains, not only in hospital wastewater but also in domestic, surface, and groundwater.

Sulfamethoxazole (SMZ or SMX) is a sulfonamide and bacteriostatic antibiotic that addresses a wide range of bacterial infections. It affects both gram negative and positive bacteria by preventing the synthesis of the physiologically active form of folic acid and the necessary cofactor in the synthesis of bacterial DNA in the bacteria.[43] Thus, a wide spectrum of bacterial strains is sensitive to sulfamethoxazole. However, the increase in shifted resistance patterns has lately limited its clinical use. However, sulfamethoxazole remains an efficient alternative to the new generation of expanded-spectrum agents, provided that resistance patterns are carefully considered. Sulfamethoxazole is also increasingly used as a cost-effective pathogen-directed therapy while decreasing or delaying the development of resistance to newer antibiotics used for empirical treatment. Thus, maintaining the resistance patterns for this vital drug is of great importance medically and environmentally through the complete degradation of its residuals from wastewater and the aquatic system.

Until now, there is no universal method for removing antibiotics from wastewater; see Figure 6.1. After being excreted with wastewater, they undergo transformations at wastewater treatment plants. This complicates the problem by producing more toxic and persistent compounds.[8] Furthermore, the concentrations of these products may be even higher than the parent compounds.[9] To date, the elimination of such compounds in the aquatic environment depends on several key factors — the inherent physicochemical properties of the antibiotic, light, temperature, pH, oxygen, and microbial communities. However, the current conventional wastewater treatment plants are not adequately designed to remove sufficient amounts and the composition of antibiotics.[10] Biodegradation and photodegradation are the most commonly applied treatment methods to eliminate antibiotics from wastewaters. In photodegradation, the contained heteroatoms, aromatic rings, and other structures in the antibiotic absorb light in a direct or indirect photolysis by reaction with the photogenerated transient species and consequently degrades upon these photochemical reactions.[11] Biodegradation depends on microbes as active agents performing their physiological pathways to degrade the antibiotics.[10,11] The latter treatment remains the most used in wastewater treatment, thanks to its sustainability, biocompatibility, safety, and cost-efficiency compared to other chemical or physical treatments. However, some of the most common biological treatments, such as activated sludge treatment, is usually incomplete, as although bioremediation by pure bacteria isolated from activated sludge may enhance antibiotics removal,[12] these bacterial strains might still develop antibiotic resistance.[13] Therefore, novel, cost-effective biodegradation technologies are still pending further research for the efficient and safe removal of antibiotics from an aqueous environment on various scales. One main issue of enhancing biodegradation efficiency is the identification of microbial strains that are involved in this process, presenting an integrated, self-sufficient solution. For example, adequate microbial strains can dually serve in Microbial Fuel Cells (MFCs) based on the same chemical principles as those of chemical oxidation and adsorption. These systems generate hydroxyl radicals and various co-species, including sulfates and hydrogen carbonates that facilitate the removal of persistent pollutants, while the microbial strains extract electrical power from the organic constituents.

Recently, microalgae-based systems have gained increased attention in pharmaceutical biodegradation, owing to their dependence on solar energy, efficient capacity for fixation of carbon dioxide, or CO_2, eco-friendliness, and being a potential feedstock for bioenergy production or other high-value products.[14] Moreover, it guarantees cost-effective remediation of various nutrients,[15] emerging contaminants (ECs),[16] and heavy metals in wastewater.[17] Microalgae also exhibit great survivability in extreme environments, making them promising candidates for enhanced wastewater treatment.

Previous studies reported the efficient removal of antibiotics from wastewater by algal strains via various methods. These are bioadsorption, bioaccumulation, and biodegradation.[18] Biodegradation occurs through rapid passive adsorption via physicochemical interactions between the cell surface and pollutants; transfer of molecules through the cell membrane; bioaccumulation, biodegradation, or all three processes. Bioadsorption is mainly achieved by ion exchange, surface complexation reactions, chelation, and micro-precipitation.[19] Thus, bioadsorption occurs when antibiotics are either absorbed into the cell wall or onto organic substances, such as extracellular polymeric substances (EPS), secreted by the microalgae into their surrounding environments.[16] Since EPS protects the cells from harsh environments,[20] microorganisms tend to secrete more EPS as an adaptive mechanism in response to antibiotic toxicity. This enables them to degrade the antibiotics efficiently.[21] EPS contributes to different functional groups such as carboxyl, amine, hydroxyl, and hydrophobic regions, thereby providing available binding sites for the adsorption of diverse organic and inorganic compounds that are integrated within the composition of a wide spectrum of antibiotics.

Numerous previous studies reported the activity of various microalgae strains in the degradation of antibiotics via bioadsorption. For example, *Chlorella sp.* achieved adsorption of 7-amino cephalosporanic acid in 10 minutes, with adsorption capacities of 4.74 mg/g, 3.09 mg/g, and 2.95 mg/g.[22] The bioadsorption capacities of microalgae are highly specific to their physical and chemical properties, such as surface chemistry and surface area.[23] The cell walls of microalgae and EPS carry negative charges due to the presence of dominant functional groups.[24] Hereafter, antibiotics with a positive charge can be adsorbed through electrostatic interactions.[24,18] In general, lipophilic compounds are prone to bioadsorption of the microalgae due to electrostatic interactions, unlike hydrophilic compounds with low bioadsorption affinities and are more persistent in growth medium.[16]

Bioadsorption can occur on both living and non-living microalgal cell surfaces, as it is a non-metabolic process. The biomass of non-living microalgae has been proved to be potent for removing antibiotics. For example, Cefalexin (at 50 mg/L) in modeled wastewater was effectively removed by non-living *Chlorella sp.*, with adsorption capacities of 129 mg/g and 63 mg/g respectively.[25] This bioadsorption by non-living microalgal biomass outweighs the use of living microalgae, as it does not have toxicity limitations, with the potential use of the biomass as sustainable algal biodiesel, and the reduction of the operational costs.[26]

As mentioned earlier, the microalgae-based bioadsorption depends on the structure of the target antibiotic, the microalgal species, and environmental conditions.[23] Employing microalgae species with a high affinity for different target antibiotics was adopted to improve microalgal adsorption technologies in bioremediation. For example, *Chlorella* genera have been identified as the most popular species for antibiotic bioremediation owing to their availability and great potential.[16] Further optimization of the bioadsorption process can be attained via optimizing the used dosage of the bio sorbent, the initial concentration of the antibiotic, the contact time, pH, and temperature, as well as stimulation of EPS excretions.[16] For example, the bioadsorption efficiency of metronidazole removal by *Chlorella vulgaris* decreased upon increasing the initial concentration of the antibiotic.[18] The pH concentration affects both ionization or dissociation of antibiotics in the aqueous media, as well as the surface charge of the bio sorbents.[27] Moreover, temperature variations have been shown to affect the rate of antibiotic adsorption on the microalgal cell surface.[17] It was also reported that the effect of the quantity and composition of EPS on antibiotics bioadsorption, as a higher proteins/polysaccharides ratio in EPS was the most potent, providing more adsorption sites due to its stronger hydrophobicity.[20]

A consequent phase of bioadsorption is bioaccumulation, which is an active metabolic process of antibiotics in living microalgal cells.[24] The transport of antibiotics across the microalgal cell membrane can occur through passive diffusion, passive-facilitated diffusion, and active uptake.[24] The first occurs when antibiotics can diffuse through the cell membrane from high to low concentration to be absorbed inside the cell. Antibiotics with low molecular weight, non-polar and lipid-soluble ones, are potent enough to pass through the cell membrane via passive diffusion due to their hydrophobicity. For example, trimethoprim and sulfamethoxazole were reported to bioaccumulate into microalgal cells via passive diffusion.[28] Changes in cell membrane permeability induced by exposure to antibiotics or stressful environmental conditions can also cause passive diffusion due to membrane depolarization or hyperpolarization.[16] Interference with the integrity of the cell membrane may also promote the passive diffusion of antibiotics. For example, adding 1% (w/v) sodium chloride significantly improved bioaccumulation of Levofloxacin by *Chlorella vulgaris* to 101 μg/g, in comparison to 34 μg/g in the control group.[24] On the other hand, passive-facilitated diffusion indicates the presence of transporter proteins to carry antibiotics across the cell membrane. The active transport process is driven by energy, where antibiotics move against a concentration gradient.[16]

Bioaccumulation is influenced by external and internal physicochemical conditions, including temperature, pH, contact time, and antibiotic concentrations. Consequently, optimizing those physicochemical parameters can affect both the rate and quantity of antibiotics accumulated by microalgae, protect the cells from any associated toxicity, and enhance the removal of the antibiotic. Recent studies reported numerous microalgal species that are tolerant to high concentrations of antibiotics and confer high bioaccumulation rates.[30] However, the accumulation of antibiotics in microalgal cells might stimulate the overproduction of reactive oxygen species (ROS).[24] Normally, ROS regulates the growth, differentiation, and proliferation of microorganisms.[31] However, increased levels of ROS induce DNA and protein denaturation, and cause cellular damage mutagenesis, and cell death due to their strong oxidative properties.[32] However, it might be possible for microalgae to deplete the accumulated antibiotics in cells via metabolism, taking accumulation as a pre-step for biodegradation.[33] For example, some microalgae species effectively depleted sulfamethazine through accumulation and subsequent intracellular biodegradation.[33] Yet, this might impose environmental hazards as antibiotics accumulated in cells can be transferred through the food chain, while also inducing the development and spread of antibiotic resistance.[34] Thus, developing systems of cell immobilization can limit these hazards by the tight control over the algal cells in algae-based wastewater treatment plants.

Thus, bioadsorption and bioaccumulation are limited in their efficiency for antibiotic removal from wastewater due to the following — bioadsorption is a reversible process, with bio-absorbed antibiotics likely to be released back into the aqueous environment;[35] and bioaccumulation can cause a greater risk of transferring antibiotics or their compounds through the food chain, causing toxicity or altering their resistance patterns. Thus, biodegradation has been proven to be the most effective and dominant mechanism for removing antibiotics from aqueous environments.[24] For example, it was reported that *Chlorella sp.* L38 achieved a 72.0% removal rate of florfenicol (159 mg/L) by biodegradation compared to bioaccumulation and bioadsorption, which only achieved a removal rate of 1.3% and 1.4% respectively.[36] Biodegradation is the breaking down of organic compounds through biotransformation that produces different metabolic intermediates,[37] or through complete mineralization to CO_2 and H_2O by pure or mixed microbial cultures.[38] Thus, biodegradation may occur through metabolic degradation, where the antibiotic is the only carbon source or electron donor/acceptor for microalgae; or the co-metabolism, in which additional organic substrates serve to both sustain biomass production and act as an electron donor for the non-growth substrate.[39]

Biodegradation of antibiotics via microalgae mainly involves extracellular and intracellular phases, where initial degradation occurs extracellularly by the EPS that act as an external digestive, surfactant, and emulsifier, to prepare the antibiotics for subsequent bioaccumulation by microalgae, and the breakdown products are digested intracellularly.[24,40] The intracellular metabolism of the antibiotic involves a two-phase enzymatic catalysis. Phase I includes oxidation, reduction, or hydrolysis reactions, catalyzed by microsomal mono-oxygenase enzymes or mixed-function oxidases.[41] Phase II includes the coupling of a xenobiotic or its metabolites that result from Phase I reactions, with large and polar compounds such as sugars and amino acids, to further increase the water solubility of the xenobiotics.[42] It was reported by Xiong et al.,[43] that sulfamethoxazole biodegradation by *Chlorella pyrenoidosa* might include Phase I reactions of oxidation and hydroxylation of the amine group and Phase II reactions of formylation and pterin-related conjugation.[43]

Microalgae-based treatment systems are either open ponds type or closed photobioreactors (PBRs) that apply suspended or *immobilized cells*.[44] Suspension cultivation of microalgae is the most adopted method, thanks to its cost-effectiveness and ability to be used in both categories.[45] The open ponds are broadly used for large-scale microalgae cultivation due to their low construction cost, low energy, and easy scale-up. However, they suffer from insufficient mixing and fluctuations in the culture conditions that hinder the biomass productivities and the removal of micropollutants in the open systems. On the other hand, closed PBRs have several advantages that outweigh open pond systems. They provide full control over culture conditions, exclude evaporation and contaminations, efficient treatment of wastewater, and biomass production.[45,46] Despite their efficiency in eliminating a wide range of contaminants, such as nutrients, heavy metals, and antibiotics, at a laboratory scale, their construction and maintenance costs have limited their large-scale application.[46] There are also the difficulties of harvesting microalgae biomass in suspended culture due to its time-consuming and energy-demanding processes. Thus, immobilization of the cell culture was proposed to solve these limitations.[47] This technique has exhibited enhanced microalgal growth rates and increased removal rates of micropollutants.[45] For instance, immobilized microalgal beads mitigated the toxicity of sulfamethoxazole and accelerated the formation of a stable microalgae-bacteria consortium, allowing for the highest antibiotic removal efficiency. Immobilization techniques are achieved either by natural passive or artificial active methods. Natural immobilization is determined by the innate characteristics of microalgal cells that can attach to a specific surface and secrete EPS to form a biofilm on solid surfaces.[40] Artificial immobilization can be conducted via various methods, including adsorption, confinement in liquid-liquid suspensions, capture with semi-permeable membranes, affinity immobilization, covalent coupling, or entrapment within polymers.[50] However, this might affect the efficiency of the biodegradation system in the long run, and add more complexity and costs for larger scales. Thus, in the current study, the natural immobilization of algal cells on customized micro-textured supports is proposed to boost the biodegradation of pollutants in algae-based wastewater treatment plants.

Thus, in the present work, biodegradation will be applied to remove sulfamethoxazole from wastewater, focusing on testing the capacity of two micro-algae strains, *Chlorella vulgaris* and *Monoraphidium conortum,* to biodegrade the antibiotic. This is achieved by analyzing the growth curve of *Chlorella vulgaris and Monoraphidium conortum* respectively, indicated by their capacity for producing chlorophyll *a* and *b*, with and without the addition of sulfamethoxazole to the growth medium. On the other hand, the two algal strains effect on the degradation rate of sulfamethoxazole is tested separately. The justification of the following methods and results is exhibited in the results and discussion section of this chapter. This is followed by a design proposal for a cell-immobilization chip-support that will host the cells of the potent algal strain for the biodegradation of sulfamethoxazole from wastewater. This micro-textured chip-support is digitally designed and 3D printed to increase algal cells attachment.

6.2. Materials and Methods

6.2.1. *Culturing the microalgae*

Two algal strains, *Chlorella vulgaris* and *Monoraphidium contortum,* were kindly provided in suspended culture by the Center of Water

Research, Universidad de Granada, Spain. Each strain was cultured in standard synthetic Combo medium in 30°C kindly provided by the Center of Water Research, at pH 8 and indirect sunlight, for 30 days. The growth curve and the biomass density of each strain were estimated in 5, 10, 15, 20, and 25 days by measuring the concentration of chlorophyll *a* and chlorophyll *b* by spectrophotometer (chlorophyll *a* at 433 nm and 666 nm, and chlorophyll *b* at 462 nm and 650 nm). This was conducted by harvesting the algal cultures (10 mL) from the cultures separately in 15 mL sterile tubes and centrifuging at 3,500 rpm for 10 minutes at 4°C.

6.2.2. *Sulfamethoxazole concentration*

On the 10th day of growth, sulfamethoxazole was introduced to each culture media of each strain separately. The antibiotic was prepared by dissolving 3 ml in 100 ml distilled water, added to 90 ml of *Chlorella vulgaris* culture and 90 ml of *Monoraphidium contortum* culture, and incubated for 20 days. A kinetic study measuring the concentration of sulfamethoxazole in the algal cultures was conducted every five days, as follows — 5, 10, 15, and 20 days of growth of the algal culture. The samples were collected from the two algal cultures separately by the centrifugation of 5 ml of each algal culture with sulfamethoxazole at 3,500 rpm for 10 minutes at 4°C. The concentration of sulfamethoxazole at each algal culture was measured separately every five days by measuring absorbance at 264 nm by spectrophotometer. This experiment was conducted in triplicate and compared to the control group without sulfamethoxazole.

6.2.3. *Sulfamethoxazole effect on the algal culture growth*

A kinetic study was conducted to measure the effect of sulfamethoxazole on each algal strain by testing their production to chlorophyll *a* and chlorophyll *b* every five days by spectrophotometer (chlorophyll *a* at 433 nm and 666 nm, and chlorophyll *b* at 462 nm and 650 nm). This experiment was conducted in triplicate and compared to the control group without sulfamethoxazole.

6.2.4. *Design and fabrication of micro-textured chips*

A micro-textured chip was designed for cell immobilization and attachment. A digital design of two micro-textured chips was conducted using

Rhinoceros 3D and 3D-printed in Polylactic acid (PLA) using the MakerBot 3D printer to fit the diameter of the potent strain, employing optimized growth media, conditions, and concentration of the pollutant. All experiments were conducted at the Catalysis Lab, Department of Inorganic Chemistry, Faculty of Science, Universidad de Granada, Spain.

6.3. Results and Discussion

6.3.1. *Chlorella vulgaris and Monoraphidium contortum growth (Culture medium and Conditions)*

In the current study, *Chlorella Vulgaris* and *Monoraphidium contortum* were employed to biodegrade sulfamethoxazole. These species were chosen for their environmental and industrial importance, availability, and easy cultivation. For example, *Chlorella vulgaris* is used for its biological and pharmacological properties as a dietary supplement that includes functional macro- and micronutrients, omega-3 polyunsaturated fatty acids, and minerals.[50] It also contains chlorophyll, intracellular proteins, carbohydrates, lipids, vitamin C, β-carotenes, and B vitamins (B1, B2, B6, and B12), which propose it as a rich dietary supplement.[57] Furthermore, *Chlorella vulgaris* is also used as a renewable energy source and in wastewater treatment.[52–54] Similarly, *Monoraphidium contortum* is used for biofuel production and in pharmaceutical industries.[55,56]

In the current study, the COMBO ATE media exhibited a high growth yield in both *Chlorella vulgaris* and *Monoraphidium contortum*. Even in the sulfamethoxazole modified media, it combatted the effect of the addition of sulfamethoxazole to both algal cultures separately. The COMBO medium includes seven major ions that simulate the composition of natural freshwater. These are Ca_2+, Mg_2+, Na+, K+, HCO_3-, SO_4 2−, and Cl−, as well as the major nutrient elements i.e., N, P, Si, B, following the methods of Guillard,[81] for the medium preparation. Trace elements play a crucial role in the metabolism of the microalgae; thus, the employed COMBO ATE media included seven trace elements that are essential for the metabolism of algae. These trace elements are Mn, Cu, Zn, Co, Fe, and Se. The COMBO ATE media was also used based on its adjustability, as it was reported to exhibit superior results in the cultivation of different microalgal species in comparison to any other artificial medium. Its composition is completely known and can be maintained among different studies.[80]

Chlorella vulgaris cultivation depends on a group of factors involving chemical and physical factors. The first is determined by the composition and concentration of the chemical species in culture medium, where the most effective are a source of carbon, nitrogen, phosphorus, silicon, and metals, such as iron, copper, zinc, vitamins, etc., particularly, the type and concentration of the carbon source and nitrogen source. The ratio between their concentrations (C/N) also affects microalgae metabolism. This, in turn, controls the quality and quantity of the biomass, lipid, carbohydrate, protein, chlorophyll, and pigments in the microalgae.[61,62] The physical factors control the cultivation environment pH, temperature, light intensity, and the intensity of aeration to the system.[63]

Since *Chlorella vulgaris* is an autotrophic microalga, the CO_2 or bicarbonate compounds are usually used as the only carbon source. CO_2 is usually introduced to the media by aeration.[64] This concentration should be balanced between the value which results in the maximum rate of cell growth and the microalgae tolerance threshold.[65,66] In the current study, the carbon source was CO_2, which was provided through phase aeration for 16 hours per day.

Nitrogen has a vital role in regulating cell growth, metabolism, protein Chlorophyll-a, Chlorophyll-b, and lipid production. Nitrogen scarcity in the culture media can cause a significant decline in the cell division rate and consequently low biomass production. Kong et al.[67] reported that the best nitrogen sources for *Chlorella vulgaris* cultivation were potassium nitrate and urea. However, the specific concentration of nitrogen source affects the lipid production intensity by the microalgae, as it was found that limiting the concentration of nitrogen leads to the production of more lipids while contradicting the biomass concentration. These aspects affect the capacity and efficiency of the microalgae in biodegrading the pollutant, as higher concentrations of biomass are required to increase the capacity of degradation. On the other hand, the amount of produced lipids controls the efficiency of degradation as it affects the cells' resistance to antibiotic toxicity as well as facilitating the previously mentioned processes of biodegradation that depend on the cell wall composition and extracellular matrix characteristics. Li et al.[68] reported that the sodium nitrate concentration of 3-20 mM achieved the highest lipid productivity in 5 mM concentration. This is congruent with the used COMBO ATE medium containing $NaNO_3$ with 1 mM concentration for both algae cultures, separately.

The addition of micronutrients including magnesium, sulfur, and iron affect the capacity of *Chlorella vulgaris* for biomass production and the chemical composition of its extracellular matrix. It was reported that the shortage in iron, magnesium, and sulfur concentrations caused inhibited photosynthesis and consequently the biomass production of *Chlorella vulgaris*. This is justified because iron is the redox catalyst in photosynthesis and facilitates the electron transport reactions within the photosynthesis process. Magnesium is an essential component of chlorophyll and facilitates microalgae cell division. The addition of sulfur enhances *Chlorella vulgaris* through cell division, protein metabolism, and fatty acid synthesis by increasing the redox enzymes. Phosphorus controls the lipid metabolism from synthesis membrane to neutral lipid storage, which affects the overall growth of the microalgae, as it is essential in the production of cellular constituents such as phospholipids, nucleotides, and nucleic acids.[69]

Monoraphidium contortum cultivation on the majority of standard common culture media was reported to be slow and challenging, which affected the adoption of treatments based on the use of this microalgae strain. Thus, the COMBO (ATE) media was used in the current study, as many previous studies reported the competence and reproducibility of this synthetic culture media in supporting the growth of a wide range of freshwater algae such as cyanobacteria and green algae.[80]

Khalil et al.[70] reported that *Chlorella vulgaris* thrived in a wide range of pH (4–10), with more favorable results achieved in the alkaline environment (pH 9 and 10). This was supported by a study that reported that upon decreasing pH concentration, the cell proliferation was consequently decreased.[71] Similarly, it was reported that the ideal pH that yields the maximum cell count in *Monoraphidium contortum* is within pH 9 and 9.5. Moreover, in *Monoraphidium contortum,* the pH concentration affects the cell size. It was reported that at pH higher than 9.5, the cells were observed to be smaller sized while increasing in diameter congruently with increased acidity of the media at pH 3.5.

Likewise, the light intensity affects the biomass productivity and the culture chemical composition, as it was proposed that light guides cell proliferation and facilitates cellular respiration and photosynthesis since it provides the required energy for facilitating the endothermic reactions for carbon metabolism while being the major factor in photosynthesis converting carbon dioxide into organic compounds, in which water and oxygen are released. In limited light intensity, the microalgae tend to produce carbon into amino acids. However, in the saturated illumination, sugar and starch production increases, and the maximum growth rate is stabilized.

However, it was reported that it is favorable to conduct phase-illumination, which employs non-continuous exposure to light, as in the case of *Chlorella vulgaris*, more cell division occurs after stopping the lighting phase, as some enzymatic mechanisms might stop during illumination.[72] Thus, employing natural daylight is favorable since it is a free-of-cost method to achieve phase-illumination. Microalgae is usually cultivated in wavelengths between 400 nm to 700 nm to be capable of photosynthesis. However, the compatible wavelength varies from one species to another.[73] In the case of *Chlorella vulgaris*, the maximum cultivation is achieved in the red-light spectrum (λ = 630 nm – 665 nm) and blue light spectrum (λ = 430 nm – 465 nm), respectively, and red light increases chlorophyll pigment. Thus, it is recommended to use both spectrums of red and blue light to achieve the maximum productivity and cell size of the *Chlorella vulgaris* biomass.

Temperature is one of the factors that most affect the various aspects of growth and fatty acid composition, enzymatic reactions, and cell membrane characteristics in various microalgal species.[74] It was reported that low temperature limits the growth and proliferation rate of the cell.[75] Thus, the optimal temperature for Chlorella vulgaris is within 30–35°C, when the maximum biomass productivity is achieved.[76] Converti et al.[61] has reported that the highest biomass production, after nine days of culturing at 30°C, was congruent with Barghbani et al.,[77] that achieved maximum biomass production after seven days of culture, in 30–35°C. Thus, in the current study, the *Chlorella vulgaris* was cultivated in 30°C for 30 days, with subsequent media renewal every 10 days. Likewise, in *M. conortum*, Fernandez et al.[74] reported that a range of 25–35°C is ideal for the growth of *Monoraphidium conortum* microalgae.

The necessity for aeration is dependent on the microalgae species, the type of culture, whether open system or photobioreactor, and the scale. Aeration is useful since it prevents the precipitation of microalgae, homogenizing the culture to expose all the cells to light and nutrients, as well as maintaining constant temperature through the culture environment and facilitating the exchange of gases within the culture. According to the culture scale, the mixing can be done by daily manual shaking for laboratory-scale cultivation in Erlenmeyer flasks and test tubes, or aerated peristaltic pump for larger scale or use the circulator arms in a large pool.[78] This was applied in the current study according to the scale of the culture used.

6.3.2. *The effect of Sulfamethoxazole on the growth of Chlorella vulgaris and Monoraphidium contortum*

In the current study, chlorophyll production was used as a biomarker to evaluate the stress of the antibiotic on the microalgae growth and metabolism, as it was reported that chlorophyll fluorescence is a sensitive parameter under environmental stress conditions that indicates the energy conversion through the photosynthesis process.[85,86] Usually, chlorophyll biosynthesis is affected by exposure to pharmaceuticals, inhibiting algal growth.[87] Moreover, a high chlorophyll content of cells acts as a protective mechanism to forage the accumulated ROS in the chloroplasts.

In the current study, it was found that sulfamethoxazole did not inhibit the growth of *Chlorella vulgaris,* unlike the growth of *Monoraphidium contortum* that was slightly inhibited by the addition of the antibiotic in comparison to the control group. This was indicated from the approximately similar production of chlorophyll *a* by each strain in the modified media compared to the control group. Figures 6.2(a) and 6.2(b) exhibit a comparison in the concentration of chlorophyll *a* by each strain separately between the control group of standard Combo ATE media and the COMO ATE modified with 3 g/100 mL of sulfamethoxazole media. The results show that the maximum production of chlorophyll *a* in the control group was achieved after 20 days of incubation in sunlight at 30°C, achieving 0.143 mg/L of chlorophyll *a* by *Chlorella vulgaris,* and 0.12 mg/L by *Monoraphidium contortum*, though declining slightly after the 20th day of growth due to the depletion of nutrients in the media. While the production of chlorophyll a by *Chlorella vulgaris* under the modified media with sulfamethoxazole in 30°C, pH 9, under sunlight, was achieved at the 25th day of growth of 0.135 mg/L and slightly increased during the following five days, reaching 0.137 mg/L, unlike the growth of *Monoraphidium contortum* that was inhibited moderately by the addition of sulfamethoxazole to the media, achieving only 0.07 mg/L at the 25th day of growth and slightly increasing during the following five days to 0.09 mg/L, which is less than the chlorophyll *a* production by *Monoraphidium contortum* in standard Combo ATE media without sulfamethoxazole.

The addition of sulfamethoxazole to the growth media divergently affected the production of chlorophyll *b* between the two algal strains because it improved the capacity of *Chlorella vulgaris* to produce

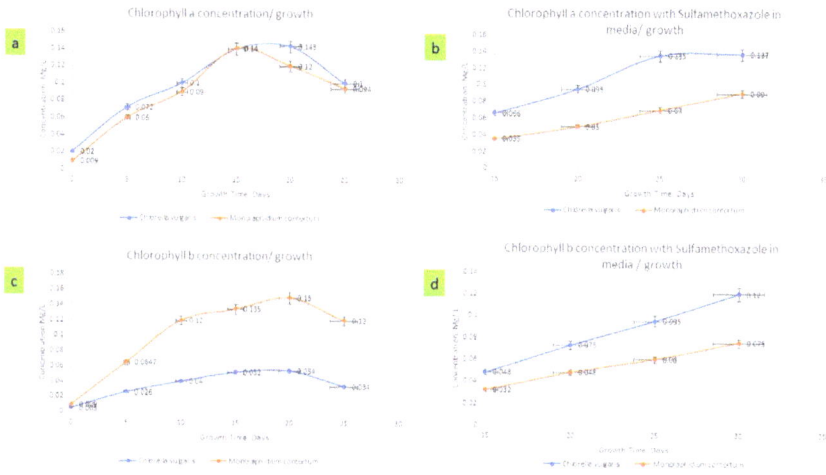

Figure 6.2. Growth and production of chlorophyll *a* and *b* by *Chlorella vulgaris* and *Monoraphidium contortum* using standard COMBO ATE medium and modified media with 3 mg/100 ml sulfamethoxazole respectively; (a) chlorophyll *a* concentration by each strain separately using standard COMBO ATE growth medium. (b) chlorophyll *a* concentration by each strain separately using modified growth medium with sulfamethoxazole. (c) chlorophyll *b* concentration by each strain separately using standard COMBO ATE growth medium. (d) chlorophyll *b* concentration by each strain separately using modified growth medium with sulfamethoxazole (by the authors).

chlorophyll *b* compared to the control group. The production levels of chlorophyll *b* by *Monoraphidium contortum* have slightly declined in the antibiotic-modified media compared to the control group. Figures 6.2(c) and 6.2(d) exhibits the increase in chlorophyll *b* levels by *chlorella* after adding 3 mg/100 ml of sulfamethoxazole to the growth medium, reaching 0.095 mg/L at the fifth day of growth while reaching only 0.034 mg/L in the control group. However, in the case of *Monoraphidium contortum*, chlorophyll *b* levels declined from 0.12 mg/L at the 25th day of growth in standard media to 0.06 mg/L in the modified media with sulfamethoxazole.

These two results prove that sulfamethoxazole increased the growth of *Chlorella vulgaris* and enhanced its ability to produce both types of chlorophyll *a* and *b*, while inhibiting the growth of *Monoraphidium contortum*, and consequently its chlorophyll *a* and *b* production.

This divergent effect of sulfamethoxazole on the two algal species is justified by the divergent biodegradation level and metabolic pathways that occur intracellular or extracellular in each algal strain. This depends on the microalgae cell wall composition and the antibiotic's chemical composition, which is the subject of degradation. It was found that the most potent algal strain in surviving the antibiotic effect is the most potent in degrading the antibiotic as well.

This could be further justified through the metabolic pathways of *Chlorella vulgaris* that are determined by its morphologic characteristics. The morphology and composition of these unicellular algae contribute to their efficiency in wastewater treatment. In the case of *Chlorella vulgaris*, its cells possess a thick cell wall of 100 nm–200 nm that provides mechanical and chemical protection for the cell and enables it to withstand chemical and mechanical stress. *Chlorella vulgaris* cells uptake carbon from their growth media by the enzyme carbonic anhydrase, which catalyzes the hydration of CO_2 to form HCO_3- and a proton. Generally, the majority of *Chlorella* species possess an outer cell wall layer and an inner cell wall layer.[58] The outer cell wall is rigid and composed of polysaccharides formed mainly by glucose and mannose. Another category of *Chlorella* species cell wall is composed of polysaccharides formed mainly by glucosamine,[59] thus, in both cases, facilitating the bioadsorption and biodegradation of different antibiotics with corresponding functional groups.

Each cell of *Chlorella vulgaris* only contains a single chloroplast composed of phospholipids. This includes two membranes — the first is permeable, while the second is selective to transport specific proteins.[57] *Chlorella vulgaris* is composed of 42–58% protein of the dry weight in its biomass; 20% of these proteins are adhered to the cell wall to facilitate the transport of the cell and support its structure, and 30% are secreted into the extracellular medium. Thus, this enhances the persistence and survivability of *Chlorella vulgaris* by maintaining high production rates even in aggressive environments, as well as playing a key role in the active transport of antibiotic break-down compounds into the algal cell.[60]

On the other hand, the antibiotic composition is the second key factor in the biodegradation reaction. Sulfamethoxazole is an isoxazole (1,2-oxazole) compound having a methyl substituent at the 5-position and a 4-aminobenzenesulfonamido group at the 3-position. It performs as an antibacterial and anti-infective that is effective against a wide range of bacterial infections by inhibiting dihydropteroate synthase, preventing the formation of dihydropteroate acid, a precursor of folic acid, which is required for bacterial growth. Thus, it works to block two consecutive steps in the biosynthesis of the nucleic acids and proteins necessary for bacterial growth and division. However, it is also considered an environmental contaminant, a xenobiotic, and a drug allergen.[82]

The reaction between *Chlorella vulgaris* and sulfamethoxazole was reported by a number of studies, as it was found that, when exposed to various sulfonamides (SAs) with different 10 mg/L, 30 mg/L, and 90 mg/L for seven days, the growth of *Chlorella vulagris* was slightly inhibited according to the examined concentrations of the antibiotic between. This was accompanied by an increased rate of proteins synthesis while soluble sugars were decreased. The oxidative stress caused by the presence of the antibiotic increased the levels of superoxide dismutase and glutathione reductase while decreasing the levels of catalase, as the antioxidant could not compensate for the levels of reactive oxygen species, which caused this oxidative damage. Similarly, it was reported that the inhibition of *Chlorella vulgaris* growth by SAs resulted from the antibiotic inhibition of folate synthesis in green algae, which implies that SAs affected the normal cellular development and folic acid synthesis in the cytoplasm, mitochondrion, and chloroplast. However, it was reported that *Chlorella vulgaris* still successfully removed up to 29% of the antibiotic.[83] These findings support results of the current study. It was reported that certain concentrations of sulfamethoxazole can perform double benefit by inducing algal growth and accelerating the removal of antibiotic contamination through the algal-mediated biodegradation process.[83] It was also reported by the same study that sulfonamides induced significantly higher levels of chlorophyll *a* and protein content production by *Chlorella vulgaris* congruently with the higher concentrations.

This reaction proves the consumption of sulfamethoxazole as a carbon source to promote algal growth, which is supported by previous studies wherein 5 mg/L and 1 mg/L levofloxacin improved the dry cell weight of *Chlorella vulgaris*.[29]

6.3.3. *Kinetic study of the degradation of SMX by Chlorella vulgaris and Monoraphidium contortum*

In the current work, a kinetic study revealed the capacity of each algal strain, *Chlorella vulgaris* and *Monoraphidium conortum*, to degrade 3 mg/100 ml sulfamethoxazole during 20 days. The following Figure 6.3 shows that the capacity of *Chlorella vulgaris* to degrade sulfamethoxazole is more than the capacity of *Monoraphidium conortum*. After the first five days of introducing sulfamethoxazole to the two algal cultures separately, *Chlorella vulgaris* biodegraded the antibiotic to 0.0655 mg/L while *Monoraphidium conortum* reached 0.077 mg/L concentration of sulfamethoxazole. The degradation rate maintained a steady

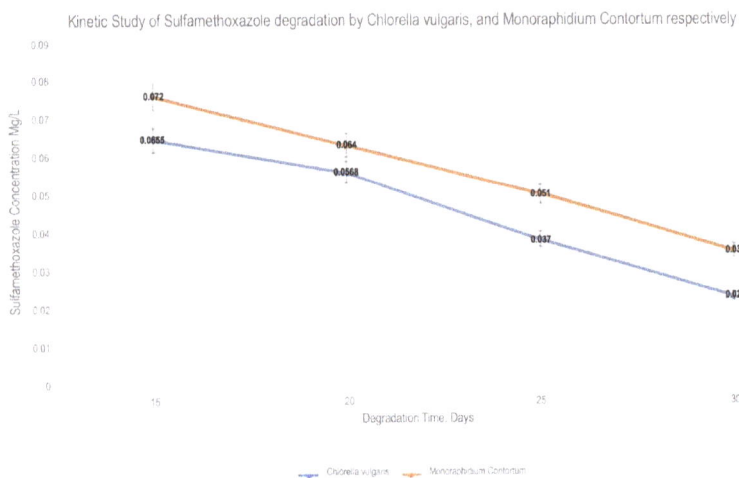

Figure 6.3.   Kinetic study of sulfamethoxazole degradation by *Chlorella vulgaris* and *Monoraphidium conortum,* respectively (by the authors).

decrease during the following 15 days, reaching 0.024 mg/L by *Chlorella vulgaris* and 0.036 mg/L by *Monoraphidium conortum*. This result implies the supremacy of *Chlorella vulgaris* over *Monoraphidium conortum* in the degradation of sulfamethoxazole, while maintaining enhanced growth behavior and production of both chlorophylls *a* and *b*. Thus, it is recommended to use *Chlorella vulgaris* for the degradation of sulfamethoxazole in wastewater treatment.

6.3.4. *Cell Immobilization 3D printed Chip support*

The reached results were used to design and optimize a microtextured supporter design to immobilize *Chlorella vulgaris* cells in a photobioreactor measuring 2 cm × 3 cm × 6 cm. Two designs were developed for a 2D chip supporter employing two different micro-textures with 50 μm resolution, which is congruent with the *Chlorella vulgaris* cell size that ranges up to 10.2 μm,[88] employing hexagonal and square channel grid respectively, to increase the angled sides of each well to increase the attachability of the cells as exhibited in Figure 6.4. The designs were printed in PLA in a micro-texture fashion by a high-resolution 3D Maker printer at the Catalysis Lab, Department of Inorganic Chemistry, Faculty of Science, University of Granada, Spain.

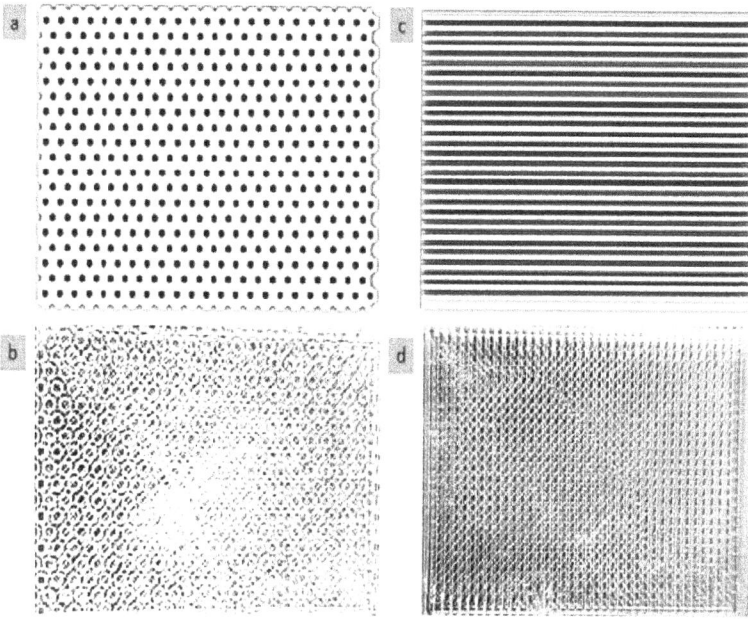

Figure 6.4. 50 μm resolution of the micro-textured chip supporter design for cell attachment and immobilization, printed in PLA. Chip dimensions. 2 cm × 3 cm. (a, b) Hexagonal grid micro-textured chip design and 3D printed piece, respectively. (c, d) Square channel grid micro-textured chip design and 3D printed piece, respectively (by the authors).

6.4. Conclusion

The current study examined the capacity of *Chlorella vulgaris* and *Monoraphidium conortum* in the biodegradation of 3 g/100 ml of sulfamethoxazole. This was conducted by employing a COMBO ATE medium, pH 9, 30°C, in sunlight and phase aeration to culture both algal strains due to its combinability and adjustability with a wide range of freshwater algal strains. Modified COMBO ATE culture media with 3 g/100 ml of sulfamethoxazole were used for both algal strains to test their ability to survive the effects of the antibiotic while also testing their capacity to digest it. The kinetic study revealed that sulfamethoxazole increased the growth of *Chlorella vulgaris* and promoted its ability to produce both types of chlorophyll *a* and *b*, achieving maximum production of chlorophyll *a* of 0.135 mg/L and chlorophyll *b* of 0.095 mg/L at the 25th day of growth, in comparison to 0.1 mg/L of chlorophyll *a* and

0.034 mg/L of chlorophyll *b* in the control group. Inhibiting the growth of *Monoraphidium conortum* achieved 0.09 mg/L of chlorophyll *a* and 0.079 of chlorophyll *b* in comparison to 0.094 mg/L of chlorophyll *a* and 0.12 of chlorophyll *b* in the control group. The results also revealed the competence of *Chlorella vulgaris* in achieving higher degradation levels of sulfamethoxazole, reaching 0.024 mg/L of sulfamethoxazole after 20 days of introducing it to the algal culture, in comparison to *Monoraphidium conortum* that degraded the antibiotic to 0.036 mg/L. These results were used to customize the digital design and digital fabrication of a 3D-printed PLA micro-textured chip support with 50 μm resolution to immobilize *Chlorella vulgaris* cells inside a pilot-scale photobioreactor. Hexagonal and square channeled supporter design was used to increase the inner edges of the 3D support for maximum cell attachment and anchorage. This supporter design could be customized for larger scales through multi-piece assembly.

Acknowledgments

This work would not have been achievable without the kind and generous support of Professor Francisco Carrasco Marín, Director of the Catalysis Laboratory, University of Granada, Faculty of Science, Department of Inorganic Chemistry, and Dr. Adriana Moral, Post-Doc at Catalysis Laboratory, University of Granada, Faculty of Science, Department of Inorganic Chemistry. Thus, we express our sincerest gratitude for their kind and generous support.

Conflict of interest: This research was conducted within a research and mobility program funded by the Erasmus+ KA-107 program.

References

1. Obimakinde, S., Fatoki, O., Opeolu, B., and Olatunji, O. (2016). Veterinary Pharmaceuticals in Aqueous Systems and Associated Effects: An Update. *Environmental Science and Pollution Research*, *24*(4), 3,274–3,297. DOI: https://doi.org/10.1007/s11356-016-7757-z.
2. Rzymski, P. et al. (2017). The Chemistry and Toxicity of Discharge Waters from Copper Mine Tailing Impoundment in the Valley of the Apuseni Mountains in Romania. *Environmental Science and Pollution Research*, *24*(26), 21,445–21,458. DOI: https://doi.org/10.1007/s11356-017-9782-y.

3. Caracciolo, A. B., Topp, E., and Grenni, P. (2015). Pharmaceuticals in the Environment: Biodegradation and Effects on Natural Microbial Communities. A Review. *Journal of Pharmaceutical and Biomedical Analysis*, *106*, 25–36. DOI: https://doi.org/10.1016/j.jpba.2014.11.040.

4. Zhang, X. X., Zhang, T., and Fang, H. H. (2009). Antibiotic Resistance Genes in Water Environment. *Applied microbiology and biotechnology*, *82*(3), 397–414. DOI: https://doi.org/10.1007/s00253-008-1829-z.

5. UN-Water (n.d.). WHO/UNICEF Joint Monitoring Program for Water Supply and Sanitation (JMP) — 2015 Update. Retrieved from: https://www.unwater.org/publications/whounicef-joint-monitoring-program-water-supply-sanitation-jmp-2015-update/.

6. Caracciolo, A. B., Topp, E., and Grenni, P. (2015). Pharmaceuticals in the Environment: Biodegradation and Effects on Natural Microbial Communities. A Review. *Journal of Pharmaceutical and Biomedical Analysis*, *106*, 25–36. DOI: https://doi.org/10.1016/j.jpba.2014.11.040.

7. Sim, W. J., Lee, J. W., Lee E. S., Shin, S. K., Hwang, S. R., and Oh, J. E. (2011). Occurrence and Distribution of Pharmaceuticals in Wastewater from Households, Livestock Farms, Hospitals and Pharmaceutical Manufactures. *Chemosphere*, *82*(2), 179–186.

8. Celiz, M. D., Perez, S., Barcelo, D., and Aga, D. S. (2009). Trace Analysis of Polar Pharmaceuticals in Wastewater by LC-MS-MS: Comparison of Membrane Bioreactor and Activated Sludge Systems. *Journal of Chromatographic Science*, *47*(1), 19–25.

9. Langford, K. and Thomas, K. V. (2011). Input of Selected Human Pharmaceuticalmetabolites into the Norwegian Aquatic Environment. *Journal of Environmental Monitoring*, *13*(2), 416–421. DOI: https://doi.org/10.1039/c0em00342e.

10. Ahmed, M. B., Zhou, J. L., Ngo, H. H., Guo, W., Johir, M. A. H., and Belhaj, D. (2017). Competitive Sorption Affinity of Sulfonamides and Chloramphenicol Antibiotics toward Functionalized Biochar for Water and wastewater treatment. *Bioresource Technology*, *238*, 306–312. DOI: https://doi.org/10.1016/j.biortech.2017.04.042.

11. Khetan, S. K. and Collins, T. J. (2007). Human Pharmaceuticals in the Aquatic Environment: A Challenge to Green Chemistry. *ChemInform*, *38*(36). DOI: https://doi.org/10.1002/chin.200736260.

12. Wang, J., Zhuan, R., and Chu, L. (2019). The Occurrence, Distribution and Degradation of Antibiotics by Ionizing Radiation: An Overview. *The Science of the total environment*, *646*, 1,385–1,397. DOI: https://doi.org/10.1016/j.scitotenv.2018.07.415.

13. Ben, Y., Fu, C., Hu, M., Liu, L., Wong, M. H., and Zheng, C. (2019). Human Health Risk Assessment of Antibiotic Resistance Associated with Antibiotic

Residues in the Environment: A Review. *Environmental Research*, *169*, 483–493. DOI: https://doi.org/10.1016/j.envres.2018.11.040.

14. Nguyen, N. V. et al. (2020). Community-level Consumption of Antibiotics According to the AWaRe (Access, Watch, Reserve) Classification in Rural Vietnam. *JAC–Antimicrobial Resistance*, *2*(3). DOI: https://doi.org/10.1093/jacamr/dlaa048.

15. Wang, J.-H., Zhang, T.-Y., Dao, G.-H., Xu, X.-Q., Wang, X.-X., and Hu H.-Y. (2017). Microalgae-based Advanced Municipal Wastewater Treatment for Reuse in Water bodies. *Applied Microbiology and Biotechnology*, *101*(7), 2,659–2,675. DOI: https://doi.org/10.1007/s00253-017-8184-x.

16. Sutherland, D. L. and Ralph, P. J. (2019). Microalgal Bioremediation of Emerging Contaminants — Opportunities and Challenges. *Water Research*, 164, no. 114921. DOI: https://doi.org/10.1016/j.watres.2019.114921.

17. Zeraatkar, A. K., Ahmadzadeh, H., Talebi, A. F., Moheimani, N. R., and McHenry, M. P. (2016). Potential Use of Algae for Heavy Metal Bioremediation, A Critical Review. *Journal of Environmental Management*, 181, 817–831. DOI: https://doi.org/10.1016/j.jenvman.2016.06.059.

18. Hena, S., Gutierrez, L., and Croué, J.-P. (2021). Removal of Pharmaceutical and Personal Care Products (PPCPs) from Wastewater using Microalgae: A Review. *Journal of Hazardous Materials*, 403, no. 124041. DOI: https://doi.org/10.1016/j.jhazmat.2020.124041.

19. Ahmed, I., Iqbal, H. M. N., and Dhama, K. (2017). Enzyme-Based Biodegradation of Hazardous Pollutants — An Overview. *Journal of Experimental Biology and Agricultural Sciences*, *5*(4), 402–411. DOI: https://doi.org/10.18006/2017.5(4).402.411.

20. Sheng, G.-P., Yu, H.-Q., and Li, X.-Y. (2010). Extracellular Polymeric Substances (EPS) of Microbial Aggregates in Biological Wastewater Treatment Systems: A Review. *Biotechnology Advances*, *28*(6), 882–894. DOI: https://doi.org/10.1016/j.biotechadv.2010.08.001.

21. Wang, J. and Wang, S. (2018). Microbial Degradation of Sulfamethoxazole in the Environment. *Applied Microbiology and Biotechnology*, *102*(8), 3,573–3,582. DOI: https://doi.org/10.1007/s00253-018-8845-4.

22. Guo, J. et al. (2020). Comparison of Oxidative Stress Induced by Clarithromycin in Two Freshwater Microalgae Raphidocelis Subcapitata and Chlorella Vulgaris. *Aquatic Toxicology*, *219*, no. 105376. DOI: https://doi.org/10.1016/j.aquatox.2019.105376.

23. Norvill, Z. N., Shilton, A., and Guieysse, B. (2016). Emerging Contaminant Degradation and Removal in Algal Wastewater Treatment Ponds: Identifying the Research Gaps. *Journal of Hazardous Materials*, *313*, 291–309. DOI: https://doi.org/10.1016/j.jhazmat.2016.03.085.

24. Xiong, J.-Q., Kurade, M. B., and Jeon, B.-H. (2018). Can Microalgae Remove Pharmaceutical Contaminants from Water? *Trends in Biotechnology*, *36*(1), 30–44. DOI: https://doi.org/10.1016/j.tibtech.2017.09.003.

25. Angulo, E., Bula, L., Mercado, I., Montaño, A., and Cubillán, N. (2018). Bioremediation of Cephalexin with non-living Chlorella sp., Biomass after Lipid Extraction. *Bioresource Technology*, *257*, 17–22. DOI: https://doi.org/10.1016/j.biortech.2018.02.079.

26. Nautiyal, P., Subramanian, K. A., and Dastidar, M. G. (2017). Experimental Investigation on Adsorption Properties of Biochar Derived from Algae Biomass Residue of Biodiesel Production. *Environmental Processes*, *4*(S1), 179–193. DOI: https://doi.org/10.1007/s40710-017-0230-2.

27. Daneshvar, E., Antikainen, L., Koutra, E., Kornaros, M., and Bhatnagar, A. (2018). Investigation on the Feasibility of Chlorella Vulgaris Cultivation in a Mixture of Pulp and Aquaculture Effluents: Treatment of Wastewater and Lipid Extraction. *Bioresource Technology*, *255*, 104–110. DOI: https://doi.org/10.1016/j.biortech.2018.01.101.

28. Xiong, Q., Hu, L.-X., Liu, Y.-S., Zhao, J.-L., He, L.-Y., and Ying, G.-G. (2021). Microalgae-based Technology for Antibiotics Removal: From Mechanisms to Application of Innovational Hybrid Systems. *Environment International*, *155*, no. 106594. DOI: https://doi.org/10.1016/j.envint.2021.106594.

29. Xiong, J. Q., Kurade, M. B., and Jeon, B. H. (2017). Biodegradation of Levofloxacin by an Acclimated Freshwater Microalga, Chlorella Vulgaris. *Chemical Engineering Journal*, *313*, 1,251–1,257. DOI: https://doi.org/10.1016/j.cej.2016.11.017.

30. Gojkovic, Z., Lindberg, R. H., Tysklind, M., and Funk, C. (2019). Northern Green Algae have the Capacity to Remove Active Pharmaceutical Ingredients. *Ecotoxicology and Environmental Safety*, *170*, 644–656. DOI: https://doi.org/10.1016/j.ecoenv.2018.12.032.

31. Zandalinas, S. I. and Mittler, R. (2018). ROS-induced ROS Release in Plant and Animal Cells. *Free Radical Biology and Medicine*, *122*, 21–27. DOI: https://doi.org/10.1016/j.freeradbiomed.2017.11.028.

32. Kumar, G., Shekh, A., Jakhu, S., Sharma, Y., Kapoor, R., and Sharma, T. R. (2020). Bioengineering of Microalgae: Recent Advances, Perspectives, and Regulatory Challenges for Industrial Application. *Frontiers in Bioengineering and Biotechnology*, *8*. DOI: https://doi.org/10.3389/fbioe.2020.00914.

33. Sun, W., Chen, L., and Wang, J. (2017). Degradation of PVA (polyvinyl alcohol) in Wastewater by Advanced Oxidation Processes. *Journal of Advanced Oxidation Technologies*, *20*(2). DOI: https://doi.org/10.1515/jaots-2017-0018.

34. Xie, W.-Y., Shen, Q., and Zhao, F. J. (2017). Antibiotics and Antibiotic Resistance from Animal Manures to Soil: A Review. *European Journal of Soil Science*, *69*(1), 181–195. DOI: https://doi.org/10.1111/ejss.12494.

35. Oberoi, A. S., Jia, Y., Zhang, H., Khanal, S. K., and Lu, H. (2019). Insights into the Fate and Removal of Antibiotics in Engineered Biological Treatment Systems: A Critical Review. *Environmental Science & Technology*, *53*(13), 7,234–7,264. DOI: https://doi.org/10.1021/acs.est.9b01131.

36. Song, C., Wei, Y., Qiu, Y., Qi, Y., Li, Y., and Kitamura, Y. (2019). Biodegradability and Mechanism of Florfenicol via Chlorella sp. UTEX1602 and L38: Experimental Study. *Bioresource Technology*, *272*, 529–534. DOI: https://doi.org/10.1016/j.biortech.2018.10.080.

37. Achermann, S., Bianco, V., Mansfeldt, C. B., Vogler, B., Kolvenbach, B. A., Corvini, P. F. X., and Fenner, K. (2018). Biotransformation of Sulfonamide Antibiotics in Activated Sludge: The Formation of Pterin-Conjugates Leads to Sustained Risk. *Environmental Science & Technology*, *52*(11), 6,265–6,274. DOI: https://doi.org/10.1021/acs.est.7b06716.

38. Alvarino, T., Nastold, P., Suarez, S., Omil, F., Corvini, P. F. X., and Bouju, H. (2016). Role of Biotransformation, Sorption and Mineralization of 14C-labelled Sulfamethoxazole under Different Redox Conditions. *Science of the Total Environment*, *542*, 706–715. DOI: https://doi.org/10.1016/j.scitotenv.2015.10.140.

39. Leng, S., Leng, L., Chen, L., Chen, J., Chen, J., and Zhou, W. (2020). The Effect of Aqueous Phase Recirculation on Hydrothermal Liquefaction/ Carbonization of Biomass: A Review. *Bioresource Technology*, *318*(124081). DOI: https://doi.org/10.1016/j.biortech.2020.124081.

40. Xiao, R. and Zheng, Y. (2016). Overview of Microalgal Extracellular Polymeric Substances (EPS) and their Applications. *Biotechnology Advances*, *34*(7), 1,225–1,244. DOI: https://doi.org/10.1016/j.biotechadv.2016.08.004.

41. Torres, M. A., Barros, M. P., Campos, S. C. G., Pinto, E., Rajamani, S., Sayre, R. T., and Colepicolo, P. (2008). Biochemical Biomarkers in Algae and Marine Pollution: A Review. *Ecotoxicology and Environmental Safety*, *71*(1), 1–15. DOI: https://doi.org/10.1016/j.ecoenv.2008.05.009.

42. Dudley, S., Sun, C., Jiang, J., and Gan, J. (2018). Metabolism of Sulfamethoxazole in Arabidopsis Thaliana Cells and Cucumber Seedlings. *Environmental Pollution*, *242*, 1,748–1,757. DOI: https://doi.org/10.1016/j.envpol.2018.07.094.

43. Xiong, Q., Liu, Y.-S., Hu, L.-X., Shi, Z.-Q., Cai, W.-W., He, L.-Y., and Ying, G.-G. (2020). Co-metabolism of Sulfamethoxazole by a Freshwater Microalga Chlorella Pyrenoidosa. *Water Research*, *175*, no. 115656. DOI: https://doi.org/10.1016/j.watres.2020.115656.

44. Hena, S., Gutierrez, L., and Croué, J.-P. (2021). Removal of Pharmaceutical and Personal Care Products (PPCPs) from Wastewater using Microalgae: A Review. *Journal of Hazardous Materials*, *403*(124041). DOI: https://doi.org/10.1016/j.jhazmat.2020.124041.

45. Gonçalves, A. L., Pires, J. C. M., and Simões, M. (2017). A Review on the Use of Microalgal Consortia for Wastewater Treatment. *Algal Research*, *24*, 403–415. DOI: https://doi.org/10.1016/j.algal.2016.11.008.

46. Vo, H. N. P. et al. (2019). A Critical Review on Designs and Applications of Microalgae-based Photobioreactors for Pollutants Treatment. *Science of the Total Environment*, *651*, 1,549–1,568. DOI: https://doi.org/10.1016/j.scitotenv.2018.09.282.

47. Ferrando, L. and Matamoros, V. (2020). Attenuation of Nitrates, Antibiotics and Pesticides from Groundwater using Immobilised Microalgae-Based Systems. *Science of the Total Environment*, *703*(134740). DOI: https://doi.org/10.1016/j.scitotenv.2019.134740.

48. Li, Y., Li, J., Pan, Y., Xiong, Z., Yao, G., Xie, R., and Lai, B. (2020). Peroxymonosulfate Activation on FeCo2S4 Modified g-C3N4 (FeCo2S4-CN): Mechanism of Singlet Oxygen Evolution for Nonradical Efficient Degradation of Sulfamethoxazole. *Chemical Engineering Journal*, *384*(123361). DOI: https://doi.org/10.1016/j.cej.2019.123361.

49. Xiao, R. and Zheng, Y. (2016). Overview of Microalgal Extracellular Polymeric Substances (EPS) and their Applications. *Biotechnology Advances*, *34*(7), 1,225–1,244. DOI: https://doi.org/10.1016/j.biotechadv.2016.08.004.

50. Eroglu, E., Smith, S. M., and Raston, C. L. (2015). Application of Various Immobilization Techniques for Algal Bioprocesses. In N. Moheimani, M. McHenry, K. de Boer, and P. Bahri (Eds.), *Biomass and Biofuels from Microalgae. Biofuel and Biorefinery Technologies*, 2, Cham: Springer, DOI: https://doi.org/10.1007/978-3-319-16640-7_2.

51. Panahi, Y., Darvishi, B., Jowzi, N., Beiraghdar, F., and Sahebkar, A. (2016). Chlorella Vulgaris: A Multifunctional Dietary Supplement with Diverse Medicinal Properties. *Current Pharmaceutical Design*, *22*(2), 164–173. DOI: https://doi.org/10.2174/1381612822666151112145226.

52. Mubashar, M. et al. (2020). Experimental Investigation of Chlorella vulgaris and *Enterobacter sp.* MN17 for Decolorization and Removal of Heavy Metals from Textile Wastewater. *Water*, *12*(11), 3,034. DOI: https://doi.org/10.3390/w12113034.

53. Otondo, A., Kokabian, B., Stuart-Dahl, S., and Gude, V. G. (2018). Energetic Evaluation of Wastewater Treatment using Microalgae, Chlorella Vulgaris. *Journal of Environmental Chemical Engineering*, *6*(2), 3,213–3,222. DOI: https://doi.org/10.1016/j.jece.2018.04.064.

54. Bifarini, M. A. S., Žitnik, M., Bulc, T. G., and Klemenčič, A. K. (2020). Treatment and Re-Use of Raw Blackwater by Chlorella vulgaris-Based System. *Water*, *12*(10), 2,660. DOI: https://doi.org/10.3390/w12102660.

55. Bogen, C. et al. (2013). Identification of Monoraphidium Contortum as a Promising Species for Liquid Biofuel Production. *Bioresource Technology*, *133*, 622–626.

56. Zhang, X., Rong, J., Chen, H., He, C., and Wang, Q. (2014). Current Status and Outlook in the Application of Microalgae in Biodiesel Production and Environmental Protection. *Frontiers in Energy Research*, *2*. DOI: https://doi.org/10.3389/fenrg.2014.00032.

57. Safi, C., Zebib, B., Merah, O., Pontalier, P.-Y., and Vaca-Garcia, C. (2014). Morphology, Composition, Production, Processing and Applications of Chlorella Vulgaris: A Review. *Renewable and Sustainable Energy Reviews*, *35*, 265–278. DOI: https://doi.org/10.1016/j.rser.2014.04.007.

58. Allard, B. and Templier, J. (2001). High Molecular Weight Lipids from the Trilaminar Outer Wall (TLS)-containing Microalgae Chlorella Emersonii, Scenedesmus Communis and Tetraedron Minimum. *Phytochemistry*, *57*(3), 459–467. DOI: https://doi.org/10.1016/s0031-9422(01)00071-1.

59. Takeda, H. (1991). Sugar Composition of the Cell Wall and the Taxonomy of Chlorella (Chlorophyceae). *Journal of Phycology*, *27*(2), 224–232. DOI: https://doi.org/10.1111/j.0022-3646.1991.00224.x.

60. Seyfabadi, J., Ramezanpour, Z., and Khoeyi, Z. A. (2010). Protein, Fatty Acid, and Pigment Content of Chlorella Vulgaris Under Different Light Regimes. *Journal of Applied Phycology*, *23*(4), 721–726. DOI: https://doi.org/10.1007/s10811-010-9569-8.

61. Converti, A., Casazza, A. A., Ortiz, E. Y., Perego, P., and Borghi, M. D. (2009). Effect of Temperature and Nitrogen Concentration on the Growth and Lipid Content of Nannochloropsis Oculata and Chlorella Vulgaris for Biodiesel Production. *Chemical Engineering and Processing: Process Intensification*, *48*(6), 1,146–1,151. DOI: https://doi.org/10.1016/j.cep.2009.03.006.

62. Belotti, G., Bravi, M., de Caprariis, B., de Filippis, P., and Scarsella, M. (2013). Effect of Nitrogen and Phosphorus Starvations on *Chlorella vulgaris* Lipids Productivity and Quality under Different Trophic Regimens for Biodiesel Production. *American Journal of Plant Sciences*, *04*(12), 44–51. DOI: https://doi.org/10.4236/ajps.2013.412a2006.

63. Daliry, S., Hallajisani, A., Mohammadi, R. J., Nouri, H., and Golzary, A. (2017). Investigation of Optimal Condition for Chlorella Vulgaris Microalgae Growth. *Global Journal of Environmental Science and Management*, *3*(2), 217–230. DOI: http://dx.doi.org/10.22034/gjesm.2017.03.02.010.

64. Hsieh, C.-H. and Wu, W.-T. (2009). Cultivation of Microalgae for Oil Production with a Cultivation Strategy of Urea Limitation. *Bioresource Technology*, *100*(17), 3,921–3,926. DOI: https://doi.org/10.1016/j.biortech.2009.03.019.

65. de Morais, M. G., da Silva Vaz, B., de Morais, E. G., and Costa, J. A. V. (2015). Biologically Active Metabolites Synthesized by Microalgae. *BioMed Research International*, pp. 1–15. DOI: https://doi.org/10.1155/2015/835761.

66. Montoya, E. Y. O., Casazza, A. A., Aliakbarian, B., Perego, P., Converti, A., and de Carvalho, J. C. (2014). Production of Chlorella Vulgaris as a Source of Essential Fatty Acids in a Tubular Photobioreactor Continuously Fed with Air Enriched with CO_2 at Different Concentrations. *Biotechnology Progress*, *30*(4), 916–922. DOI: https://doi.org/10.1002/btpr.1885.

67. Kong, W., Song, H., Cao, Y., Yang, H., Hua, S., and Xia, C. (2011). The Characteristics of Biomass Production, Lipid Accumulation and Chlorophyll Biosynthesis of Chlorella Vulgaris Under Mixotrophic Cultivation. *African Journal of Biotechnology*, *10*(55), 11,620–11,630.

68. Li, Y., Huang, J., Sandmann, G., and Chen, F. (2008). Glucose Sensing and the Mitochondrial Alternative Pathway are Involved in the Regulation of Astaxanthin Biosynthesis in the Dark-grown Chlorella Zofingiensis (Chlorophyceae). *Planta*, *228*(5), 735–743.

69. Ermis, H., Guven-Gulhan, U., Cakir, T., and Altinbas, M. (2020). Effect of Iron and Magnesium Addition on Population Dynamics and High Value Product of Microalgae Grown in Anaerobic Liquid Digestate. *Scientific Reports*, *10*(1), p. 3510. DOI: https://doi.org/10.1038/s41598-020-60622-1.

70. Khalil, Z. I., Asker, M. M. S., El-Sayed, S., and Kobbia, I. A. (2009). Effect of pH on Growth and Biochemical Responses of Dunaliella Bardawil and Chlorella Ellipsoidea. *World Journal of Microbiology and Biotechnology*, *26*(7), 1,225–1,231. DOI: https://doi.org/10.1007/s11274-009-0292-z.

71. Yan, C., and Zheng, Z. (2013). Performance of Photoperiod and Light Intensity on Biogas Upgrade and Biogas Effluent Nutrient Reduction by the Microalgae Chlorella sp. *Bioresource Technology*, *139*, 292–299.

72. Hultberg, M., Jönsson, H. L., Bergstrand, K. J., and Carlsson, A. S. (2014). Impact of Light Quality on Biomass Production and Fatty Acid Content in the Microalga Chlorella Vulgaris. *Bioresource Technology*, *159*, 465–467.

73. Blair, M. F., Kokabian, B., and Gude, V. G. (2014). Light and Growth Medium Effect on Chlorella Vulgaris Biomass Production. *Journal of Environmental Chemical Engineering*, *2*(1), 665–674.

74. Zeng, X., Guo, X., Su, G., Danquah, M. K., Chen, X. D., Lin, L., and Lu, Y. (2016). Harvesting of microalgal biomass. In Name of Editor/s? (Eds.), *Algae Biotechnology*, Cham: Springer, pp. 77–89.
75. Nishida, I., and Murata, N. (1996). Chilling Sensitivity in Plants and Cyanobacteria: the Crucial Contribution of Membrane Lipids. *Annual Review of Plant Biology*, *47*(1), 541–568.
76. Chinnasamy, S., Ramakrishnan, B., Bhatnagar, A., and Das, K. C. (2009). Biomass Production Potential of a Wastewater Alga Chlorella Vulgaris ARC 1 under Elevated Levels of CO_2 and Temperature. *International Journal of Molecular Sciences*, *10*(2), 518–532.
77. Barghbani, R., Rezaei, K., and Javanshir, A. (2012). Investigating the Effects of Several Parameters on the Growth of Chlorella Vulgaris using Taguchi's Experimental Approach. *International Journal of Biotechnology for Wellness Industries*, *1*(2), 128–133.
78. Morris, H. J., Almarales, A., Carrillo, O., and Bermúdez, R. C. (2008). Utilisation of Chlorella Vulgaris Cell Biomass for the Production of Enzymatic Protein Hydrolysates. *Bioresource Technology*, *99*(16), 7,723–7,729. DOI: https://doi.org/10.1016/j.biortech.2008.01.080.
79. Fernández, M. B., Tossi, V., Lamattina, L., and Cassia, R. (2016). A Comprehensive Phylogeny Reveals Functional Conservation of the UV-B Photoreceptor UVR8 from Green Algae to Higher Plants. *Frontiers in Plant Science*, *7*, 1,698.
80. Kilham, S. S., Kreeger, D. A., Lynn, S. G., Goulden, C. E., and Herrera, L. (1998). COMBO: A Defined Freshwater Culture Medium for Algae and Zooplankton. *Hydrobiologia*, *377*(1), 147–159.
81. Guillard, R. R. (1975). Culture of phytoplankton for feeding marine invertebrates. In Name of Editor/s? (Eds.). *Culture of Marine Invertebrate Animals*, Boston: Springer, 1975, pp. 29–60.
82. CHEBI: 9332 — sulfamethoxazole (2017). Retrieved from: http://www.ebi.ac.uk/chebi/searchId.do?chebiId=CHEBI:9332.
83. Chen, S. et al. (2020). Sulfonamides-induced Oxidative Stress in Freshwater Microalga Chlorella Vulgaris: Evaluation of Growth, Photosynthesis, Antioxidants, Ultrastructure, and Nucleic Acids. *Scientific Reports*, *10*(1). DOI: https://doi.org/10.1038/s41598-020-65219-2.
84. Xiong, J.-Q., Kurade, M. B., Patil, D. V., Jang, M., Paeng, K.-J., and Jeon, B.-H. (2017). Biodegradation and Metabolic Fate of Levofloxacin via a Freshwater Green Alga, Scenedesmus Obliquus in Synthetic Saline Wastewater. *Algal Research*, *25*, 54–61. DOI: https://doi.org/10.1016/j.algal.2017.04.012.
85. Xiao, Y., Huang, Q., Chen, L., and Li, P. (2010). Growth and Photosynthesis responses of Phaeodactylum Tricornutum to Dissolved Organic Matter from

Salt Marsh Plant and Sediment. *Journal of Environmental Sciences*, *22*(8), 1,239–1,245.

86. Perales-Vela, H. V., García, R. V., Gómez-Juárez, E. A., Salcedo-Álvarez, M. O., and Cañizares-Villanueva, R. O. (2016). Streptomycin Affects the Growth and Photochemical Activity of the Alga Chlorella Vulgaris. *Ecotoxicology and Environmental Safety*, *132*, 311–317.

87. Nie, X., Wang, X., Chen, J., Zitko, V., and An, T. (2008). Response of the Freshwater Alga Chlorella Vulgaris to Trichloroisocyanuric Acid and Ciprofloxacin. *Environmental Toxicology and Chemistry: An International Journal*, *27*(1), 168–173.

88. Vander Wiel, J. B. et al. (2017). Characterization of Chlorella Vulgaris and Chlorella Protothecoides using Multi-Pixel Photon Counters in a 3D Focusing Optofluidic System. *RSC Advances*, *7*(8), 4,402–4,408. DOI: https://doi.org/10.1039/c6ra25837a.

Chapter 7

Assessing Environmental Skincare Products: A Proposed Framework Using SWARA, ARAS, and COPRAS Methods

Berrak Aksakal*,¶, Zeliha Mahmat†, Figen Balo‡, and Lutfu S. Sua§

*Department of Public Health Services, Elazig Provincial Health Directorate

†Independent Researcher

‡Department of METE, Firat University, Elâzığ, Turkey

§Department of Management and Marketing, Southern University and A&M College, Baton Rouge, USA

¶berrak.aksakal@yahoo.com

Abstract

Skincare is one of the most important issues in the process of adapting to the changing and developing society of today's women and men. In addition to being an area where women show more interest, a certain segment of men, if not most, care about skincare as well. In this chapter, while examining the products of some brands used for skincare, the authors has tried to find the most harmless of them. Step-wise Weight Assessment Ratio Analysis (SWARA), Additive Ratio Assessment (ARAS), and Complex Proportional Assessment (COPRAS) were used to select the most harmless skincare product. While SWARA was used to find the criteria weights, ARAS and COPRAS rules were applied to determine such a product, and the two methods were compared. Through the combination of these methods, skincare products of various brands were evaluated and determined to be the most harmless one. Pairwise combination of the three quantitative methods is a unique approach developed for the purpose of this study and it offers an objective assessment of various decision alternatives.

Keywords: Skincare, ARAS, COPRAS, SWARA, public health, environmental effect, natural product.

7.1. Introduction

Skincare is one of the most important issues in the process of adapting to the changing and developing society of today's women and men. Of course, this situation differs from country to country, from society to society. How skincare habits are formed is also an important issue because not all skincare habits are correct. Regarding this issue, Gokdemir et al. conducted a study taking into account the Turkish society; they made an evaluation about the level of knowledge of people about skincare.[1] Within the scope of their studies, they examined the patients who visited the dermatology outpatient clinic where they worked from October 2006 to May 2007 and conducted a survey by asking some questions about skincare.[1] As a result of the study, it was found that the knowledge of the Turkish population about skincare products and skincare was insufficient.[1] At the same time, it should be noted that products used for skincare have an important place among skincare habits. In addition, many of the various preservatives and different chemical agents added for the purpose of increasing the quality, performance, and effectiveness of the products mentioned earlier create different toxicological effects in the human body by displaying different health risks, ranging from a mild skin sensitivity described as eczema to a serious life-threatening hypersensitivity or fatal poisoning called anaphylaxis. However, harmful chemicals can create particularly harmful effects for some periods of the human life. It has been proven that pregnancy, infancy, and adolescence periods are among them. For these reasons, the indiscriminate use of skincare products has become an important problem threatening public health in recent years.[2,3]

The prominent products and harmful effects in skincare products can be summarized as follows. Moisturizing products, which are frequently used skincare products, are classified according to the type of active and protective additives they contain. It can cause eczematous conditions such as irritant dermatitis, allergic contact dermatitis, photo-allergic reaction, hair follicle inflammatory conditions called folliculitis, a type of acne called cosmetic acne, sweat gland blockage, and systemic poisoning.[4] It is thought that paraben, which is used in many different cosmetic products, including moisturizers, in order to protect the substance from associated products by preventing the microbial activities in the product and to prevent the deterioration of the substance, may be associated with breast cancer due to its estrogenic activity.[5] On the other hand, studies have shown that parabens negatively affect the production of hormones and impair genital system functions.[6]

There are studies in which the benzophenones (oxybenzone) which are in skincare products used as sun cream has been shown to have hormone-disrupting effects.[7] On the other hand, the Environmental Working Group (EWG) states that oxybenzone and octinoxate can cause hormone disruptions and skin allergies, homosalate has hormone-disrupting effects, and avobenzone, mexoryl SX, octisalate, and octocrylene can cause skin allergies.[8] In addition, there is a possibility that lead in some sun creams can cause toxicity as it is a heavy metal. Since lead causes toxicity with neurological system effects, it can cause speech and learning difficulties, and behavioral problems. It can also cause miscarriages when used during pregnancy. It can cause infertility in both sexes and delay the onset of puberty in girls.

Another remarkable skincare product in terms of its harmful effects are products that lighten the skin. Although there are some restrictions on the use of mercury which is considered a heavy metal in recent years, it can still be used in mixtures to some extent. It is known that mercury has toxic effects on the nervous system, reproductive system, immune system, and respiratory system.[9] In a study where four cases were examined due to the use of mercury-containing whitening creams for two to six months, it was found that the cases caused a kidney pathology called "minimal change disease of the kidney", and it was shown that blood mercury levels returned to normal in one to seven months, urinary mercury levels returned to normal in nine to 16 months and urinary proteinuria got better in one to nine months after the use of the creams was terminated and the necessary treatments were applied.[10] In another study on mercury-related hyperpigmentation on the face, high mercury levels in the blood and urine and also neuropsychiatric findings were found in a 42-year-old female patient due to the long-term use of a cream containing a high mercury content (17.5%) as a bleaching agent.[11] Considering that creams containing high levels of mercury are not packaged products, but are drugs specially prepared by pharmacies, it is of great importance to inform pharmacists of the toxicity of cosmetic products, because the content of these products can irritate the skin and cause undesirable conditions such as eczema and allergies. Again, it is of great importance for human health that skincare products are natural and do not contain harmful chemicals. Despite this, it is impossible to talk about the fact that skincare products contain 100% naturalness and do not contain chemicals, even if they are beneficial.

There are studies by Gunes and Kurtoglu on skincare that focus on the skincare and properties of newborn babies.[12] Tatli and Gurel also have studies on newborn baby skincare and the physiological properties of skin.[13] Sarikaya and Altunisik examined consumer attitudes and focused on preferences that guide consumer preferences by doing a study on personal care products.[14]

In this chapter, skincare products were evaluated and criterion weights were determined by the SWARA method. Literature studies on the SWARA method are as follows. Ozdagoglu et al. conducted a study on the evaluation of alternative macroelisa equipment used in a university hospital.[15] They used SWARA and Weighted Aggregated Sum Product Assessment (WASPAS) methods in their study.[5] Chen et al. studied the relationship between landslide-related criteria by using the SWARA method to model in their studies on landslide susceptibility.[16] Ghorabaee et al. evaluated SWARA in a blurry environment in the evaluation of construction equipment in terms of sustainability in their study.[17]

Another Multi-Criteria Decision Making (MCDV) method used in the article is ARAS. Literature studies using the ARAS method are as follows. Yildirim et al. applied the ARAS method by considering the airline companies in their studies on personnel selection.[18] Ercan and Kundakci made a selection for a pattern program by using Operational Competitiveness Rating (OCRA) and ARAS methods in a textile business.[19] Buyukozkan and Gocer conducted a study examining potential suppliers by applying the ARAS method in an intuitive fuzzy environment in a digital supply chain.[20] Using Analytic Hierarchical Process (AHP) and ARAS methods, Tamošaitienė et al. chose the most suitable supplier for their work on problems related to the supply chain in the construction industry.[21] Iordache et al. used the ARAS method to select the most suitable location among the four alternatives by conducting a case study of salt caves in

Romania.[22] Zavadskas et al. used AHP and ARAS methods in the selection of the construction site for a deep-sea port in their study.[23]

The third method discussed in the article is COPRAS. Literature studies with COPRAS are as follows. Yazdani et al. used COPRAS and the Decision Making Trial and Evaluation Laboratory (DEMATEL) methods in the selection and ranking of green supplier alternatives.[24] Ghorabaee et al. evaluated airlines with multiple service quality using WASPAS, COPRAS, TOPSIS, and Evaluation based on Distance from Average Solution (EDAS) methods.[25] Nguyen et al. used fuzzy Analytic Network Process (ANP) and COPRAS-G method expressed with gray relations in the alternative choice of machine tools.[26]

Examples of studies in which SWARA and ARAS methods are used together can be given as follows. In his study, Ozbek evaluated the financial structures of Factoring companies in Borsa Istanbul (BIST) using MCDV methods such as SWARA, ARAS, TOPSIS, and Multi-Objective Optimization on the Basis of Ratio Analysis (MOORA).[27] Ozbek and Engur selected five different universities about Student Affairs Automation (OIO) used in universities and performed performance evaluation using ARAS, SWARA, and EDAS methods.[28] As a result of the analysis, they determined which alternative was the most suitable SWB in their studies.[28] Silver et al. evaluated the financial performance of firms which traded in the construction field using ARAS and SWARA methods.[29]

The studies in which SWARA and COPRAS methods are used together are as follows. Cakir and Karabiyik chose the most suitable one among different cloud storage service providers by using SWARA and COPRAS methods in their studies.[30] Zolfani and Zavadskas conducted a study on the sustainable development of buildings in rural areas in Iran by SWARA and COPRAS methods.[31] In their study on the retrofitting of public buildings, Volvačiovas et al. focused on nine different alternatives, determined their criterion weights with the SWARA method, and compared the most suitable alternatives with TOPSIS and COPRAS methods.[32] In their study on environmental sustainability, Zolfani et al. handled the evaluation of projects related to hotel construction using SWARA and COPRAS methods.[33] In their work on reverse logistics, which is an important issue for many companies, Zarbakhshnia et al. found the most suitable third-party reverse logistics providers by applying SWARA and COPRAS in a fuzzy environment.[34] Blue et al. used Stepwise Weight Assessment Ratio Analysis with Grey Relations

(SWARA-G) and COmplex PRoportional ASsessment of alternatives with Grey relations (COPRAS-G) methods with gray numbers while evaluating the transportation scenarios of the bus rapid transit line simulated with ARENA 14.[35] Yucenur et al. have presented the most suitable city selection method for a new facility in addition to the existing facilities in Turkey.[36] They have utilized the SWARA method when determining the criterion weights and twelve criteria were evaluated.[36] They tried to find out which of the three alternative cities was the most suitable with the COPRAS method.[36]

The aim of the study is defined in the second part of the chapter. In the third, fourth, and fifth sections, SWARA, ARAS, and COPRAS methods of the problem are explained and their application steps are given. In the sixth section, the application was made. Then the conclusion section is given.

7.2. The Objective of the Study

People have different lifestyles and working conditions affecting the level of pollution their skin is exposed to during the day. Although pollution rate is different, daily skincare is important for every person. Skincare is required in terms of both cleaning the skin and preventing the formation of unwanted problems on the skin. There are different skincare products that are used while performing this care. What also piques curiosity is how harmless the products used are to health. In this study, skincare products of some brands were considered within the framework of certain criteria and the authors try to determine the most harmless skincare product. In the study, MCDV methods were applied by taking into account the results of the survey conducted on The Good Shopping Guide[a] which products will be evaluated according to which criteria. While applying the methods, evaluation was made with 10 alternatives and 10 criteria. Since the criterion weights do not have equal importance, criterion weights were found by the SWARA method and these weights were used in the ARAS and COPRAS methods to determine the most harmless skincare product. Then, the ordering of the alternatives in both methods was compared.

---

[a]Ethical Shopping — The Good Shopping Guide (2001). Retrieved from: https://the goodshoppingguide.com/

7.3. SWARA Method

While the SWARA method helps determine the criterion weights, the easy inclusion of expert opinions into the method has made its use widespread.[37] In addition, in the studies conducted with the questionnaire, the correct answers of the members participating in the questionnaire increase the consistency of the method and members have the opportunity to freely evaluate the criteria since they do not need any other scale.[38] The steps of the five-step method are given as follows.[39]

Step 1: Criteria are ranked from the most important to the least important with expert opinion.

Step 2: The relative importance level (s_j) is determined for all criteria. So jth criterion ($j + 1$). It is written in comparison with the criterion.

Step 3: k_j coefficient is calculated.

$$k_j = \begin{cases} 1 & j = 1 \\ sj + 1 & j > 1 \end{cases} \tag{7.1}$$

Step 4: w_j variable is calculated (x_{j-1} is equal to w_{j-1}).

$$w_j = \begin{cases} 1 & j = 1 \\ \dfrac{x_{j-1}}{k_j} & j > 1 \end{cases} \tag{7.2}$$

Step 5: Relative weight (q_j) is calculated. The sum of the criteria weights must be equal to '1'.

$$q_j = \frac{w_j}{\sum_{k=1}^{n} w_k} \tag{7.3}$$

7.4. ARAS Method

The ARAS method was founded by Zavadskas and Turskis.[40] In the method, the value ratios of the utility functions of the alternatives and the values of the utility functions of the optimum alternatives are compared.[41] The steps of the method are as follows.[40]

Step 1: A decision matrix is created.

$$X = \begin{bmatrix} x_{01} & \cdots & x_{0n} \\ \vdots & \ddots & \vdots \\ x_{m1} & \cdots & x_{mn} \end{bmatrix} i = 0,1,\ldots,m; \ j = 1,2,\ldots,n \tag{7.4}$$

Maximization $\hspace{3cm} x_{0j} = \max_i x_{ij} \hspace{2cm}$ (7.5)

Minimization $\hspace{3.2cm} x_{0j} = \min_i x_{ij} \hspace{2cm}$ (7.6)

Step 2: The decision matrix is normalized. The normalization process consists of different stages according to the criteria being maximized and minimized. While the normalization process of the criteria with maximization direction in Equation (7.7) is done in one step, the normalization process of the criteria that is in the direction of minimization in Equation (7.8) is done in two stages.

$$\overline{x}_{ij} = \frac{x_{ij}}{\sum_{i=0}^{m} x_{ij}} \tag{7.7}$$

$$x_{ij} = \frac{1}{x_{ij}^*}; \hspace{1cm} \overline{x}_{ij} = \frac{x_{ij}}{\sum_{i=0}^{m} x_{ij}} \tag{7.8}$$

Step 3: The weighted matrix is created. The normalized matrix is multiplied by the criteria weights. The point to note here is that $\sum_{j=1}^{n} w_j = 1$. It should also be $0 < W_i < 1$.

$$\widehat{x}_{ij} = \overline{x}_{ij} \cdot w_j; \ i = 0,\ldots,m \tag{7.9}$$

$$\widehat{X} = \begin{bmatrix} \widehat{x}_{01} & \cdots & \widehat{x}_{0n} \\ \vdots & \ddots & \vdots \\ \widehat{x}_{m1} & \cdots & \widehat{x}_{mn} \end{bmatrix} i = 0,1,\ldots,m; \ j = 1,2,\ldots,n \tag{7.10}$$

Step 4: The optimum function (S_i) is calculated. Using the weighted matrix in Equation (7.10), the optimum function value of each alternative is determined.

$$S_i = \sum_{j=1}^{n} \widehat{x_{ij}}; \qquad i = 0,\ldots,m: \qquad j = 1,\ldots,n \qquad (7.11)$$

Step 5: The degree of benefit (K_i) is calculated and the ranking is made. K_i is determined by the ratio between the optimum value of the best decision option (S_0) and S_i.

$$K_i = \frac{S_i}{S_0}; \quad i = 0,\ldots,m \qquad (7.12)$$

7.5. COPRAS Method

The COPRAS method was developed at the Vilnius Gediminas Technical University in 1996 by Zavadskas and Kaklauskas.[42] The steps of the method are as follows.[43]

Step 1: Decision matrix is created.

$$X = \begin{bmatrix} x_{11} & \cdots & x_{1m} \\ \vdots & \ddots & \vdots \\ x_{n1} & \cdots & x_{nm} \end{bmatrix} \qquad (7.13)$$

Step 2: Decision matrix is standardized. This process is obtained by multiplying the values normalized by Equation (7.14) by the criterion weights.

$$d_{ij} = \frac{x_{ij} q_i}{\sum_{j=1}^{n} x_{ij}}, \quad i = \overline{1,m}; \ j = \overline{1,n} \qquad (7.14)$$

$$q_i = \sum_{j=1}^{n} d_{ij}, \quad i = \overline{1,m}; \quad j = \overline{1,n} \qquad (7.15)$$

The q_j values given in Equation (7.15) express the weight of each criterion.

Step 3: The weighted and normalized values are collected. S_{-i} is calculated according to the minimization-oriented criteria given in Equation (7.16), and while the value is as small as possible, S_{+i} is calculated according to the maximization-oriented criteria and the value is requested to be as large as it is.

$$S_{+j} = \sum_{i=1}^{m} d_{+ij}; \quad S_{-j} = \sum_{i=1}^{m} d_{-ij}, \quad i = \overline{1,m}; \quad j = \overline{1,n} \quad (7.16)$$

Step 4: The relative significance (Q_j) is calculated for each alternative.

$$Q_j = S_{+j} + \frac{S_{-min} \sum_{j=1}^{n} S_{-j}}{S_{-j} \sum_{j=1}^{n} \frac{S_{-min}}{S_{-j}}}, \quad j = \overline{1,n} \quad (7.17)$$

Step 5: The degree of utility (N_j) is calculated for the alternatives. The degree of utility is calculated with the help of Equation (7.18) and is the best option with a value of '100'.

$$N_j = \left(\frac{Q_j}{Q_{max}} \right) x100\% \quad (7.18)$$

7.6. Application

The study is aimed at determining the most harmless skincare product. For this, the most harmless among the 10 alternative skincare products was determined, according to the 10 criteria. These criteria are shown in Table 7.1 with maximization and minimization aspects.

In addition, the initial decision matrix to be used in ARAS and COPRAS is given in Table 7.2. Criteria weights to be determined in the SWARA method will be used in the ARAS and COPRAS methods.

Table 7.1.   Criteria.

Criterion	Symbol	Kind
Environmental report	K1	MAX
Organic	K2	MAX
Nuclear power	K3	MIN
Animal welfare	K4	MAX
Vegetarian	K5	MAX
Armaments	K6	MAX
Irresponsible marketing	K7	MIN
Political donations	K8	MAX
Public record criticisms	K9	MIN
Ethical accreditation	K10	MAX

Table 7.2. Initial decision matrix.

Criterion/ Brand	K1 MAX	K2 MAX	K3 MIN	K4 MAX	K5 MAX	K6 MAX	K7 MIN	K8 MAX	K9 MIN	K10 MAX
Brand 1	9	9	3	3	7	9	3	9	3	7
Brand 2	5	1	1	1	5	5	1	5	3	3
Brand 3	9	9	3	7	7	9	3	9	3	7
Brand 4	3	1	1	1	3	5	1	5	3	3
Brand 5	5	1	1	1	3	5	1	5	3	3
Brand 6	7	3	3	9	9	9	3	9	3	9
Brand 7	5	1	1	1	3	5	1	5	5	3
Brand 8	9	3	3	9	9	9	3	9	3	9
Brand 9	5	1	1	1	3	5	1	3	5	3
Brand 10	9	9	3	9	9	9	3	7	3	7

Table 7.3. Calculation of parameters.

Criterion	Order of importance	Sj	Kj	Wj	Qj
Environmental report	K3		1	1	0,2227
Organic	K6	0,2	1,2	0,8333	0,1855
Nuclear power	K2	0,3	1,3	0,6410	0,1427
Animal welfare	K5	0,3	1,3	0,4931	0,1098
Vegetarian	K4	0,2	1,2	0,4109	0,0915
Armaments	K9	0,1	1,1	0,3735	0,0832
Irresponsible marketing	K1	0,5	1,5	0,2490	0,0554
Political donations	K10	0,1	1,1	0,2264	0,0504
Public record criticisms	K7	0,4	1,4	0,1617	0,0360
Ethical accreditation	K8	0,6	1,6	0,1010	0,0225
SUM				**4,49**	**1**

7.7. SWARA Application

The k_j, w_j, and q_j values were calculated by taking the expert opinion on the relative importance level of the criteria. Criteria weights found at the end of the operations are given in Table 7.3.

As seen in Table 7.3, Nuclear Power (K3) appears as the most important criterion, while Armaments (K6) take the second place, and Organic

(K2) takes the third place. The least important criterion weight was found to be Political Donations (K8). At the same time, the sum of criterion weights gives '1' as indicated at the bottom right. These values were used to find the most suitable alternative in the ARAS and COPRAS methods, and the results of the two methods were compared.

7.8. ARAS Application

The values are normalized by using the values in Table 7.2 and Equations 7.7 and 7.8 and multiplied by the criterion weights (Table 7.3) found in the SWARA method. The resulting matrix is used for sorting and is written in Equations 7.11 and 7.12. The ranking is given in Table 7.4.

At the end of the values found and the ranking, Brand 10 appears as the most harmless skincare product. Brand 3 is in second place and Brand 1 is in third place. The product with the worst ranking value was found to be Brand 9. Brand 7 is followed by Brand 4.

7.9. COPRAS Application

The equation was placed in Equation 7.14 by using the initial decision matrix (Table 7.2) values. Normalized process and weighting process has

Table 7.4. Optimum Function, Benefit Rating, and Ranking.

	Si	Ki	
Optimum	0,17083	1	Ranking
Brand 1	0,09612	0,5626	3
Brand 2	0,07069	0,4138	6
Brand 3	0,10329	0,6046	2
Brand 4	0,06593	0,3859	8
Brand 5	0,06741	0,3946	7
Brand 6	0,09206	0,5388	5
Brand 7	0,06558	0,3839	9
Brand 8	0,09354	0,5475	4
Brand 9	0,06498	0,3804	10
Brand 10	0,10956	0,6413	1

Table 7.5.   Performance values.

	S+i	S-i	S-min	S-i-sum	(S-min/ Si-)	Sum(S-min/Si-)	Qi	Qmax	Ni	Rank
Brand 1	0,0946	0,0461	0,0203	0,3419	0,4394	6,808	0,1167	0,1329	87,825	3
Brand 2	0,0374	0,0203			1		0,0876		65,918	6
Brand 3	0,1033	0,0461			0,4394		0,1254		94,384	2
Brand 4	0,0319	0,0203			1		0,0821		61,803	8
Brand 5	0,0336	0,0203			1		0,0838		63,068	7
Brand 6	0,0891	0,0461			0,4394		0,1112		83,69	5
Brand 7	0,0336	0,0252			0,8056		0,074		55,72	9
Brand 8	0,0908	0,0461			0,4394		0,1129		84,955	4
Brand 9	0,0329	0,0252			0,8056		0,0734		55,206	10
Brand 10	0,1108	0,0461			0,4394		0,1329		100	1

been done. The results were placed in Equations (7.16–7.18), and the ranking process was made. Results are shown in Table 7.5.

According to the COPRAS method, the most harmless skincare product was Brand 10, while Brand 3 ranked second and Brand 1 ranked third. Brand 9 had the lowest ranking value, followed by Brand 7 and Brand 4.

7.10. Conclusion

When it comes to health, the first thing that comes to mind is that there is no biological internal disorder. However, health is not only regarding any internal discomfort or conditions but it also encompasses the health of the skin that covers our body like a shield. Although human skin is a self-renewable layer, undesirable results may occur if it is not well cared for and cleaned regularly. For this, they must be cleaned with skincare products. It is desired that these products that help to clean are also harmless in terms of human health. However, although it is not a 100% harmless product, the most harmless one can be chosen.

In this study, skincare products of some brands were evaluated and determined as the most harmless. Multi-Criteria Decision Making methods were used for this. In order to choose the most suitable alternative, criterion weights were found by the SWARA method. These weights found were used in ARAS and COPRAS methods. According to the SWARA method, the criterion with the highest weight was Nuclear Power

(K3), followed by Armaments (K6), and Organic (K2). The criterion with the least weight is Political Donations (K8). The results obtained from the ARAS and COPRAS methods were compatible with each other, and Brand 10 was found to be the most harmless skincare product. Brand 3 is in second place and Brand 1 is in third place. The worst result was found to be Brand 9. Brand 7 followed by Brand 4.

Since skincare is necessary for today's conditions, consumers should research the content of skincare products and pay attention when buying products. By providing training to doctors, pharmacists, and skin beauticians, it can be ensured that awareness on the subject becomes more widespread in society. Manufacturer companies should also consider human health as the most important factor. Mass media can be used to make people more aware of the issue. Other studies can be done to enlighten consumers regarding skincare. Further studies can be done on the ingredients of skincare products. After all, every ingredient in skincare products may not be healthy for human skin.

References

1. Assessment of knowledge about skin care among Turkish people. Turkderm-Turk Arch Dermatol Venereol (2008). *42*(2): 60–63.
2. Bilal, M. and Iqbal, H. M. N. (2019). An Insight into Toxicity and Human-Health-Related Adverse Consequences of Cosmeceuticals: A Review. *Science of the Total Environment, 670*(20), 555–568. DOI: 10.1016/j.scitotenv.2019.03.261.
3. Özkan, K., Danacı, M. Ö., and Çetin, Z. (2019). Interaction of Pregnancy Physiology and Cosmetic Products with Teratogen Effect and Personal Cleaning and Care Products. *H.Ü. Sağlık Bilimleri Fakültesi Dergisi, 6*(3), DOI: 10.21020/husbfd.606508.
4. Daye M., and Mevlitoğlu, M. (2011). Humidifiers. *Selçuk Üniv Tıp Derg, 27*(2), 124–127.
5. Gülle, S. (2019). Method development for trace level paraben analysis in cosmetic products. Master Thesis. Cumhuriyet University, Institute of Health Sciences, Sivas.
6. Tan, A. S. and Tüysüz, M. (2013). Use of Preservatives and Preservative Effectiveness Tests in Cosmetic Products. *ANKEM Derg, 27*(2), 83–91. DOI:10.5222/ankem.2013.083.
7. Suzuki, T., Kitamura, S., Khota, R., Sugihara, K., Fujimoto, N., and Ohta, S. (2005). Estrogenic and Antiandrogenic Activities of 17 Benzophenone

Derivatives used as UV Stabilizers and Sunscreens. *Toxicology and Applied Pharmacology, 203*, 9–17.

8. EWG Guide to Sunscreen: "The Trouble with Ingredients in Sunscreens". Access: https://www.ewg.org/sunscreen/report/thetrouble-with-sunscreen-chemicals/.

9. Alam, M. F., Akhter, M., Mazumder, B., Ferdous, A., Hossain, M. D., Dafader, N. C., and AtiqueUllah, A. K. M. (2019). Assessment of Some Heavy Metals in Selected Cosmetics Commonly used in Bangladesh and Human Health Risk. *Journal of Analytical Science and Technology, 10*(2), 2–8. DOI: 10.11867s40543-018-0162-0.

10. Tang, H. L., Mak, Y. F., and Chu, K.H. (2013). Minimal Change Disease caused by Exposure to Mercury-Containing Skin Lightening Cream: A Report of 4 Cases. *Clinical Nephrology, 79*, 326–329.

11. Çağlar, A. B. and Saral, S. (2014). Toxicity Concerns in Cosmetology. *Turkish Journal of Dermatology, 4*: 248–251.

12. Güneş, T. and Kurtoğlu, S. (2005). Yenidoğan Derisinin Özellikleri ve Bakımı. *Turkiye Klinikleri Pediatric Sciences — Special Topics, 1*(4), 1–4.

13. Tatlı, M. M. and Gürel, M. S. (2002). Newborn Skin Care. *Turkiye Klinikleri Journal of Pediatrics, 11*(2), 108–112.

14. Sarıkaya, N. and Altunışık, R. (2011). KişiselBakım Olgusu ve Kişisel Bakım Ürünlerine Yönelik Tüketici Tutum ve Tercihlerini Etkileyen Faktörler Üzerine Bir Araştırma. *Eskişehir Osmangazi Üniversitesi İktisadi ve İdari Bilimler Dergisi, 6*(2), 389–413.

15. Özdağoğlu, A., Keleş, M. K., and Eren, F. Y. (2019). Evaluation of Makroelisa Equipment Alternatives in a University Hospital by WASPAS and SWARA Methods. *Suleyman Demirel University Journal of Faculty of Economics & Administrative Sciences, 24*(2), 319–331.

16. Chen, W., Panahi, M., Tsangaratos, P., Shahabi, H., Ilia, I., Panahi, S., and Ahmad, B. B. (2019). Applying Population-Based Evolutionary Algorithms and a Neuro-Fuzzy System for Modeling Landslide Susceptibility. *Catena, 172*, 212–231.

17. Ghorabaee, M. K., Amiri, M., Zavadskas, E. K., and Antucheviciene, J. (2018). A New Hybrid Fuzzy MCDM Approach for Evaluation of Construction Equipment with Sustainability Considerations. *Archives of Civil and Mechanical Engineering, 18*(1), 32–49.

18. Yıldırım, B. I., Uysal, F., and Ilgaz, A. (2019). Personnel Selection in Airline Companies: An Application with ARAS Method. *Süleyman Demirel Üniversitesi Sosyal Bilimler Enstitüsü Dergisi, 2*(33), 219–231.

19. Ercan, E. and Kundakçı, N. (2017). Comparison of ARAS and OCRA Methods in Sen Program Selection in a Textile Company. *Afyon Kocatepe Üniversitesi Sosyal Bilimler Dergisi, 19*(1), 83–105.

20. Büyüközkan, G. and Göçer, F. (2018). An Extension of ARAS Methodology Under Interval Valued Intuitionistic Fuzzy Environment for Digital Supply Chain. *Applied Soft Computing, 69,* 634–654.
21. Tamošaitienė, J., Zavadskas, E. K., Šileikaitė, I., and Turskis, Z. (2017). A Novel Hybrid MCDM Approach for Complicated Supply Chain Management Problems in Construction. *Procedia Engineering, 172,* 1,137–1,145.
22. Iordache, M., Schitea, D., Deveci, M., Akyurt, İ. Z., and Iordache, I. (2019). An Integrated ARAS and Interval Type-2 Hesitant Fuzzy Sets Method for Underground Site Selection: Seasonal Hydrogen Storage in Salt Caverns. *Journal of Petroleum Science and Engineering, 175,* 1,088–1,098.
23. Zavadskas, E. K., Turskis, Z., and Bagočius, V. (2015). Multi-Criteria Selection of a Deep-Water Port in the Eastern Baltic Sea. *Applied Soft Computing, 26,* 180–192.
24. Yazdani, M., Chatterjee, P., Zavadskas, E. K., and Zolfani, S. H. (2017). Integrated QFD–MCDM Framework for Green Supplier Selection. *Journal of Cleaner Production, 142,* 3,728–3,740.
25. Ghorabaee, M. K., Amiri, M., Zavadskas, E. K., Turskis, Z., and Antucheviciene, J. (2017). A New Hybrid Simulation-Based Assignment Approach for Evaluating Airlines with Multiple Service Quality Criteria. *Journal of Air Transport Management, 63,* 45–60.
26. Nguyen, H. T., Dawal, S. Z. M., Nukman, Y., and Aoyama, H. (2014). A Hybrid Approach for Fuzzy Multi-Attribute Decision making in Machine Tool Selection with Consideration of the Interactions of Attributes. *Expert Systems with Applications, 41*(6), 3,078–3,090.
27. Ozbek, A. (2018). Evaluation of Financial Structures of Factoring Companies Traded on BIST by Multi-Criteria Decision Making Methods. *Yönetim ve Ekonomi: Celal Bayar Üniversitesi İktisadi ve İdari Bilimler Fakültesi Dergisi, 25*(1), 29–53.
28. Ozbek, A. and Engür, M. (2019). Student Affairs Automation Selection with Multi-Criteria Decision Making Methods. *Afyon Kocatepe Üniversitesi İktisadi ve İdari Bilimler Fakültesi Dergisi, 21*(1), 1–18.
29. Gümüş, U. T., Öziç, H. C., and Sezer, D. (Evaluation of Financial Performance of Businesses Traded in Construction and Public Works Sector on BIST with SWARA and ARAS Methods. *OPUS UluslararasıToplum Araştırmaları Dergisi, 10*(17), 835–858.
30. Cakir, E. and Karabıyık, B. K. (2017). Evaluation of Cloud Storage Service Providers using the Integrated SWARA–COPRAS Method. *Bilişim Teknolojileri Dergisi, 10*(4), 417–434.
31. Zolfani, S. H. and Zavadskas, E. K. (2013). Sustainable Development of Rural Areas' Building Structures Based on Local Climate. *Procedia Engineering, 57,* 1,295–1,301.

32. Volvačiovas, R., Turskis, Z., Aviža, D., and Mikštienė, R. (2013). Multi-attribute Selection of Public Buildings Retrofits Strategy. *Procedia Engineering*, *57*, 1,236–1,241.
33. Zolfani, S. H., Pourhossein, M., Yazdani, M., and Zavadskas, E. K. (2018). Evaluating Construction Projects of Hotels Based on Environmental Sustainability with MCDM Framework. *Alexandria Engineering Journal*, *57*(1), 357–365.
34. Zarbakhshnia, N., Soleimani, H., and Ghaderi, H. (2018). Sustainable Third-Party Reverse Logistics Provider Evaluation and Selection using Fuzzy SWARA and Developed Fuzzy COPRAS in the Presence of Risk Criteria. *Applied Soft Computing*, *65*, 307–319.
35. Mavi, R. K., Zarbakhshnia, N., and Khazraei, A. (2018). Bus Rapid Transit (BRT): A Simulation and Multi Criteria Decision Making (MCDM) Approach. *Transport Policy*, *72*, 187–197.
36. Yucenur, G. N., Çaylak, Ş., Gönül, G., and Postalcıoğlu, M. (2020). An Integrated Solution with SWARA & COPRAS Methods in Renewable Energy Production: City Selection for Biogas Facility. *Renewable Energy*, *145*, 2,587–2,597.
37. Keršuliene, V., Zavadskas, E. K., and Turskis, Z. (2010). Selection of Rational Dispute Resolution Method by Applying New Stepwise Weight Assessment Ratio Analysis (SWARA). *Journal of Business Economics and Management*, *11*(2): 243–258.
38. Stanujkic, D., Karabasevic, D., and Zavadskas, E. K. (2015). A Framework for the Selection of a Packaging Design Based on the SWARA Method. *Inzinerine Ekonomika-Engineering Economics*, *26*(2), 181–187.
39. Cakir, E. (2017). Determination of Criterion Weights by the SWARA–Copeland Method: Application in a Manufacturing Plant. *Adnan Menderes Üniversitesi Sosyal Bilimler Enstitüsü Dergisi*, *4*(1), 42–56.
40. Zavadskas, E. K. and Turskis, Z. (2010). A New Additive Ratio Assessment (ARAS) Method in Multicriteria Decision-making. *Technological and Economic Development of Economy*, *16*(2), 159–172.
41. Shariati, S., Yazdani-Chamzini, A., Salsani, A., and Tamosaitiene, J. (2014). Proposing a New Model for Waste Dump Site Selection: Case Study of Ayerma Phosphate Mine. *Inzinerine Ekonomika-Engineering Economics*, *25*(4), 410–419.
42. Zavadskas, E. K. and Kaklauskas, A. (1996). Systemotechnical evaluation of buildings. *Pastatų sistemo techninisįvertinimas*. Vilnius: Technika, p. 280.
43. Zavadskas, E. K., Kaklauskas, A., Banaitis, A., and Kvederyte, N. (2004). Housing Credit Access Model: The Case for Lithuania. *European Journal of Operational Research*, *155*(2), 335–352.

Chapter 8

Step-Wise Weight Appraisal Ratio Analysis in the Assessment of Biosurfactant

Figen Balo*,‡ and Lutfu S. Sua†

*Dept. of METE, Engineering Faculty, Firat University, Turkey
†Department of Management and Marketing, Southern University, USA
‡figenbalo@gmail.com

Abstract

Using and consuming healthy and natural products has been of concern for many of us. We want all the products we eat, drink, wear, and use to be natural or healthy, but over the years, it has become either very difficult to find natural products or it can be very expensive. Cleaning our body and kitchen utensils with natural products has also recently emerged as an important issue. In this study, we evaluate washing liquids from the viewpoint of human and environmental health. While making the evaluation, nine important criteria were taken into consideration and six alternative washing liquids were selected to find out which one is more suitable. Multi Criteria Decision Making Methodologies (MCDM) were utilized while performing the ranking process. Criteria weights were found through the Step-wise Weight Appraisal Ratio Analysis (SWARA) methodology. The alternatives' ranking was determined through the Technique for Order Preference by Similarity to Ideal Solution (TOPSIS) methodology.

Keywords: SWARA, TOPSIS, biosurfactant, MCDM, weight assessment.

8.1. Introduction

People have to eat in order to continue their vital activities and survive. For this purpose, they carry out eating activities at home, at work, or outside. Kitchen utensils used for food should be clean and disinfected with cleaners that will not harm human health. At the same time, among the liquids used for cleaning, the least harmful to the environment should be selected. Cowden et al. stated in their study that 16% of food poisoning incidents in England and Wales could be related to washing liquids.[1] In this chapter, a study has been done on washing liquids. Washing liquids were evaluated from the viewpoint of human and environment health and the study tries to find the most suitable one. The most suitable washing fluid from the viewpoint of human and environment health was determined using Step-wise Weight Appraisal Ratio Analysis (SWARA) and Technique for Order Preference by Similarity to Ideal Solution (TOPSIS) methodologies, which are among the Multi-Criteria Decision Making (MCDM) methodologies. Criterion weights were obtained by the SWARA methodology, and these weights were used in the selection of the most proper option in the TOPSIS method.

Mattick et al. studied the survival of food-borne pathogens during dishwashing at home, and then their transition to kitchen surfaces, food and dishwashing sponges.[2] Carpentier and Taché conducted a study on hazard control and taking basic precautions in microbiological situations.[3]

Some of the studies using the SWARA method are as follows. Ghenai et al. conducted studies on the evaluation of indicators in renewable energy systems in terms of sustainability.[4] The evaluation of the criteria was made using the SWARA method and the Additive Ratio Assessment (ARAS) method was used to find the most suitable alternative.[4] Aghdaie et al. determined the cluster characteristics by defining the synergies related to MCDM and data mining, while weighting was made using the SWARA method and the ranking of the clusters from the best to the worst was done using the VIseKriterijumska Optimizacija I Kompromisno

Resenje (VIKOR) method.[5] Dehnavi et al. developed a hybrid model to evaluate landslide sensitive areas using Adaptive Neuro-Fuzzy Inference System (ANFIS) and SWARA based on Geographical Information System.[6] Zavadskas et al. created the evaluation model theoretically by using Multi-Objective Optimization by Ratio Analysis Plus Full Multiplicative Form (MULTIMOORA) and SWARA methodologies in the selection of elements and materials in housing constructions.[7] Ighravwe and Oke discussed how to choose the appropriate maintenance strategy based on the sustainability criteria for public structures utilizing SWARA, Weighted Aggregated Sum Product Assessment (WASPAS), ARAS, and Fuzzy Axiomatic Design (FAD).[8] Nezhad et al. researched on the sector of nanotechnology in Iran to prioritize the planning of high technology sectors with SWARA and WASPAS methods.[9] Stanujkic et al. discussed making a selection for packaging design using the SWARA method.[10] Karabašević et al. have made the ranking of companies by ARAS and SWARA methodologies based on corporate social responsibility criteria.[11] Işık and Adalı used SWARA and Operational Competitiveness Rating (OCRA) as an integrated decision-making method in hotel selection.[12] Shukla et al. used SWARA and Preference Ranking Organization Methodology for Enrichment Evaluations (Promethee) methodologies for the Enterprise Resource Planning (ERP) application's selection that aims to achieve success.[13] Yazdani et al. used MCDM methods while choosing materials, determined the criterion weights with SWARA, and also found the order of the alternatives by comparing them with WASPAS and Multi-Objective Optimization on the basis of Ratio Analysis (MOORA).[14]

Some of the studies conducted with the TOPSIS method are as follows. Boran et al. have done an intuitively fuzzy multi-criteria study that suggests making a group decision in problems encountered in supplier selection using the TOPSIS method.[15] Kaya and Kahraman conducted a study that chose the most suitable energy alternative for energy planning using fuzzy AHP and fuzzy TOPSIS methods.[16] Işıklar and Büyüközkan determined the desired features while choosing a mobile phone through a questionnaire, and by finding criteria weights with AHP method, alternatives were sorted from the finest to the worst using the TOPSIS method.[17] In their study based on the Panchet dam in India, Bid and Siddique evaluated alternatives with TOPSIS and WASPAS methods after humanitarian risk assessment through Delphi survey.[18] Sadeghzadeh and Salehi presented a mathematical analysis to advance strategic technologies for the fuel cell with the TOPSIS method and a survey study in the automotive industry.[19] Demireli made the performance evaluation of some state-owned banks in Turkey by utilizing the TOPSIS methodology.[20] Uygurtürk and Korkmaz made the financial performance evaluation of enterprises operating in Turkey in the field of metal industry using the TOPSIS method.[21] Ertuğrul and Özçil evaluated the criteria that consumers consider when choosing air conditioners, listed the alternatives with TOPSIS and VIKOR methods, and made a study by finding a result that the TOPSIS method was more reliable.[22] Akyüz et al. evaluated the economic performance of the Joint Stock Company, which is traded in the ceramics sector at the stock exchange in Istanbul, with the TOPSIS method based on a 10-year period.[23] Supçiller and Çarpraz determined the criteria weights with AHP and made the alternative order utilizing the TOPSIS methodology, based on a problem related to supplier selection.[24] Ertuğrul and Öztaş showed the consistency of the results of the two methods by listing the alternatives with the Complex Proportional Assessment (COPRAS) and TOPSIS methods in determining the most ideal retirement plan for individuals considering individual retirement.[25] Efe and Kurt evaluated eight criteria and 10 candidates by applying fuzzy TOPSIS and fuzzy AHP methodologies for the personnel selection in the port operation.[26] Uzun and Kazan considered 12 criteria in order to determine the most ideal main engine in the project they dealt with in shipbuilding, these criteria weights were defined by AHP methodology, and the criteria weights found were used for seven alternative machines in PROMETHEE and TOPSIS methodologies.[27] Yayar and Baykara handled efficiency and effectiveness issues in their study on participation banks in Turkey and evaluated their work in 2005–2011 with the TOPSIS methodology.[28]

The aim of the work is stated in the second part of this chapter. In the following sections, the SWARA and TOPSIS methods to be used in this chapter are explained and their implementation steps are given. The definition of the problem is made and the methods are applied. The results of the applied methods are emphasized and suggestions are made about what else can be done about the subject.

8.2. Methodology

Criteria weights are evaluated in the SWARA methodology according to available survey results. Determining which of the selected washing liquids is the most suitable alternative is accomplished by using TOPSIS methodology. During the research, the selection of washing liquids to compare and the criteria to be evaluated were taken from The Good Shopping Guide website,[a] and the criterion weights were assessed in the SWARA methodology according to the outcomes of the survey on the site. In this study which is conducted in light of expert opinions, the researchers tried to determine which of the selected washing liquids was the most suitable by using TOPSIS method.

The questionnaire was presented to the experts by preparing an evaluation table with the four main criteria and sub-criteria determined as follows. Evaluation was made on the basis of the brand company. Alternatives were rated in line with the results of the survey evaluated by experts including company owners, chemical engineers, and biologists involved in the manufacturing of biosurfactants. The survey was supported with the information of the current official documents of the products. The evaluated criteria along with their explanations are provided as follows.

8.2.1. *Environmental criteria*

- *Environmental report*: The environmental reporting quality of a company producing biosurfactants is important to obtain information about the compliance of biosurfactants with ecological standards. A good environmental report is important in terms of seeing the clear effects of products, as opposed to vague statements about their environmental impact. For this purpose, companies that participated in the survey but could not publish an environmental report were given a lower rating. An average rating is given to the companies with insufficient reports, while the best ratings are given to the companies that have published a tangible, company-wide product environmental impact report in the past two years.

  The products of companies that produce in a way that does not harm people, the environment, or animals are shown with an average rating. The highest rating is given to the products manufactured by companies with an official environmental policy or report. These companies are the ones that have proven their products with an official report, offering organic, vegetarian, i.e., plant-based, and environmentally-friendly alternatives.[29]

- *Toxic chemical policy*: Biosurfactants may contain chemicals related to many environmental and health problems. Some of the common health problems caused by certain chemicals in these products are

---

[a] https://thegoodshoppingguide.com/

developmental damage, asthma, and cancer. It is also necessary to take into account the harmful effects on the ecological environment. Some of the possible harms are that chemicals can stay in the environment indefinitely, contaminate rivers and streams, and possibly enter the food chain. Ratings are created based on the chemical use policies of biosurfactant companies or whether they use certain chemicals (formaldehyde, phthalates, parabens, and triclosan) during production. The highest rating is given if the manufacturing company does not use any of the chemicals mentioned above or has a policy in this direction. Companies that use only one of these four chemicals are given a medium rating. The lowest rating was applied to the companies that manufacture using all the chemicals mentioned or that do not have any policy in this regard.[30,31]

- *Nuclear*: Nuclear waste has remained dangerous for decades. Nuclear power is a target for environmental and social advocacy because of the polluting properties of radioactive waste and its link to nuclear weapons production. The nuclear industry is an industry relying on the necessity of being an electricity producer that does not produce greenhouse gases as a way of combating the climate change. However, environmentalists argue that it is more effective in supporting a sustainable future by using more energy efficient production, and more renewable and cleaner energy sources such as wind, solar, and wave energy. For this reason, it is important to consider the nuclear criterion in environmental studies. Experts in the nuclear industry emphasize that the evaluation of the effects of the product and its production stages as well as the product wastes in the manufacture of a certain product is important in terms of radioactive waste management. In addition, it is important to evaluate this criterion in terms of human and environmental health. The company where the product is manufactured is included in the lowest rating if listed in the Nuclear Industry Association, World Nuclear Association, World Nuclear Transport Institute, or International Nuclear Engineering Purchasing Guide.[32]

- *Animal Welfare*: At the end of the 20th century, more than 100 million animals in the world are being subjected to experimental testing. Rats, mice, cats, guinea pigs, fish, birds, dogs, and rabbits are often used in these tests, but many more animals are being used. Testing products such as biosurfactant is a very small part of animal testing. However, in many countries, it is a legal requirement to test all new chemical products on animals. In order for a product to be registered, it must be standardized with a series of tests. Alternative non-animal tests are also available, such as the use of skin scraps, cell, and tissue cultures, clinical studies, and computer generation for all standard irritation and toxicity tests. That is why the highest rating is given to the companies that do not test on animals. If a company's product does not meet one of the criteria mentioned, a medium rating is applied. Similarly, companies that do not have an explicit policy to test on animals, or do not make any statement on this matter, they are given medium rating. Companies that test animals are rated low.[33]

- *Vegetarian/vegan*: For a product to be evaluated positively within the framework of the vegan criterion, the product should not contain animal fibers, additives of animal origin, animal ingredients, eggs, bee products, milk, or dairy products. It should not contain slaughter by-products and human-derived substances. There should be no cross contamination in the production process. If there is contact with non-vegetarian products on the production line, proper cleaning must be applied before vegetarian production begins. Necessary procedures must be in place to avoid packaging errors. This criterion supports companies producing the investigated product to use vegetarian-based raw materials.[34]

- *Armaments*: This represents the involvement of the company producing the product in the supply or production of conventional or nuclear weapons, including tanks, ships, aircraft, and armored vehicles. A rating was made taking into account these activities.

- *Political donations*: This criterion is included to assess whether companies fund political parties. In some countries, such as Germany, corporate financing is quite reasonably prohibited by law. A medium-rating indicates that the company and its employees have donated more than USD12,500 to the political organizations. A low rating indicates donations of more than USD 65,000.[35]

- *Public record criticisms*: The wide range of criticisms covered by these criteria relate mainly to the environment and human rights. The low rating indicates multiple serious criticisms received from organizations working on this topic worldwide over the past five years.

- *Ethical accreditation*: While there are many single-issue certification bodies that provide standards for organic products, fair trade, cruelty-free, or energy efficiency; ethical product, ethical brand, and ethical company logos span the full spectrum of ethical concerns. Products, brands, or companies that has received Ethical Accreditation are only those associated with strong business ethics and a strong code of ethics. In addition, the product was provided with the production of environmentally friendly products with standard business ethics. They have the privilege of carrying Ethical Brand, Ethical Company, and Ethical Product logos.[36]

8.2.2. *SWARA methodology*

In addition to being a widely used and simple method among criterion weighting methods in recent years, the SWARA method also provides convenience in terms of directly integrating expert opinions into the application.[37] Steps of the method are as follows.[38]

Step 1: The ranking beginning from the most significant criterion is made by considering the expert opinions.

Step 2: The comparative significance (s_j) of the average value is defined.[37] Comparison is made about how superior it is according to the criteria.

Step 3: After calculating the s_j criteria, the k_j criteria are calculated.

$$k_j = \begin{cases} 1 & j=1 \\ sj+1 & j>1 \end{cases} \tag{8.1}$$

Step 4: w_j is calculated.

$$w_j = \begin{cases} 1 & j=1 \\ \dfrac{x_{j-1}}{k_j} & j>1 \end{cases} \tag{8.2}$$

Step 5: q_j criteria weights are calculated.

$$q_j = \frac{w_j}{\sum_{k=1}^{n} w_k} \tag{8.3}$$

8.2.3. *TOPSIS methodology*

The TOPSIS methodology was developed by Hwang and Yoon.[38] The two basic points on which this methodology is based are Negative Ideal Solution (NIS) and Positive Ideal Solution (PIS). The best decision option in the method is found by evaluating the shortest distance of all alternatives to the PIS and the furthest range to the NIS.[39] Its steps are as follows.[40]

Step 1: Decision matrix is created.

$$A_{ij} = \begin{bmatrix} a_{11} & \cdots & a_{1n} \\ \vdots & \ddots & \vdots \\ a_{m1} & \cdots & a_{mn} \end{bmatrix}$$

Step 2: The standard decision matrix is created with the formula given in equation 8.4.

$$r_{ij} = \frac{a_{ij}}{\sqrt{\sum_{i=1}^{m} a_{ij}^2}} \quad i = 1,\ldots,m \qquad j = 1,\ldots,n \tag{8.4}$$

The matrix R_{ij} formed at the end of the normalization process is obtained.

$$R_{ij} = \begin{bmatrix} r_{11} & \cdots & r_{1n} \\ \vdots & \ddots & \vdots \\ r_{m1} & \cdots & r_{mn} \end{bmatrix}$$

Step 3: Weighted standard decision matrix (V_{ij}) is constituted. While doing this, the criterion weights are found by another method and multiplied by the values in the R_{ij} matrix. Also, the sum of criterion weights must be $\sum_{j=1}^{n} w_j = 1$.

$$V_{ij} = \begin{bmatrix} w_1 r_{11} & \cdots & w_n r_{1n} \\ \vdots & \ddots & \vdots \\ w_1 r_{m1} & \cdots & w_n r_{mn} \end{bmatrix}$$

Step 4: Negatif Ideal (A⁻) and Ideal (A*) resolutions are constituted. While creating these solutions, the V_{ij} matrix is used. While PIS (A*) criteria values that should be maximized consist of the largest, NIS (A⁻) values consist of the smallest. While PİÇ(A*) values of the criteria that should be minimized consist of the smallest, NIS (A⁻) values consist of the largest. This situation is shown in Equations (8.5–8.6).

$$A^* = \{V_1^*, V_2^*, \dots, V_n^*\} \text{(Largest in Max case, smallest in Min case)} \quad (8.5)$$

$$A^- = \{V_1^-, V_2^-, \dots, V_n^-\} \text{(Smallest in Max case, largest in Min case)} \quad (8.6)$$

Step 5: Measures $(S_i^* \text{ ve } S_i^-)$ which are the distances to PIS and NIS are calculated.

$$S_i^+ = \sqrt{\sum_{j=1}^{n}(v_{ij} - v_j^+)^2} \qquad i = 1, 2, \dots, m \qquad (8.7)$$

$$S_i^- = \sqrt{\sum_{j=1}^{n}(v_{ij} - v_j^-)^2} \qquad i = 1, 2, \dots, m \qquad (8.8)$$

Step 6: Using $(C_i^*), S_i^* \text{ ve } S_i^-$ proximity to the ideals solution values, the alternative that is closest to the PIS is found to be the best solution as shown in Equation (8.9).

$$C_i^+ = \frac{S_i^-}{S_i^- + S_i^+} \qquad i = 1, 2, \dots, m \qquad (8.9)$$

C_i^* value in Equation (8.9), is within the range of $0 \le C_i^* \le 1$ and if this value is equivalent to '1', then the alternative is at the PIS point, if it is equivalent to '0', then the alternative is at the NIS point.

8.3. Application

Six washing liquids were selected as the number of alternatives and an evaluation was made according to nine criteria. Criterion weights were determined using the SWARA method because it is a straightforward method to apply. The most suitable alternative was selected using the TOPSIS method. The initial decision matrix is given in Table 8.1.

Table 8.1.   Initial decision matrix.

Alternatives/ Criteria	Max	Min	Min	Max	Max	Max	Max	Min	Max
	Environmental report (K1)	Toxic chemicals policy (K2)	Nuclear (K3)	Animal welfare (K4)	Vegetarian/ Vegan (K5)	Armaments (K6)	Political donations (K7)	Public record criticisms (K8)	Ethical accrediation (K9)
Brand 1	9	3	3	7	7	9	9	9	7
Brand 2	3	5	1	3	3	5	5	5	3
Brand 3	3	5	1	3	3	5	5	5	3
Brand 4	5	3	1	1	3	5	1	1	3
Brand 5	9	3	3	9	9	9	9	9	7
Brand 6	7	3	3	9	9	9	9	9	7

Table 8.2.   Criteria weights.

Criteria	Order of importance	Sj	Kj	Qj	Wj
Environmental report (K1)	K3		1	1	0,2005
Toxic chemicals policy (K2)	K2	0,1	1,1	0,9090	0,1822
Nuclear (K3)	K6	0,2	1,2	0,7575	0,1519
Animal welfare (K4)	K5	0,5	1,5	0,5050	0,1012
Vegetarian/Vegan (K5)	K4	0,1	1,1	0,4591	0,0920
Arnaments (K6)	K9	0,15	1,15	0,3992	0,0800
Political donations (K7)	K1	0,1	1,1	0,3629	0,0727
Public record criticisms (K8)	K8	0,1	1,1	0,3299	0,0661
Ethical accrediation (K9)	K7	0,25	1,25	0,2639	0,0529
TOTAL				**4,986**	**1**

8.3.1. SWARA Application

Criterion weights are shown in Table 8.2 using those given in Equations 8.1–8.3. $\sum_{j=1}^{n} w_j = 1$ is shown at the bottom-right of the table.

8.3.2. TOPSIS Application

In order to rank the alternatives, an initial decision matrix is created starting from Equation (8.4). Normalization process is performed in line with the values shown in the last two rows in Table 8.3.

PIS and NIS values are found by multiplying the criterion weights found in the SWARA method with the normalized values. By using these values in Equations 8.7–8.8, positive and negative ideal separation measures are obtained. Finally, the relative proximity to the ideal resolution is calculated and displayed in Table 8.4.

According to Table 8.4, the most suitable washing liquid is Bio D (Brand 5) with 0.556 ($C^* = 1$ refers that the alternative is at the PIC point), while ECOS (Brand 6) with 0.549 takes the second place, followed by Attitude-Bio with 0.519. Spectra (Brand 1) takes the third place. The remaining ranking is Brand 4 with 0.491, Brand 2 with 0.490, and Brand 3 with 0.457.

Table 8.3.   Normalized values.

	Max	Min	Min	Max	Max	Max	Max	Min	Max
Alternatives/ Criteria	Environmental report (K1)	Toxic chemicals policy (K2)	Nuclear (K3)	Animal welfare (K4)	Vegetarian/ Vegan (K5)	Armaments (K6)	Political donations (K7)	Public record criticisms (K8)	Ethical accrediation (K9)
Brand 1	9	3	3	7	7	9	9	9	7
Brand 2	3	5	1	3	3	5	5	5	3
Brand 3	3	5	1	3	3	5	5	5	3
Brand 4	5	3	1	1	3	5	1	1	3
Brand 5	9	3	3	9	9	9	9	9	7
Brand 6	7	3	3	9	9	9	9	9	7
Sum of square	254	86	30	230	238	318	294	78	174
Square root	15,937	9,273	5,477	15,165	15,427	17,832	17,146	8,831	13,190

Table 8.4.   Solution values.

	S_i^*		S_i^-		C_i^*	Ranking	
S_1^*	0,7684	S_1^-	0,8305	C_1^*	0,5193	Brand 1	3
S_2^*	0,8412	S_2^-	0,8098	C_2^*	0,4904	Brand 2	5
S_3^*	0,8930	S_3^-	0,7524	C_3^*	0,4572	Brand 3	6
S_4^*	0,8661	S_4^-	0,8360	C_4^*	0,4911	Brand 4	4
S_5^*	0,7473	S_5^-	0,9376	C_5^*	0,5564	Brand 5	1
S_6^*	0,7529	S_6^-	0,9151	C_6^*	0,5486	Brand 6	2

8.4. Conclusion

Eating and drinking plays a significant role in almost every aspect of life. Besides these, it is of great importance that they are healthy and hygienic. While the cleaning of food products is important, the hygiene of kitchen materials is just as important. When cleaning, instead of using random washing liquids, one should consciously prefer washing liquids that cause the least harm to the nature and the environment. In this study, washing liquids were evaluated from the viewpoint of human and environment health. While making the evaluation, nine important criteria were taken into consideration and six alternative washing liquids were selected to find out which one is most suitable. Environmental report, toxic chemical policy, and nuclear are evaluated as primary environmental criteria. According to environmental radiation protection standards, it is evaluated with the nuclear criterion whether it emits radioactive material to the environment during its production.

Multi Criteria Decision Making Techniques were utilized while performing the ranking process. Criterion weights were found through the SWARA methodology. The sorting of the alternatives was determined through the TOPSIS methodology. According to the outputs, the most significant criterion was found to be Nuclear (K3) by weight. After the criterion weights were calculated, the alternatives were listed. The most ideal alternative is Bio D (Brand 5).

Public studies are needed to raise people's awareness on this issue. At the same time, the more consumers read and research on this subject, the more producers and manufacturers will give more importance to nature and human health.

References

1. Cowden, J. M., Wall, P. G., Adak, G., Evans, H., Le, S. B., and Ross, D. (1995). Outbreaks of Foodborne Infectious Intestinal Disease in England and Wales: 1992 and 1993. *Communicable Disease Report. CDR review*, *5*(8), R109-17.
2. Mattick, K., Durham, K., Domingue, G., Jørgensen, F., Sen, M., Schaffner, D., and Humphrey, T. (2003). The Survival of Foodborne Pathogens during Domestic Washing-up and Subsequent Transfer onto Washing-up Sponges, Kitchen Surfaces and Food. *International Journal of Food Microbiology*, *85*(3), 213–226.
3. Taché, J. and Carpentier, B. (2014). Hygiene in the Home Kitchen: Changes in Behaviour and Impact of Key Microbiological Hazard Control Measures. *Food Control*, *35*(1), 392–400.
4. Ghenai, C., Albawab, M., and Bettayeb, M. (2020). Sustainability Indicators for Renewable Energy Systems using Multi-Criteria Decision-Making Model and Extended SWARA/ARAS Hybrid Method. *Renewable Energy*, *146*, 580–597.
5. Aghdaie, M. H., Zolfani, S. H., and Zavadskas, E. K. (2014). Synergies of Data Mining and Multiple Attribute Decision Making. *Procedia-Social and Behavioral Sciences*, *110*, 767–776.
6. Dehnavi, A., Aghdam, I. N., Pradhan, B., and Varzandeh, M. H. M. (2015). A New Hybrid Model Using Step-Wise Weight Assessment Ratio Analysis (SWARA) Technique and Adaptive Neuro-Fuzzy Inference System (ANFIS) for Regional Landslide Hazard Assessment in Iran. *Catena*, *135*, 122–148.
7. Zavadskas, E. K., Bausys, R., Juodagalviene, B., and Garnyte-Sapranavicien, I. (2017). Model for Residential House Element and Material Selection by Neutrosophic MULTIMOORA Method. *Engineering Applications of Artificial Intelligence*, *64*, 315–324.
8. Ighravwe, D. E., and Oke, S. A. (2019). A Multi-Criteria Decision-Making Framework for Selecting a Suitable Maintenance Strategy for Public Buildings using Sustainability Criteria. *Journal of Building Engineering*, *24*, DOI: 10.1016/j.jobe.2019.100753.
9. Nezhad, M. R. G., Zolfani, S. H., Moztarzadeh, F., Zavadskas, E. K., and Bahrami, M. (2015). Planning the Priority of High Tech Industries Based on SWARA–WASPAS Methodology: The case of the nanotechnology

industry in Iran. *Economic Research-Ekonomska Istrazivanja*, *28*(1), 1,111–1,137.

10. Stanujkic, D., Karabasevic, D., and Zavadskas, E. K. (2015). A Framework for the Selection of a Packaging Design Based on the SWARA Method. *Inzinerine Ekonomika-Engineering Economics*, *26*(2), 181–187.

11. Karabašević, D., Paunkovic, J., and Stanujkić, D. (2016). Ranking of Companies According to the Indicators of Corporate Social Responsibility Based on SWARA and ARAS Methods. *Serbian Journal of Management*, *11*(1), 43–53.

12. Işık, A.T. and Adalı, E. A. (2016). A New Integrated Decision Making Approach Based on SWARA and OCRA Methods for the Hotel Selection Problem. *International Journal of Advanced Operations Management*, *8*(2), 140–151.

13. Shukla, S., Mishra, P. K., Jain, R., and Yadav, H. C. (2016). An Integrated Decision Making Approach for ERP System Selection Using SWARA and PROMETHEE Method. *International Journal of Intelligent Enterprise*, *3*(2), 120–147.

14. Yazdani, M., Zavadskas, E. K., Ignatius, J., and Abad, M. D. (2016). Sensitivity Analysis in MADM Methods: Application of Material Selection. *Engineering Economics*, *27*(4), 382–391.

15. Boran, F. E., Genç, S., Kurt, M., and Akay, D. (2009). A Multi-Criteria Intuitionistic Fuzzy Group Decision Making for Supplier Selection with TOPSIS Method. *Expert Systems with Applications*, *36*(8), 11,363–11,368.

16. Kaya, T. and Kahraman, C. Multicriteria Decision Making in Energy Planning Using A Modified Fuzzy TOPSIS Methodology. (2011). *Expert Systems with Applications*, *38*(6), 6,577–6,585.

17. Işıklar, G. and Büyüközkan, G. (2007). Using a Multi-Criteria Decision Making Approach to Evaluate Mobile Phone Alternatives. *Computer Standards & Interfaces*, *29*(2), 265–274.

18. Bid, S. and Siddique, G. (2019). Human Risk Assessment of Panchet Dam in India Using TOPSIS and WASPAS Multi-Criteria Decision-Making (MCDM) Methods. *Heliyon*, *5*(6), 251–265.

19. Sadeghzadeh, K. and Salehi, M. B. (2011). Mathematical Analysis of Fuel Cell Strategic Technologies Development Solutions in the Automotive Industry by the TOPSIS Multi-Criteria Decision Making Method. *International Journal of Hydrogen Energy*, *36*(20), 13,272–13,280.

20. Demireli, E. (2010). TOPSIS Multi-Criteria Decision Making System: An Application on Public Banks in Turkey. *Girişimcilik ve Kalkınma Dergisi*, *5*(1), 1–13.

21. Uygurtürk, H. and Korkmaz, T. (2012). Determination of Financial Performance by TOPSIS Multi-Criteria Decision Making Method: An

Application on Basic Metal Industry Enterprises. *Eskişehir Osmangazi Üniversitesi İktisadi ve İdari Bilimler Dergisi*, *7*(2), 95–115.

22. Ertuğrul, İ. and Özçil, A. (2014). Air Conditioner Selection with TOPSIS and VIKOR Methods in Multi-Criteria Decision Making. *Çankırı Karatekin Üniversitesi İİBF Dergisi*, *4*(1), 267–282.

23. Akyüz, Y., Bozdoğan, T., and Hantekin, E. (2011). Evaluation of Financial Performance with TOPSIS Method and an Application. *Afyon Kocatepe Üniversitesi İktisadi ve İdari Bilimler Fakültesi Dergisi*, *13*(1), 73–92.

24. Supçiller, A. and Çapraz, O. (2011). Supplier Selection Application Based on AHP–TOPSIS Method. *Ekonometri ve İstatistik e-Dergisi*, *13*, 1–22.

25. Ertuğrul, İ. and Öztaş, T. (2016). Application of Decision Making Methods in Choosing an Individual Pension Plan: COPRAS and TOPSIS Example. *Çankırı Karatekin Üniversitesi Sosyal Bilimler Enstitüsü Dergisi*, *7*(2), 165–186.

26. Efe, B. and Kurt, M. (2018). Personnel Selection Practice in a Port Operator. *Karaelmas Science and Engineering Journal*, *8*(2), pp. 417–427.

27. Uzun, S., and Kazan, H. (2016). Comparison of Multi-Criteria Decision Making Methods AHP TOPSIS and PROMETHEE: Main Engine Selection Application in Shipbuilding. *Journal of Transportation and Logistics*, *1*(1), 99–113.

28. Yayar, R. and Baykara, H. V. (2012). An Implementation upon Efficiency and Productivity of Participation Banks with TOPSIS Method. *Business and Economics Research Journal*, *3*(4), 21.

29. Płociniczak, M. P., Płaza, G. A., Piotrowska-Seget, Z., and Cameotra, S. S. (2011). Environmental Applications of Biosurfactants: Recent Advances, *International Journal of Molecular Sciences*, *12*(1): 633–654.

30. Edwards, K. R., Lepo, J. E., and Lewis, M. A. (2003). Toxicity Comparison of Biosurfactants and Synthetic Surfactantsused in Oil Spill Remediation to Two Estuarine Species, *Marine Pollution Bulletin*, *46*, 1,309–1,316.

31. Malkapuram, S. T., Sharma, V., Gumfekar, S. P., Sonawane, S. H., Sonawane, S., Boczkaj, G., and Seepana, M. M. (2021). A Review on Recent Advances in the Application of Biosurfactants in Wastewater Treatment. *Sustainable Energy Technologies and Assessments*, *48*, 101576.

32. Gebregiorgis, A., Vaccaria, M., Shiv, P., and Rtimic, S. (2021). Preparation, Characterization and Application of Biosurfactant in Various Industries: A Critical Review on Progress, Challenges and Perspectives. *Environmental Technology & Innovation*, *24*, 102090.

33. Vatsa, P., Sanchez, L., Clement, C., Baillieul, F., and Dorey, S. (2010). Rhamnolipid Biosurfactants as New Players in Animal and Plant Defense against Microbes. *International Journal of Molecular Sciences*, *11*(12): 5,095–5,108.

34. Sobrınho, H. B. S., Luna, J. M., Rufıno, R. D., Porto, A. L. F., and Sarubbo, L. A. (2013). Biosurfactants: Classification, Properties and Environmental Applications, *Biotechnology*, Vol. 11: Bioremediation, 1–29.

35. Shabani, M. H., Seyyed Mohammad Mousavi, A. J., and Abdi-Khanghah, M. (2020). Comparison of Produced Biosurfactants Performance in in-situ and ex-situ MEOR: Micromodel Study. *Energy Sources, Part A: Recovery, Utilization, and Environmental Effects*, DOI: 10.1080/15567036.2020.1810826.

36. Gayathiri, E., Prakash, P., Karmegam, N., Varjani, S., Awasthi, M. K., and Ravindran, B. (2022). Biosurfactants: Potential and Eco-Friendly Material for Sustainable Agriculture and Environmental Safety: A Review. *Agronomy*, *12*, 662.

37. Keršuliene, V., Zavadskas, E. K., and Turskis, Z. (2010). Selection of Rational Dispute Resolution Method by Applying New Stepwise Weight Assessment Ratio Analysis (SWARA). *Journal of Business Economics and Management*, *11*(2): 243–258.

38. Çakır, E. (2017). Determination of Criterion Weights by the SWARA–Copeland Method: Application in a Manufacturing Plant. *Adnan Menderes Üniversitesi Sosyal Bilimler Enstitüsü Dergisi*, *4*(1), 42–56.

39. Hwang, C. L. and Yoon, K. (1981). *Multiple Attribute Decision Making Methods and Application, A State-of-the-Art Survey*, Berlin, Heidelberg, New York.

40. Dumanoğlu, S. and Ergül, N. (2010). Financial Performance Measurement of Technology Companies Traded in the ISE. *Muhasebe ve Finansman Dergisi*, *48*, 101–111.

Chapter 9

Conceptualizations of Resilient Community for Social Futures

Eija Meriläinen

Human geography, Örebro University, Örebro, Sweden

Institute for Risk and Disaster Reduction & Institute for Global Health, University College London, London, United Kingdom

Hanken School of Economics, Helsinki, Finland

eijamerilainen@fastmail.com

Abstract

The concepts "resilience" and "community" appear frequently in research on disasters associated with major society-shaping phenomena such as climate change or urbanization. The chapter seeks to explicate how social science research conceptualizes the *resilient* community, building on a descriptive overview of disaster studies and climate change adaptation literature. Furthermore, the chapter explores how different conceptualizations of the resilient community frame social futures amid disasters. Three key conceptualizations of resilient community arise from the literature: (i) resilient community of belonging, (ii) resilient community of practice, and (iii) resilient community as an object of governance. Patterns arising across these conceptualizations resonate with previous critiques on how a focus on resilience and communities can serve to suppress the ideas and practices of the social, in line with neoliberal politics. Yet the conceptualizations are not univocally compatible with neoliberalism and can point to a variety of social futures. Conceptualizations of the resilient community can draw attention to the self-organizing of communities, as well as illustrate how a civil society mobilizes around a cause. Ideally, the cause is in line with the interests of the disaster-affected people, who typically are also marginalized in and across societies.

Keywords: Community; community resilience; conceptualization.

9.1. Introduction

Resilience has become both the ideal and the governance framework of the day to address the complexity and uncertainty of life.[1,2] The concept and discourse of resilience has had significant purchase in research, policy, and practice on major society-shaping phenomena such as climate change and urbanization. Resilience serves as an aspirational psychological trait,[3,4] an organizing principle for communities,[5,6] and a mode of international intervention.[7,8] While resilience can refer to ways in which people and communities weather everyday struggles,[9] it is associated particularly with shocks that come seemingly unannounced. Whether the origins of those shocks are natural or political, resilience is called for (resilience as something that is requested, e.g., "call for patience") to prepare for, respond to, and recover from them.[10–12] Resilience implies that there is a way to live with the disasters associated with major phenomena, treated as if they were inevitable forces of either nature or "development" [cf. 13].

Resilience highlights the capacity of communities to bounce back or forward to a "new normal" after a shock, typically with scant external support,[14–17] and it tends to emphasize how affected people exhibit bottom-up organizing, self-reliance, and agency.[18,19] Some consider the increased prevalence of resilience as a paradigmatic shift, drawing the focus away from people's vulnerabilities to their agency and capacities, while others perceive vulnerability, capacity, and resilience as complimentary concepts and approaches to governance.[16,20–23] For the proponents, resilience carries the hope of doing more with less; that is, performing public services or interventions with fewer resources than before, while additionally empowering the civil society.[7,24] Meanwhile, the critics of resilience thinking problematize how responsibility over governance activities and outcomes is increasingly placed on smaller spatial scales, both through responsibilizing disaster-affected communities and individuals, and through decentralizing formal disaster governance.[19,25–27] This is often done without appropriate redistribution of resources across scales. As communities affected are typically those marginalized in and across societies in the first place,[28] this is likely to perpetuate social inequalities.

Community as a concept is often present in resilience research, policy, and practice. The impacts of phenomena such as climate change or disasters are not neatly restricted within "jurisdictional, organizational and other forms of boundaries",[29] and the community scale may seem increasingly powerless to address issues evolving in a transnational context. However, it is particularly in times and situations of uncertainty that community has been argued to become both an important entity[30] and a concept. While resilience is discussed on a variety of scales — for example global, national, regional, city-level, and individual — the relevance of local community scales tends to be foregrounded. The resilience of a community is interpreted to be more than that of the individuals making it, as the "whole is more than the sum of its parts".[17] More resilience on the local community level is further thought to lead to a more resilient national community.[31] Community is taken for something intrinsically good,[32] and both academics and practitioners refer to it to emphasize how people-centered and participatory their approach to development or disaster-related work is, even when that might not be entirely justified.[33]

The ways in which (resilient) community is depicted influences what kinds of sociality are recognized and supported in practice, as well as what kinds of politics are put forth.[6] A central concern identified by previous research is that resilience is likely to be expected from marginalized "communities",[9,26,34] while the state and other organizations work to anticipate and pre-empt the adverse disasters facing those "facilitated" seen as "valued life".[28,35] For those facilitated, the future appears as a shock-proofed and open-ended set of possible trajectories branching out from the present.[36] For the marginalized "communities" of whom resilience is expected, meanwhile, the future is assumed immediate and precarious, tethered to disasters and recovery from them.[37]

Previous literature has explored community resilience[6,32] and conceptualizations of community within related fields.[33,38] This chapter aims is to make explicit the diverse conceptualizations of *resilient* community in research studying disasters. A *resilient* community appears as an entity of ever-appropriate proportions for navigating the uncertainties of major society-shaping phenomena, but critical scrutiny is in order. Methodologically, the chapter builds on a descriptive overview of social science research deploying resilient community in the context of disasters. The work is informed particularly by literature on community resilience from the fields of disaster studies and climate change adaptation. Three conceptualizations of resilient community arise from the literature, and they are entitled (i) resilient community of belonging, (ii) resilient community of practice, and (iii) resilient community as an object of governance. The conceptualizations are not mutually exclusive, and the typology elaborates patters of meaning that arise within each of the three conceptualizations, as well as across them.

The contribution of this chapter is in explicating how social science research conceptualizes the *resilient* community. The chapter explores also how the conceptualizations frame the *social* futures amid various disasters. Research and media tend to portray people's futures as personal and private, even if accounts of shared and common future have recently started dotting the media narratives amid major society-shaping phenomena.[39] Yet, Urry[39] sees both accounts as problematic. Futures are surely not private, but neither are they commons, as inequalities persist under capitalism and corporations have disproportionate power over how futures unfold. An emphasis on *social* futures can help to shift the interrogation away from the privatized realm of individuals and communities, while not assuming a singular common future either. Furthermore,

> "within contemporary disorganized capitalism, futures thinking is a major way of bringing the state and civil society back in from the cold. Moreover, if we focus upon social futures, this forces a transcendence of both markets and technologies. 'Social futures' — authorize the participation of range of many relevant actors including states and civil society in making futures."[39]

Thus, in exploring the *social* futures framed by the conceptualizations of resilient community, the attention is drawn particularly to how relations between states and civil societies are portrayed and shaped through community resilience.

9.2. State, (Civil) Society, and the Resilient Community Under Neoliberalism

Society can be defined as "the organization of human beings into forms that transcend the individual person, bringing them into relationship with one another that possess some measure of coherence, stability and, indeed, identifiable 'reality'".[40] Society — or social — is the connecting tissue between personal life and the state that brings citizens together beyond their differences and inequalities both conceptually, and in practice.[41] In this chapter, "social" implies a broad understanding of social connections, while "society" refers to the collective social realm that associates with a given state.

The origins of the notion of *civil* society are often traced back to ancient Greek cities, where the civil society and the state were deeply entwined.[42] However, in modern discourse, the state and civil society are typically portrayed as separate from one another. In liberal political thought, civil society comprises most simplistically the "non-market, non-state sphere of voluntary public activity" undertaken by individuals with curbed self-interest.[43] The civil society appears as the welcomed counter-balance to the state's excesses.[43] These connotations flooded the Anglophone political thought particularly in the 1980s from Latin America and Eastern Europe, where the notion of civil society was associated with the people's opposition to oppressive states.[44] Yet civil society, seen to arise from various and competing interests, is not automatically on the side of social justice. The conceptualizations of civil society are often premised on the fictional idea of economic equality,[43] and despite forming the majority, the marginalized can remain powerless also in the civil society.[44] Some streams of Marxist thought share the idea of civil society being separate from the state with liberals, but in contrast, regard the civil society as a problematic realm of private (economic) interest.[43,44] Gramsci, however, sees hegemony mutually reinforcing the civil society and the state.[44] Despite the wide variety of (modern) understandings of civil society, not limited to the above, in this chapter, the notion of civil society resonates most with the restricted liberal one, as it best corresponds with the views articulated across the reviewed articles.

A state can be thought of as a unitary entity, as well as an "ensemble — of institutions, organizations, and interactions involved in the exercise of political leadership and in the implementation of decisions that are, in principle, collectively binding on its political subjects" according to Jessop.[45] In the Weberian vein, the state could be defined through (i) its state apparatus exercising authority and power over the people within it, and states outside it, (ii) a core territory where the power of the state apparatus remains relatively uncontested, and (iii) a stable population of political subjects bound by the state's authority.[45] However, it is helpful to approach and scrutinize the state as a social relation in itself, yet the details of this perspective are not discussed here due to limitations in space.[45]

The state, even when it formally carries the trappings of democracy, is not unequivocally "good", or on the side of the "people." States have tendencies of centralizing power and remain at risk of co-optation by the elites. Yet, a state is one of the few political entities in the capitalist nation-state system that can be held accountable to the civil society.[41] Furthermore, states play a crucial role in balancing the scales in a society, which is necessary for fostering political equality than underlies democracy.[41] The balancing of scales can mean, for instance, redistributing surplus value in a society equitably, providing basic services such as education and healthcare, as well as securing formal political participation through voting rights and rights to stand as a candidate.

Capitalism could perhaps best be framed as an "institutionalized social order", in that it — much like feudalism — shapes life much beyond the economic or political system.[46] Contemporary social realms and states operate under neoliberal capitalism to varying degrees. Neoliberal ideology holds that the freedom of market and entrepreneurship are the key source of societal well-being,[47,78] while neoliberalism could be considered "an always mutating project of state-facilitated market rule".[49] Neoliberalism has featured both a "roll-back" of the state from providing social welfare and regulating markets, and a subsequent "roll-out" with the state involving increasingly "technocratic economic management and invasive social policies" that foreground the economic interests of a narrow set of actors.[50] In the process, the state has not disappeared; instead, it has been repurposed and harnessed to facilitate the markets more explicitly. States are only part of governance, where private and non-state actors have increasing power over the provision of common good, and profit-driven logic shapes the states themselves.[51]

In order to fend off threats towards the power of markets and those enriched by them, neoliberal thinkers and political figures have since the 1970s striven to suppress the idea of a collective social realm, or the "social".[41] Neoliberalism has sought to render the idea of the social inexistent in both discourse and practice, in order to delegitimize collective claims towards the state arising from the people.[41] Neoliberalism is also implicated in how the bottom-up idea of the *civil* society has been watered down, as privatized proxies such as non-governmental organizations (NGOs) have taken over the space of civil society.[52]

Beyond emphasizing that the social does not exist, neoliberal ideology assumes that states are not responsible for responding to the supposedly inevitable and impersonal forces of either nature or "development".[13] Thus, responding and adapting major society-shaping phenomena becomes the responsibility of individuals and communities. Further, actors with actual economic and political power over the root causes of phenomena (e.g., climate change or urbanization) are framed as legitimately, or at very least inevitably, operating beyond the scale of the state in neoliberal discourse.[13] Even those sympathetic to holding the states responsible might be skeptical about the states' power over markets and corporations. However, under neoliberalism, states are very much complicit in reinforcing the power of markets and "encasing" it from democratic politics[53] and precisely for the reason states should be held accountable for and by the civil society.

The ways in which "community resilience" is typically evoked and analyzed fit very well into the broader neoliberal politics and discourses,[6] with

- the major phenomena shaping the present and future societies being framed as natural or inevitable, coming from "outside" of the given social unit,
- the responsibility to deal with the adverse consequences of the phenomena being pushed to the sub-national scales and "communities", as the states are framed as unequipped or sub-optimal to deal with the disasters, even if
- the states play a central role in maintaining the existing societal system granting economic and political power to private actors such as corporations, while
- non-state actors such as corporations frame the adverse disasters connected to their activities (e.g., greenhouse gasses contributing to climate change) as externalities that they should not be responsible for, and yet
- non-state actors such as corporations are celebrated for technical fixes, they might provide for communities to deal with various disasters.

The ways in which community resilience is portrayed and enacted upon in the context of neoliberalism are often problematic to the welfare of the marginalized "communities". However, research that evokes ideas of resilient communities does not necessary play for the neoliberal team. Rather, some of the conceptualizations of the resilient community can be more radical and help re-imagine more just social futures. How the relations between civil societies and states are framed is a central part of the analytical scrutiny directed at conceptualizations of resilient community in this chapter.

9.3. Methodology: Drawing Out Three Conceptualizations of Resilient Community

The chapter builds on a descriptive overview and qualitative content analysis that were conducted in June 2018 in Scopus. The search was restricted to articles discussing "community resilience" in its title, abstract, or keywords. To narrow the topic down to disaster-related community resilience, concepts of "hazard" and "disaster" were added to the keyword search. The focus was on disasters connected to natural hazards. For this reason concepts that were likelier to encompass also armed conflicts were not included, including terms such as "emergency", "crisis", and "catastrophe". Furthermore, the search was limited to articles within the social sciences. It ranged all time periods until the present, resulting in 244 articles to be analyzed. Research on disaster risk reduction and climate change adaptation were dominant in the reviewed literature, but other issues such as (unequal) processes related to urbanization, economic systems, and health were also prevalent themes. Hence, insights from the reviewed literature are relevant to social science research on shocks (e.g., natural hazard induced disasters) and changes (e.g., how climate change shapes lives and livelihoods) more broadly.

The analysis focused on abstracts and article sections discussing the conceptualization of the community in the 244 articles selected. The qualitative content analysis went beyond the manifest (explicit and visible) content into the latent (implicit and underlying) content.[54] The coding scheme was developed iteratively, and it resulted ultimately in three categories and 16 related codes. The number of codes in the final scheme was concise, informing the core typology and the structure of the chapter.

The three categories of conceptualizations that emerged from the analysis frame the discussion. The three key types of resilient community identified from the literature are (i) resilient community of belonging, (ii) resilient community of practice, and (iii) resilient community as an object of governance. The following three sections will each introduce one of these perspectives. The perspectives are not mutually exclusive in that individual articles may draw elements from two or three categories, even if the elements may not always amount to a coherent whole. The typology is, however, helpful at elaborating patters of meaning that arise within the three conceptualizations and across them. The three conceptualizations resonate to a degree with the three conceptualizations identified by Räsänen et al.,[38] developed in the context of disaster risk reduction policy and practice. Here, however, the focus is on how resilient community is conceptualized in academic research. Attention is paid particularly to how certain key ideas and emphases appear either within or across the articles.

In the following sections, references from the reviewed literature are used to illustrate the key ideas. As individual articles reviewed may evoke elements from more than one conceptualization of resilient community, also a handful of reviewed articles are cited in the context of more than one resilient community category. In some cases, central references appearing in the reviewed articles are referred to directly. Furthermore, some conceptual or theoretical references beyond community resilience literature are used to make sense of the literature in the broader context of social science research. This layer of interpretation is what contributes to making this a *descriptive* review.

9.4. Resilient Community of Belonging

This section explores *resilient community of belonging*, the first of the three conceptualizations outlined in this chapter. Yuval-Davis[55] has argued that "belonging" can be understood on three analytical levels, namely on the levels of

(i) social locations, which refer to people's positionalities within a society,
(ii) identifications of people and their emotional attachments, as well as
(iii) value systems that people use to define the boundaries of belonging.

A "community of belonging" is to a great extent about who is excluded from, and who is included into, a community.[55] A "social fabric" ties together the people that comprise the community and is a source of agency and strength. Particularly, people that are thought to live off the nature or are in the margins of a society are considered to rely on a resilient community of belonging.

The broad strokes of Yuval-Davis'[55] outline of a community of belonging are apt for describing the resilient community of belonging. In disaster-related community resilience literature, a community of belonging exists irrespective of all hazards. Resilient community of belonging is an entity living through the phases of a disaster, and disasters over time. The community of belonging typically relies on attachments to and in place, and exhibits a temporal continuity. The following two sub-sections introduce the community of belonging in relation to these key aspects, namely, with respect to how a resilient community of belonging is defined as (i) attachments to and in place, and as (ii) continuity of attachments over time.

9.4.1. *Resilient community of belonging as attachments to and in place*

In their highly influential article on community resilience, Norris et al.[17] define a typical community as one that "has geographic boundaries and a shared fate." This is a prevalent conceptualization of a community which assumes that people's main attachments to other humans and the non-human world are bound within a relatively restricted geographical area.[56] The dominance of the place-based understanding of community in resilience literature has been traced back to the influence of environmental and other non-social sciences.[31] Beyond emphasizing place, many authors refer to "local" communities when they want to highlight the geographically delimited nature of the people's main attachments.

Many authors on community resilience recognize that

- a locality can contain several, often overlapping, communities and groups that are differently affected by a hazard, such as different genders,[57] ethnic groups,[58] and religious groups,[59] and that
- the main attachments and social ties of people may stretch beyond a restricted locality.

However, the place-based community of belonging persists in community resilience literature. It allows for researchers to frame a geographically bounded community that faces a disaster as a whole. Leitch and Bohensky[15] argue that place-attachment in resilience discourse tends to be an aspiration projected on the disaster-affected people from the outside. The aspired ideal is for individuals to remain locked in place, rather than leave the disaster-affected place and "community".[15] This is particularly problematic when the very place and connections fostering the community are made more vulnerable by political and economic processes beyond the place,[60] some which may be enacted in the name of increasing resilience.

A typical justification for defining a community through attachments to and in place is that the specific community is argued to be either isolated or place-based[23] enough from the actors and processes taking place at scales beyond the local. This is particularly the case with populations living on small islands or in remote areas.[61-63] Rural communities, for instance, present a large sub-group of resilient communities of belonging. Rural communities are thought to have lives and livelihoods that depend on local nature, making them both more vulnerable to external shocks such as those connected to climate change, but also overall more resilient than other people.[64-66] The idea of living off the elements, such as land[64] or water,[65] seemingly resonates with the ecological metaphor of resilience.

Often, rural communities of belonging are not conceptualized through what they are, but in relation to what they are not — urban. Rural and urban communities might even be placed on a continuum, as illustrated in the work of Rapaport et al.,[67] where the resilience of communities is discussed in urban, sub-urban, and rural contexts. Their finding, echoed or implied by many other articles, is that rural communities are the most resilient ones due to their social capacities, while urban communities are the least resilient. This type of research might, unwittingly, reproduce the contested idea that life in large and densely populated cities is somehow essentially different from idealized communal rural existence — "with urban life" characterized by the break-down of social structures resulting in mobility, instability, and insecurity.[68]

The romanticizing of contemporary rural communities is also problematized within community resilience research. As McManus et al.[69] point out, in many parts of the "developed" world, the rural decline accompanied with new communication technologies and working practices have led to a loss of community in rural areas. The sense of locality that tied farmers to their local rural towns is withering away along with it.[69] While cities are at the heart of capitalist processes which may contribute to the alienation of communities[70] and urban life heavily entwined with infrastructure,[71] the processes of globalization, capitalism, and climate change shape communities' resilience beyond the simplistic urban/rural or local/global dichotomies.[23]

Yet, when particularly urban regions and people are diagnosed to lack the characteristics and dimensions of *resilient communities of belonging*, articles might prescribe community creation as the solution to the perceived lack of urban resilience.[67,72] Often, a neighborhood becomes the proxy for a community.[73] Notably, in marginalized, low-income, and informal settlements, neighborhood-level attachments and social bonds are portrayed as an important source of community resilience.[74,75] The emphasis on a community bound together in place may have problematic implications for social and societal mobility. While some authors see resilience and resourcefulness of individual members of the community as a contribution to community level resilience,[74] others recognize a potential tension between individual resilience and community level resilience, since resilient individuals might not remain as members of a community in the long-term.[76] The implication of the latter perspective is that for the sake of resilience of a community, the boundaries of the community of belonging should be enforced.

While the understanding of a tightly knit and often exclusionary community is left more implicit in some articles, concepts such as "social cohesion" and "social fabric," among others, serve to make this explicit. Social cohesion is shown to lead to higher resilience of a community,[77] whereas "social fabric" tends to refer to communities' internal social capacities as opposed to "hard" infrastructure and systems that might be externally imposed.[29,78] This often implies relatively clear boundaries between those who are included into, or excluded from, the community.[55]

9.4.2. *Resilient community of belonging as continuity of attachments*

The resilient community of belonging typically has an emphasis on the continuity of connections over time.[79] The attachments of people to one another and to a place are expected to have evolved over time, potentially over generations. Not only does a resilient community of belonging live through a single disaster as one whole, over time, the community may have lived through and with several disasters.[80,81]

Concepts such as traditional society or institutions,[16,82] indigeneity,[80,83] and tribal community[84] often signal the attachments over time in community resilience literature. The members of a community of belonging are thought to have developed traditions and cultural practices that support their ways of living in a place and with one another over extended periods of time. Traditional and indigenous knowledges and practices are seen as central to community resilience, but insufficiently integrated into formal disaster governance systems.[85] That said, resilience literature has also been criticized for romanticizing traditional or indigenous communities,[86] especially when it comes to dealing with the increasingly devastating impacts of climate change[87] and extractivism.[58,88]

The idea of (bounded) place is central to the resilient community of belonging perspective, but it is not necessary. Another way to express the continuity of attachments characterizing a community of belonging is to highlight the shared identity. Cultural and historical connections are seen as the source of people's capacity to survive and adapt,[89,90] and reinforcing identities is thought to increase resilience.[91]

9.5. Resilient Community of Practice

The *resilient community of practice* conceptualization highlights the ways in which a disaster brings about a community and galvanizes people to action. Wenger[92] defines communities of practice as groups of people sharing a "concern or a passion for something they do and learn how to do it better as they interact regularly", and their key feature and purpose relates to social learning.[93] In the reviewed articles, *resilient community of practice* is often discussed in the aftermath of a disaster, with particularly disaster-affected people acting on a shared concern. The existence and legitimacy of *resilient community of practice* depends on the success in self-organizing and leveraging social networks for the purpose of responding to and recovering from a disaster.

In articles deploying the *resilient community of practice* conceptualization, much of the action described is likely to occur within a limited geographical area where a disaster has unfolded. However, the *resilient community of practice* conceptualization carries no assumption that people coming together in place because of a disaster would have previously considered themselves a "community" — or been considered by others as such. While a *resilient community of practice* may have central nodes that coincide with geographical impacts of a disaster and the affected people, the boundaries of the community are blurred and membership is ambiguous, with the social network re-configuring over time. In the following subsection, the resilient community of practice is explored as a social network. The second subsection builds upon the social network perspective, elaborating how the community of practice spans out from the core, rather than being limited to it.

9.5.1. *Resilient community of practice as a self-organizing social network*

According to the *resilient community of practice* conceptualization, the core community consists of people who are affected by a disaster and/or are doing something to respond to and recover from it. Disaster-affected people are not merely victims, but first and foremost survivors,[65] the agentic subjects of disaster governance. They are thought to be able to help themselves and self-organize a significant part of the disaster response and recovery.[12,94] Community resilience is described as endogenous and autonomous.[95,96] The emphasis on bottom-up organizing is shared with the *resilient community of belonging* conceptualization, but in literature deploying the resilient community of practice, self-organization is what (re)produces the community, rather than being something that a pre-existing entity does. A "community of practice" configures around a shared concern (a disaster) and is performed through self-organization.

The *resilient community of practice* is not a pre-existing, static, nor well-bounded entity.[97,98] Neither is it, however, a random collection of individuals. Rather, the view of sociality put forth by the *resilient community of practice* conceptualization can be understood through a social network perspective.[99] People might be members of several social networks, or communities, simultaneously, but a major disaster is likely to re-configure those social networks and ties to respond to the gravity of the shock.[100] These social networks are likely to evolve as the time passes on from the moment of a shock.[100] Identities of community members are not fixed, and the common fate of experiencing a disaster can also forge new community identities.[101,102] The "community of practice" perspective is particularly prevalent in community resilience literature discussing urban and densely populated places.

9.5.2. *Resilient community of practice and connections spanning out from the core*

The *resilient community of belonging* conceptualization emphasized the resilience arising from within a fairly bounded entity that exhibits continuity over time. In the case of *resilient community of practice*, there are no firm boundaries between those that belong to the community and those that do not. Furthermore, the community of practice is emergent, and its existence is related to the shared cause. If the concern disappears, the community of practice dissolves. The social network perspective is useful for imagining how a community might span out from the disaster-affected people. The affected people can be seen as the central nodes of the social network, the initial core of the resilient community of practice.

The strength of this perspective in the era of globalization is that it does not assume the community to exist in a disconnected periphery, nor does it assume that:

- all preparedness for, response to, and recovery from a disaster takes place within a clearly delimited community, and
- all disasters, such as those related to climate-change, would be entirely external to a community.

Rather, both disasters and community resilience evolve on multiple scales between "local" and "global".[23] As (re)configurations of social networks are expected, this type of conceptualization allows also for mobility, whether within a society or across space. Disaster and climate change-related threats can disperse social networks further across space, so assuming community to remain bounded in order to assess its resilience may be problematic. For instance, small islands can serve as ideal types a *community of belonging*, but sea level rise associated with climate change may even lead to the relocation of the entire population.[61,103]

A hazard putting the disaster in motion can be assumed to be spatially delimited. People directly affected are then likely to share an approximate locality with one another. This resonates with Wisner and Kelman's[79] definition of a community as "people living in a locality and their extended networks elsewhere in space and time." The core of the resilient community of practice is likely to comprise the people in a certain (disaster-affected) place, who act as the central nodes in a social network spanning beyond the place.

A *resilient community of practice* centers around a disaster, but a disaster does not entirely re-shape social relations. An important basis for a resilient community of practice is formed by pre-existing social networks of people and places affected, supported by "community infrastructure" such as neighborhood centers, community service agencies, and tribal organizations.[104,105] A rich set of voluntary organizing models,[56] social coordination abilities,[106] and leadership capacities[107] within community organizations and associations prior to a disaster is seen as central to community resilience. Furthermore, diaspora groups play an important role in mobilizing relief across places.[108]

Resilient community of practice responding to a disaster is emergent. This means, firstly, that pre-existing social networks reconfigure as a result of a disaster, with actors assuming new roles and relationships to respond to and recover from a disaster. Secondly, actors that are, prior to the disaster, not engaged with the core of the social network will become engaged after and because of the disaster. These actors include organizations dedicated to disaster response and recovery,[109] and they can be voluntary and non-profit, commercial, or governmental. In this way, actors that are "external" to the disaster-affected people before a disaster become a part of the community of practice. Community organizations and infrastructure can act as links between "external" actors, attracting resources and coordinating action.[104] Some of these community organizations are online-based.[12]

9.6. Resilient Community as an Object of Governance

The third conceptualization of community deployed in the reviewed articles on disaster-related community resilience is labelled here as *resilient community as an object of governance*. This perspective is in line with Bulley's[32] observation that a community is produced and governed through resilience. This perspective exhibits two key features that the subsections will explore in further detail: (i) communities are defined through their relation to an organization that governs them top-down, and (ii) communities are perceived as collections of individuals. This approach does not assume strong relationships between members of a community. Rather, the central relationship is that between the organization and individual community members, with the former defining the terms of the relationship.

9.6.1. *Resilient community as an object governed by organizations*

Research deploying the *resilient community as an object of governance* conceptualization typically aims at influencing policy and practice on how organizations and institutions govern disasters and communities. The organizations depicted encompass governmental institutions, NGOs, commercial entities, and research institutes. Some of these entities are involved in governance in the status quo, irrespective of any hazards, while others are specialized in disaster risk reduction and disaster governance.

The *resilient community as an object of governance* perspective tends to portray the community in relation to the organizations and institutions, as "them" to be governed. A "community" does not actively govern itself or its parts, but is governed by institutions that are conceived to exist on higher scales. For example, Oktari et al.[110] recognize that schools form a subsystem of a community. Yet, when it comes to governing and influencing this part of the community, the power is described to rest on higher scales and in the hands of experts of the national and regional school system.[110] Typically, the community is depicted as separate from the governing actors, even if some exceptions explore governing actors as community members. For instance, Sciulli et al.'s[111] research is on the resilience of local councils, while Sugino et al.'s[112] research explores the mixed roles of nurses in case of a disaster.

Governmental or public institutions discussed in the articles include the local and national governments, as well as entities such as the military, libraries, and schools. Particularly, when research addresses governmental organizations, a geographically delimited and place-based approach to community resilience becomes prominent.[113,114] Yet, in comparison to the community of belonging perspective, the *resilient community as an object of governance* perspective does not assume community to be a place-based social *entity*. It instead focuses on individuals within its geo-graphically determined jurisdiction that are acknowledged as citizens, stakeholders, or otherwise legitimate users of its institutions.

The role of the transnational and national NGOs is also seen as sig-nificant when it comes to building "community resilience".[115,116] These entities shape disaster governance through producing resilience discourse,[117,118] material resilience building activities[119,120] and gate-keeping access to resources.[121] Similarly, commercial actors are seen as central to community resilience. For instance, in the context of the United States the "Whole Community" approach involves all sectors and resources of a community "whether belonging to the local govern-ment, non-profit organizations, private business, or even individuals and neighborhoods", with the private and commercial sector playing a major role.[122]

Commercial organizations can be defined as central disaster govern-ance actors in relation to which the "community" is defined. However, beyond this, private businesses are framed as community members in their own right in some cases.[123–125] For example, Orhan[124] views busi-nesses as part of a community and argues that building the resilience of businesses is needed to improve the resilience of the community, or soci-ety, as a whole. However, Carpenter[126] argues that the private sector can also negatively influence the resilience of a community — for instance, problematic real estate development practices of private businesses can inhibit or disrupt the social networks that resilience stems from. Seeing private businesses as community members whose well-being is an objec-tive of resilience building can, then, be questionable.

9.6.2. *Resilient community as individuals to be governed*

The articles that adopt the *resilient community as an object of governance* perspective depict — often implicitly — a community as a collection of individuals. "Community" is the unit or scale of governance on which resilience is built, assessed,[122] and compared against the resilience of other "communities".[127] These "community" units are defined primarily through the individuals' relation to an organization, but they can addi-tionally have other features. For instance, neighborhoods are an example of geographically defined units that do not necessarily fully correspond-ence with a jurisdiction of an organization. Resilient communities are compounded from the individuals[128,129] or households[130] within these community units.

The *resilient community as an object of governance* perspective is characterized by the deployment of the concept "community," but with little or no analytical scrutiny of the term. Research deploying this conceptualization of resilient community may not in fact focus on a "community" as a whole in the first place. Rather, the spotlight can be on a vulnerable target demographic within the broader community. Improving the resilience of the vulnerable demographic is implied to improve the resilience of the community as a whole. Such vulnerable demographics are, for instance, the children[131] the elderly,[132,133] the disabled,[134] the women,[87] or the small-scale farmers.[64,115] Disaster education for vulnerable children is expected to reduce the vulnerability of a community as awareness, knowledge, skills, and agency instilled in children are thought to spill over to the community.[135,136]

Governments at different scales are key actors involved in resilience building in resilient communities (as objects of governance). The resilient communities that governments are involved with comprise individual citizens. In this context, resilient communities are not depicted as a collective civil society, and the conceptualization of citizenship across the articles tends to be rather passive. The "citizen" does not actively shape disaster governance but is governed as a citizen-stakeholder. Similarly, as in other policy initiatives coming top-down, the citizen-stakeholders are not political animals that govern themselves or demand rights, but rather the "general citizens" in the name of whom plans are drafted and action is taken, while the actual citizens remain absent from decision-making and policy processes.[137] Also the stakeholders of disaster governance led by commercial and non-governmental organizations are typically framed as disconnected and apolitical individuals, following the *resilient community as an object of governance* perspective.

A concrete way in which both governmental and non-governmental organizations and institutions frame a resilient community is as users and targets of infrastructures, policies, and technologies reconstructed or introduced in a disaster governance context. Physical,[138] information,[139] transportation,[140] and energy/electric[141] infrastructures are built top-down for individual users of the infrastructure. Frameworks, policies, and technologies are also tools of intervention for institutions and organizations to develop community resilience top-down. This is often done with the support of "experts". While there are examples where infrastructure was (re)constructed by community members themselves,[142] this is mostly seen as undesirable according to this perspective.

9.7. Summarizing Discussion

This section summarizes and discusses the three conceptualizations of resilient community arising from the literature, namely: (i) resilient community of belonging, (ii) resilient community of practice, and (iii) resilient community as an object of governance (Table 9.1).

The *resilient community of belonging* is an entity that persists continuously across spans of time, including through disasters. A resilient community of belonging is bound together in place by a "social fabric" and exists irrespective of shocks. Particularly marginalized people, living either in the rural areas or in urban margins, are thought to rely on a community of belonging when facing disruptions. This conceptualization of a resilient community brings welcomed attention to the agency, strength, and social fabric that arise from long-term social connections in and to place. Yet, if the very society-shaping phenomena that communities are expected to be resilient amid disperse them in space, the *resilient*

Table 9.1.   Three conceptualizations of resilient community.

	Resilient community of belonging	Resilient community of practice	Resilient community as an object of governance
Consists of	Marginalized people	Disaster-affected core community and their social networks	People defined in relation to an organization or institution
Sociality	A bounded entity in place	Social networks spanning out from disaster-affected people	Individuals, likely in place
Relationship to other actors	Limited engagement	Collaborating with various actors	Governed top-down by other actors
Role of disasters	Exists independent of disasters	Exists because of a disaster	Exist independent of disasters

community of belonging perspective may problematically insist on temporal continuity, with the future akin to the present.[60] The perspective may further inadvertently frame all societal mobility of community members as a problem for the community resilience, as "resilient" individuals may cease to be part of the marginalized community.[76]

The conceptualization of the *resilient community of belonging* can also be problematic for social politics striving to address the root causes of the society-shaping phenomena by holding states and other powerful actors accountable. The focus on a bounded and temporally continuous marginalized community carries the further danger of conceptually re-enforcing the boundaries between the "community" and the society. The claims towards the state are weakened when they seem to arise from outside of the (civil) society. Marginalized people and groups might justifiably see the state rather as a problem than a solution. Furthermore, for a specific community of belonging, major society-shaping phenomena might be appropriately treated as driven by external forces. Yet, by focusing only on the resilience of the community, the community's futures remain precarious, and the root phenomena that produce adverse impacts of disasters remain. The phenomena such as climate change or urbanization continue to be framed as external to communities, but also to the state and other actors having power over them.

The *resilient community of practice* view, meanwhile, observes how a disaster re-configures people's existing social networks. People and actors come together due to a disruption, not in spite thereof. The *resilient community of belonging* and *resilient community of practice* perspectives both highlight the self-organizing capacity of people associated with a community. However, a key difference lies in that a *resilient community of practice* perspective comes into being as a result of self-organization, while for *resilient community of belonging*, self-organizing is something a pre-existing community exhibits. While the core of the *resilient community of practice* perspective lies in people affected by a disaster or striving to address one, the community is not bounded. This conceptualization of a resilient community, seen as a re-configuration of social networks around a cause, is very much compatible with the idea of a civil society mobilizing around a cause. The *social* futures are emerging from action, and not just for a core community. Rather, past relations are re-configured and new ones are forged, shaping the trajectories that both the community and society might take.

A key source of both strength and weakness that the *resilient community of practice* perspective shares with the *resilient community of belonging* perspective relates to the emphasis on bottom-up action. Under neoliberal capitalism, the "institutionalized social order" of the day,[46] the chances for autonomous self-organization are heavily restricted by states and other actors, for instance, through the reinforced protection of private property. Actors "external" to the community can to a degree support the community claims, for instance, through providing resources and coordination.[104] Yet, addressing the major society-shaping phenomena in the long-term would likely require a broader front than one mobilized around a specific affected "core" community. While a resilient community of practice has its limitations, its openness is likely to keep the idea of the social realm alive.

The third and final conceptualization, labelled here as the *resilient community as an object of governance* perspective, differs from the first two perspectives in that it treats the community as a collection of individuals to be governed. The strength of this perspective is that it can take stock of the society's various parts, with the community becoming a demographic sample of people with differing disaster vulnerability profiles. People are portrayed as users of infrastructure, stakeholders in political processes, and individuals who can reduce their own and their community's vulnerability. Yet, rather than being political animals, community members are framed as individual citizen-stakeholders defined by their associations to governing organizations and institutions. Communities are not seen to form a social body that can jointly shape politics; rather, they are only targets of politics.

The *resilient community as an object of governance* conceptualization also frames the governing authorities — the state in particular — as separate from the resilient communities. This weakens the legitimacy of claims toward organizations and institutions arising from "communities," treated as arbitrary units defined by these very actors. This conceptualization of a resilient community appears primarily concerned with the future of actors such as the state, rather than the well-being and futures of the community or the social. The lack of accountability can be even more glaring when a resilient community is defined in relation to a technology or infrastructure provided by non-state actors. While non-state actors might be celebrated for technical fixes to supporting the resilience of communities, they are unlikely to be held accountable for the outcomes.

9.8. Conclusion

The chapter set out to make explicit the diverse ways in which the *resilient community* is conceptualized in research on disasters associated with major society-shaping phenomena such as climate change or urbanization. The chapter explored also how the conceptualizations frame the *social futures* amid various disasters. Particular attention was paid to how the relationship between the state and civil society was portrayed. By evoking a "resilient community," research shapes what kinds of sociality are recognized and supported in governance, as well as what kinds of politics are put forth [cf. 6].

The three conceptualizations of the resilient community outlined in this chapter are not mutually exclusive, and individual pieces of research may potentially draw from several of the categories. The key contribution of the typology, then, is not to file individual pieces of research under a particular category based on their conceptualization of resilient community. Rather, the typology illustrates both how shared patterns arise across the three conceptualizations, as well as how each conceptualization has their own key emphases and understandings of social futures. Between the conceptualizations, the implied social and collective politics can even be oppositional.

The patterns that are shared across the perspectives resonate with the previous critical perspectives on community resilience under neoliberalism,[6] that find that a focus on communities (and individuals) serving to obscure the idea of the social and society.[26,40] This, in turn, may steer the attention away from collective politics needed to address the major society-shaping phenomena that underlie the disasters. When no "society" is thought to exist, the legitimacy of the civil societies' claims in holding states accountable is severely weakened. Yet, even if broad strokes of resilient community framings resonate with neoliberal politics, the diversity between the three conceptualizations illustrates that the "resilient community" can also command more radical and just imaginaries of social futures.

The emphasis on bottom-up dynamics embedded in the *resilient community of belonging* and *resilient community of practice* perspectives can serve to highlight the agency of disaster-affected people. However, these perspectives may also hold communities responsible for their resilience without questioning whether they have the power or resources within and across societies to enact the actions and changes necessary.[19,25-27] The adverse implications of the *resilient community of belonging* perspective, for instance, can be that members of marginalized communities owe it not only to their collective resilience and well-being, but to the state, to remain continuously in place.

The *resilient community as an object of governance* perspective, meanwhile, views people as individuals rather than collectives. These individuals are defined not in relation to one another, but in relation to the state or another organization. While the *resilient community of belonging* and *resilient community as an object of governance* perspectives have differing understandings of sociality on the level of community, neither connects these communities to a broader notion of the social. Disasters faced by individuals or bounded communities appear singular, rather than a part of a society's internal politics. This may further allow states to treat climate change, urbanization, and the crisis-tendencies of the economic system as if they were inevitable forces of either nature or "development" [cf. 13], despite often being very much complicit in crafting and fueling these major society-shaping phenomena. Social futures amid various disasters are implied to remain firmly in the control of states, as well as NGOs and commercial entities.

Conceptualizations of resilient community can serve to isolate the interests of individuals and communities from the idea of a society. However, they can also illustrate how a civil society mobilizes around a cause. Ideally, the cause is in line with the interests of the disaster-people, who typically are also marginalized in and across societies. The *resilient community of practice* perspective shows how a social network can center and configure on a core community's concern. The core community is not bounded but engages external actors to respond to and recover from a disaster. This framing conveys an understanding of a social realm that can come together and enact collective politics. The futures that unfold from a disaster are not pre-defined by past status quo or the state alone.

To conclude, the chapter joins the chorus in calling attention to how "the notion of resilient community" is conceptualized and deployed in neoliberal politics[6,18,31] that aim at suppressing the ideas and practices of the social. Yet, a resilient community is not, in all its conceptualizations, univocally compatible with neoliberalism, and ideas of more just social futures can arise in the context of these conceptualizations.

Acknowledgments

I would like to thank the three reviewers for their constructive and on-the-point reviews that made a valuable contribution to the chapter. Additionally, I wish to thank the following people for their support generally and/or their help with former incarnations of this chapter: Jeff Hearn, Ilan Kelman, Nikodemus Solitander, Jaakko Aspara, Yewondwossen Tesfaye Gemenchu, Linda Annala, Minchul Sohn, and Wojciech Piotrowicz.

This research was supported by Tore Browaldhs Stiftelse [grant number B21-0005] and Belmont Forum through the UK's Natural Environment Research Council (NERC) [grant number NE/T013656/1].

References

1. Davoudi, S. (2016). Resilience and governmentality of unknowns, in *Governmentality After Neoliberalism*, M. Bevir, Ed. London: Routledge.
2. Duffield, M. (2011). Total War as Environmental Terror: Linking Liberalism, Resilience, and the Bunker. *South Atlantic Quarterly*, *110*(3), 757–769. DOI: 10.1215/00382876-1275779.
3. Bonanno, G. A., Galea, S., Bucciarelli, A., and Vlahov, D. (2006). Psychological Resilience After Disaster: New York City in the Aftermath of the September 11th Terrorist Attack. *Psychological Science*, *17*(3), 181–186. DOI: 10.1111/j.1467-9280.2006.01682.x.
4. Konnikova, M. (2016, February 12). How people learn to become resilient. *The New Yorker*, Retrieved from: http://www.newyorker.com/science/maria-konnikova/the-secret-formula-for-resilience.
5. Towe, V. L., Chandra, A., Acosta, J. D., Chari, R., Uscher-Pines, L., and Sellers, C. (2015). *Community Resilience: Learn and Tell Toolkit.* RAND Corporation.
6. Zebrowski, C. and Sage, D. (2017). Organising community resilience: An examination of the forms of sociality promoted in community resilience programmes. *Resilience*, *5*(1), 44–60. DOI: 10.1080/21693293.2016.1228158.
7. Pugh, J. (2014). Resilience, complexity and post-liberalism. *Area*, *46*(3), 313–319. DOI: 10.1111/area.12118.
8. Vitale, R. (2017). *Building Resilience Through Iterative Processes: Mainstreaming Ancestral Knowledge, Social Movements, and the Making of Sustainable Programming in Bolivia.* Oxfam. DOI: 10.21201/2017.0322.

9. Diprose, K. (2014). Resilience is Futile: The Cultivation of Resilience is not An Answer to Austerity and Poverty. *Soundings: A Journal of Politics and Culture*, *58*(1), 44–56.

10. Paton, D. and Johnston, D. (2001). Disasters and Communities: Vulnerability, Resilience and Preparedness. *Disaster Prevention and Management*, *10*(4), 270–277. DOI: 10.1108/EUM0000000005930.

11. Pendall, R., Foster, K. A., and Cowell, M. (2010). Resilience and Regions: Building Understanding of the Metaphor. *Cambridge Journal of Regions, Economy and Society, 3*(1), 71–84. DOI: 10.1093/cjres/rsp028.

12. Vallance, S., and Carlton, S. (2015). First to Respond, Last to Leave: Communities' Roles and Resilience Across the '4Rs'. *International Journal of Disaster Risk Reduction*, *14*(1), 27–36. DOI: 10.1016/j.ijdrr.2014.10.010.

13. Jessop, B. (2002). Liberalism, Neoliberalism, and Urban Governance: A State–Theoretical Perspective. *Antipode*, *34*(3), 452–472. DOI: 10.1111/1467-8330.00250.

14. Houston, J. B., Spialek, M. L., Cox, J., Greenwood, M. M., and First, J. (2015). The Centrality of Communication and Media in Fostering Community Resilience: A Framework for Assessment and Intervention. *American Behavioral Scientist*, *59*(2), 270–283. DOI: 10.1177/0002764214548563.

15. Leitch, A. M., and Bohensky, E. L. (2014). Return to 'A New Normal': Discourses of Resilience to Natural Disasters in Australian Newspapers 2006–2010. *Global Environmental Change*, *26*, 14–26. DOI: 10.1016/j.gloenvcha.2014.03.006.

16. Manyena, S. B. (2006). The Concept of Resilience Revisited. *Disasters*, *30*(4), 434–450. DOI: 10.1111/j.0361-3666.2006.00331.x.

17. Norris, F. H., Stevens, S. P., Pfefferbaum, B., Wyche, K. F., and Pfefferbaum, R. L. (2008). Community Resilience as a Metaphor, Theory, Set of Capacities, and Strategy for Disaster Readiness. *American Journal of Community Psychology*, *41*(1–2), 127–150. DOI: 10.1007/s10464-007-9156-6.

18. Grove, K. (2014). Agency, Affect, and the Immunological Politics of Disaster Resilience. *Environment and Planning D*, *32*(2), 240–256. DOI: 10.1068/d4813.

19. Sletto, B. and Nygren, A. (2015). Unsettling Neoliberal Rationalities: Engaged Ethnography and the Meanings of Responsibility in the Dominican Republic and Mexico. *International Journal of Urban and Regional Research*, *39*(5), 965–983. DOI: 10.1111/1468-2427.12315.

20. Gaillard, J. C. (2010). Vulnerability, Capacity and Resilience: Perspectives for Climate and Development Policy. *Journal of International Development*, *22*(2), 218–232. DOI: 10.1002/jid.1675.

21. Sudmeier-Rieux, K. I. (2014). Resilience — An Emerging Paradigm of Danger or of Hope? *Disaster Prevention and Management*, *23*(1), 67–80. DOI: 10.1108/DPM-12-2012-0143.

22. Turner II, B. L. (2010). Vulnerability and Resilience: Coalescing or Paralleling Approaches for Sustainability Science? *Global Environmental Change*, *20*(4), 570–576. DOI: 10.1016/j.gloenvcha.2010.07.003.

23. Wilson, G. A. (2012). Community Resilience, Globalization, and Transitional Pathways of Decision-Making. *Geoforum*, *43*(6), 1,218–1,231. DOI: 10.1016/j.geoforum.2012.03.008.

24. Bankoff, G. (2019). Remaking the World in Our Own Image: Vulnerability, Resilience and Adaptation as Historical Discourses. *Disasters*, *43*(2), 221–239. DOI: 10.1111/disa.12312.

25. Blackburn, S. (2014). The Politics of Scale and Disaster Risk Governance: Barriers to Decentralisation in Portland, Jamaica. *Geoforum*, *52*(Supplement C), 101–112. DOI: 10.1016/j.geoforum.2013.12.013.

26. Chandler, D. and Reid, J. (2016). *The Neoliberal Subject: Resilience, Adaptation and Vulnerability*. London; New York: Rowman & Littlefield International.

27. Coaffee, J. (2013). Rescaling and Responsibilising the Politics of Urban Resilience: From National Security to Local Place-Making. *Politics*, *33*(4), 240–252. DOI: 10.1111/1467-9256.12011.

28. Collins, T. W. (2010). Marginalization, Facilitation, and the Production of Unequal Risk: The 2006 Paso del Norte Floods. *Antipode*, *42*(2), 258–288. DOI: 10.1111/j.1467-8330.2009.00755.x.

29. O'Sullivan, T. L., Kuziemsky, C. E., Toal-Sullivan, D., and Corneil, W. (2013). Unraveling the Complexities of Disaster Management: A Framework for Critical Social Infrastructure to Promote Population Health and Resilience. *Social Science & Medicine*, *93*(0), 238–246. DOI: 10.1016/j. socscimed.2012.07.040.

30. Farole, T., Rodríguez-Pose, A., and Storper, M. (2011). Human Geography and the Institutions that Underlie Economic Growth. *Progress in Human Geography*, *35*(1), 58–80. DOI: http://dx.doi.org/10.1177/0309132510 372005.

31. Mulligan, M., Steele, W., Rickards, L., and Fünfgeld, H. (2016). Keywords in Planning: What Do We mean by 'Community Resilience'? *International Planning Studies*, *21*(4), 348–361. DOI: 10.1080/13563475.2016.1155974.

32. Bulley, D. (2013). Producing and Governing Community (through) Resilience. *Politics*, *33*(4), 265–275. DOI: 10.1111/1467-9256.12025.

33. Titz, A., Cannon, T., Krüger, F., Titz, A., Cannon, T., and Krüger, F. (2018). Uncovering 'Community': Challenging an Elusive Concept in Development and Disaster Related Work. *Societies*, *8*(71), 1–28. DOI: 10.3390/ soc8030071.

34. MacKinnon, D. and Derickson, K. D. (2012). From Resilience to Resourcefulness: A Critique of Resilience Policy and Activism. *Progress in Human Geography*, *37*(2), 253–270. DOI: 10.1177/0309132512454775.

35. Anderson, B. (2010). Preemption, Precaution, Preparedness: Anticipatory Action and Future Geographies. *Progress in Human Geography, 34*(6), 777–798. DOI: 10.1177/0309132510362600.
36. Anderson, B., Grove, K., Rickards, L., and Kearnes, M. (2019). Slow Emergencies: Temporality and the Racialized Biopolitics of Emergency Governance. *Progress in Human Geography*, p. 0309132519849263. DOI: 10.1177/0309132519849263.
37. Bracke, S. (2016). Bouncing Back: Vulnerability and resistance in times of resilience. In *Vulnerability in Resistance*, Durham, N.C.: Duke University Press, 52–75.
38. Räsänen, A., Lein, H., Bird, D., and Setten, G. (2020). Conceptualizing Community in Disaster Risk Management. *International Journal of Disaster Risk Reduction*, no. 45, p. 101,485.
39. Urry, J. (2016). *What is the Future?* John Wiley & Sons.
40. Philo, C. (2009). Society. In D. Gregory, R. Johnston, G. Pratt, M. Watts, and S. Whatmore (Eds.), *The Dictionary of Human Geography* (5th edition), Malden, MA: Wiley-Blackwell, pp. 701–703.
41. Brown, W. (2019). *In the Ruins of Neoliberalism: The Rise of Antidemocratic Politics in the West.* New York: Columbia University Press.
42. Ehrenberg, J. (2017). Civil Society and the Classical Heritage. In *Civil Society* (2nd Ed.). NYU Press, 13–39. Retrieve from: https://www.jstor.org/stable/j.ctt1bj4rhv.4
43. Ehrenberg, J. (2017). Conclusion: Pessimism, Activism, and Political Revival. In *Civil Society* (2nd Ed.). NYU Press, 271–300. Retrieved from: https://www.jstor.org/stable/j.ctt1bj4rhv.12.
44. Buttigieg, J. A. (2005). The Contemporary Discourse on Civil Society: A Gramscian Critique. *Boundary 2, 32*(1), 33–52. DOI: 10.1215/01903659-32-1-33.
45. Jessop, B. *The State: Past, Present, Future.* Oxford, United Kingdom: Polity Press, 2016. Retrieved from: http://ebookcentral.proquest.com/lib/ucl/detail.action?docID=4306419
46. Fraser, N., and Jaeggi, R. (2018). *Capitalism: A Conversation in Critical Theory*, 1st edition. Medford, MA: Polity.
47. Chomsky, N. (1999). *Profit Over People: Neoliberalism and Global Order.* Seven Stories Press.
48. Harvey, D. (2005). *A Brief History of Neoliberalism.* OUP Oxford.
49. Peck, J. and Theodore, N. (2019). Still Neoliberalism?, *South Atlantic Quarterly, 118*(2), 245–265. DOI: 10.1215/00382876-7381122.
50. Peck, J. and Tickell, A. (2002). Neoliberalizing Space. *Antipode, 34*(3), 380–404. DOI: 10.1111/1467-8330.00247.
51. Ferguson, J. (2010). The Uses of Neoliberalism. *Antipode, 41*(s1), 166–184. DOI: 10.1111/j.1467-8330.2009.00721.x.

52. Kaldor, M. (2003). The Idea of Global Civil Society. *International Affairs*, *79*(3), 583–593. DOI: 10.1111/1468-2346.00324.
53. Slobodian, Q. (2018). *Globalists: The End of Empire and the Birth of Neoliberalism*. Cambridge, Massachusetts: Harvard University Press.
54. Graneheim, U. H. and Lundman, B. (2004). Qualitative Content Analysis in Nursing Research: Concepts, Procedures and Measures to Achieve Trustworthiness. *Nurse Education Today*, *24*(2), 105–112. DOI: 10.1016/j.nedt.2003.10.001.
55. Yuval-Davis, N. (2006). Belonging and the Politics of Belonging. *Patterns of Prejudice*, *40*(3), 197–214. DOI: 10.1080/00313220600769331.
56. Dutta, S. (2017). Creating in the Crucibles of Nature's Fury: Associational Diversity and Local Social Entrepreneurship after Natural Disasters in California, 1991–2010. *Administrative Science Quarterly*, *62*(3), 443–483. DOI: 10.1177/0001839216668172.
57. Proudley, M. (2013). Place Matters. *Australian Journal of Emergency Management*, *28*(2), 11–16.
58. Colten, C. E., Grimsmore, A. A., and Simms, J. R. Z. (2015). Oil Spills and Community Resilience: Uneven Impacts and Protection in Historical Perspective. *Geographical Review*, *105*(4), 391–407. DOI: 10.1111/j.1931-0846.2015.12085.x.
59. Guarnacci, U. (2016). Joining the Dots: Social Networks and Community Resilience in Post-Conflict, Post-Disaster Indonesia. *International Journal of Disaster Risk Reduction*, 16, 180–191. DOI: 10.1016/j.ijdrr.2016.03.001.
60. Lewis, J. and Kelman, I. (2010). Places, People and Perpetuity: Community Capacities in Ecologies of Catastrophe. *ACME: An International Journal for Critical Geographies*, *9*(2), Art. no. 2.
61. Chacowry, A., McEwen, L. J., and Lynch, K. (2018). Recovery and Resilience of Communities in Flood Risk Zones in a Small Island Developing State: A Case Study from a Suburban Settlement of Port Louis, Mauritius. *International Journal of Disaster Risk Reduction*, 28, 826–838. DOI: 10.1016/j.ijdrr.2018.03.019.
62. Eriksson, H., Albert, J., Albert, S., Warren, R., Pakoa, K., and Andrew, N. (2017). The Role of Fish and Fisheries in Recovering from Natural Hazards: Lessons learned from Vanuatu. *Environmental Science and Policy*, 76, 50–58. DOI: 10.1016/j.envsci.2017.06.012.
63. Ferdinand, I., O'Brien, G., O'Keefe, P., and Jayawickrama, J. (2012). The Double Bind of Poverty and Community Disaster Risk Reduction: A Case Study from the Caribbean. *International Journal of Disaster Risk Reduction*, *2*(0), 84–94. DOI: 10.1016/j.ijdrr.2012.09.003.
64. Mathews, J. A., Kruger, L., and Wentink, G. J. (2018). Climate-smart Aagriculture for Sustainable Agricultural Sectors: The Case of Mooifontein.

Jamba: Journal of Disaster Risk Studies, *10*(1), p. a492. DOI: 10.4102/jamba.v10i1.492.

65. Moreno, J., Lara, A., and Torres, M. (2019). Community Resilience in Response to the 2010 Tsunami in Chile: The Survival of a Small-Scale Fishing Community. *International Journal of Disaster Risk Reduction*, 33, 376–384. DOI: 10.1016/j.ijdrr.2018.10.024.

66. Himes-Cornell, A. and Kasperski, S. (2015). Assessing Climate Change Vulnerability in Alaska's Fishing Communities. *Fisheries Research*, 162, 1–11. DOI: 10.1016/j.fishres.2014.09.010.

67. Rapaport, C., Hornik-Lurie, T., Cohen, O., Lahad, M., Leykin, D., and Aharonson-Daniel, L. (2018). The Relationship between Community Type and Community Resilience. *International Journal of Disaster Risk Reduction*, 31, 470–477. DOI: 10.1016/j.ijdrr.2018.05.020.

68. Wirth, L. (1938). Urbanism as a Way of Life. *American Journal of Sociology*, *44*(1), 1–24. DOI: 10.1086/217913.

69. McManus, P., et al. (2012). Rural Community and Rural Resilience: What is Important to Farmers in Keeping their Country Towns Alive? *Journal of Rural Studies*, *28*(1), 20–29. DOI: 10.1016/j.jrurstud.2011.09.003.

70. Harvey, D. (1989). *The Urban Experience*. Baltimore, Md: Johns Hopkins University Press.

71. Graham, S. and McFarlane, C. (2014). *Infrastructural Lives: Urban Infrastructure in Context*. Routledge.

72. González-Muzzio, C. (2013). The Place and the Role of Social Capital in Post-Disaster Community Resilience. Approaches using a Case Study After the Earthquake of 27/F [{*El rol del lugar y el capital social en la resiliencia comunitaria posdesastre. Aproximaciones mediante un estudio de caso después del terremoto del 27/F*]. *Eure*, *39*(117), 25–48. DOI: 10.4067/S0250-71612013000200002.

73. Kontokosta, C. E. and Malik, A. (2018). The Resilience to Emergencies and Disasters Index: Applying Big Data to Benchmark and Validate Neighborhood Resilience Capacity. *Sustainable Cities and Society*, 36, 272–285. DOI: 10.1016/j.scs.2017.10.025.

74. Harte, E. W., Childs, I. R., and Hastings, P. A. (2009). Imizamo Yethu: A Case Study of Community Resilience to Fire Hazard in an Informal Settlement Cape Town, South Africa. *Geographical Research*, *47*(2), 142–154. DOI: 10.1111/j.1745-5871.2008.00561.x.

75. Usamah, M., Handmer, J., Mitchell, D., and Ahmed, I. (2014). Can the Vulnerable be Resilient? Co-existence of Vulnerability and Disaster Resilience: Informal Settlements in the Philippines. *International Journal of Disaster Risk Reduction*, 10, Part A, 178–189. DOI: 10.1016/j.ijdrr.2014.08.007.

76. Chelleri, L., Waters, J. J., Olazabal, M., and Minucci, G. (2015). Resilience trade-offs: Addressing Multiple Scales and Temporal Aspects of Urban Resilience. *Environment and Urbanization, 27*(1), 181–198. DOI: 10.1177/0956247814550780.

77. Patel, R. B. and Gleason, K. M. (2018). The Association between Social Cohesion and Community Resilience in Two Urban Slums of Port au Prince, Haiti. *International Journal of Disaster Risk Reduction,* 27, 161–167. DOI: 10.1016/j.ijdrr.2017.10.003.

78. Khew, Y. T. J., et al. (2015). Assessment of Social Perception on the Contribution of Hard-Infrastructure for Tsunami Mitigation to Coastal Community Resilience After the 2010 Tsunami: Greater Concepcion Area, Chile. *International Journal of Disaster Risk Reduction,* 13, 324–333. DOI: 10.1016/j.ijdrr.2015.07.013.

79. Wisner, B. and I. Kelman. (2015). Community Resilience to Disasters. In J. D. Wright (Ed.), *International Encyclopedia of the Social & Behavioral Sciences* (2nd Edition), Oxford: Elsevier, pp. 354–360. Retrieved from: http://www.sciencedirect.com/science/article/pii/B9780080970868280197.

80. Rahman, A., Sakurai, A., and Munadi, K. (2018). The Analysis of the Development of the Smong Story on the 1907 and 2004 Indian Ocean Tsunamis in Strengthening the Simeulue Island Community's Resilience. *International Journal of Disaster Risk Reduction,* 29, 13–23. DOI: 10.1016/j.ijdrr.2017.07.015.

81. Cho, S. E., Won, S., and Kim, S. (2016). Living in Harmony with Disaster: Exploring Volcanic Hazard Vulnerability in Indonesia. *Sustainability (Switzerland), 8*(9). DOI: 10.3390/su8090848.

82. Gaillard, J.-C. and Le Masson, V. (2007). Traditional Societies' Response to Volcanic Hazards in the Philippines: Implications for Community-Based Disaster Recovery. *Mountain Research and Development, 27*(4), 313–317. DOI: 10.1659/mrd.0949.

83. Morley, P., Russell-Smith, J., Sangha, K. K., Sutton, S., and Sithole, B. (2016). Evaluating Resilience in Two Remote Indigenous Australian Communities. *Australian Journal of Emergency Management, 31*(4), 44–50.

84. Fan, M.-F. (2015). Disaster Governance and Community Resilience: Reflections on Typhoon Morakot in Taiwan. *Journal of Environmental Planning and Management, 58*(1), 24–38. DOI: 10.1080/09640568.2013.839444.

85. Mercer, J., Kelman, I., Taranis, L., and Suchet-Pearson, S. (2010). Framework for Integrating Indigenous and Scientific Knowledge for Disaster Risk Reduction. *Disasters, 34*(1), 214–239. DOI: 10.1111/j.1467-7717.2009.01126.x.

86. Brown, K. (2014). Global Environmental Change I: A Social Turn for Resilience? *Progress in Human Geography*, *38*(1), 107–117. DOI: 10.1177/0309132513498837.

87. Rakib, M. A., Islam, S., Nikolaos, I., Bodrud-Doza, M., and Bhuiyan, M. A. H. (2017). Flood Vulnerability, Local Perception and Gender Role Judgment using Multivariate Analysis: A Problem-Based 'Participatory Action to Future Skill Management' to Cope with Flood Impacts. *Weather and Climate Extremes*, 18, 29–43. DOI: 10.1016/j.wace.2017.10.002.

88. Bambrick, H. (2018). Resource Extractivism, Health and Climate Change in Small Islands. *International Journal of Climate Change Strategies and Management*, *10*(2), 272–288. DOI: 10.1108/IJCCSM-03-2017-0068.

89. Gundersen, V., Kaltenborn, B. P., and Williams, D. R. (2016). A Bridge over Troubled Water: A Contextual Analysis of Social Vulnerability to Climate Change in a Riverine Landscape in South-east Norway. *Norsk Geografisk Tidsskrift*, *70*(4), 216–229. DOI: 10.1080/00291951.2016.1194317.

90. Rahmayati, Y., Parnell, M., and Himmayani, V. (2017). Understanding Community-led Resilience: The Jakarta Floods Experience. *Australian Journal of Emergency Management*, *32*(4), 58–66.

91. Boeri, A., Longo, D., Gianfrate, V., and Lorenzo, V. (2017). Resilient Communities. Social Infrastructures for Sustainable Growth of Urban Areas. A Case Study. *International Journal of Sustainable Development and Planning*, *12*(2), 227–237. DOI: 10.2495/SDP-V12-N2-227-237.

92. Wenger, E. (2011). Communities of Practice: A Brief Introduction. *STEP Leadership Workshop, University of Oregon*. Retrieved from: https://scholarsbank.uoregon.edu/xmlui/handle/1794/11736.

93. Wenger, E. (2000). Communities of Practice and Social Learning Systems, Communities of Practice and Social Learning Systems. *Organization*, *7*(2), 225–246, DOI: 10.1177/135050840072002.

94. Matthies, A. (2017). Community-Based Disaster Risk Management in the Philippines: Achievements and Challenges of the Purok System. *Austrian Journal of South-East Asian Studies*, *10*(1), 101–108. DOI: 10.14764/10.ASEAS-2017.1-7.

95. Fois, F. and Forino, G. (2014). The Self-Built Ecovillage in L'aquila, Italy: Community Resilience as a Grassroots Response to Environmental Shock. *Disasters*, *38*(4), 719–739. DOI: 10.1111/disa.12080.

96. Imperiale, A. J. and Vanclay, F. (2016). Experiencing Local Community Resilience in Action: Learning from Post-Disaster Communities. *Journal of Rural Studies*, 47, 204–219, DOI: 10.1016/j.jrurstud.2016.08.002.

97. Barrios, R. E. (2014). 'Here, I'm not at ease': Anthropological Perspectives on Community Resilience. *Disasters*, *38*(2), 329–350. DOI: 10.1111/disa.12044.

98. Barrios, R. E. (2017). *Governing Affect: Neoliberalism and Disaster Reconstruction.* Lincoln and London: University of Nebraska Press.
99. B. Pfefferbaum, R. L. Van Horn, and R. L. Pfefferbaum. (2017). A Conceptual Framework to Enhance Community Resilience Using Social Capital. *Clinical Social Work Journal, 45*(2), 102–110. DOI: 10.1007/s10615-015-0556-z.
100. S. Misra, R. Goswami, T. Mondal, and R. Jana. (2017). Social Networks in the Context of Community Response to Disaster: Study of a Cyclone-Affected Community in Coastal West Bengal, India. *International Journal of Disaster Risk Reduction, 22,* 281–296. DOI: 10.1016/j.ijdrr.2017.02.017.
101. Ntontis, E., Drury, J., Amlôt, R., Rubin, G. J., and Williams, R. (2018). Emergent Social Identities in a Flood: Implications for Community Psychosocial Resilience. *Journal of Community and Applied Social Psychology, 28*(1), 3–14. DOI: 10.1002/casp.2329.
102. Goldstein, B. E. (2008). Skunkworks in the Embers of the Cedar Fire: Enhancing Resilience in the Aftermath of Disaster. *Human Ecology, 36*(1), 15–28. DOI: 10.1007/s10745-007-9133-6.
103. Locke, J. T. (2009). Climate Change-Induced Migration in the Pacific Region: Sudden Crisis and Long-Term Developments. *The Geographical Journal, 175*(3), 171–180. DOI: 10.1111/j.1475-4959.2008.00317.x.
104. Redshaw, S., Ingham, A. P. V., Hicks, P. J., and Millynn, J. (2017). Emergency Preparedness through Community Sector Engagement in the Blue Mountains. *Australian Journal of Emergency Management, 32*(2), 35–40.
105. Thornley, L., Ball, J., Signal, L., Lawson-Te Aho, K., and Rawson, E. (2015). Building community resilience: Learning from the Canterbury earthquakes. *Kotuitui, 10*(1), 23–35. DOI: 10.1080/1177083X.2014.934846.
106. Grube, L. and Storr, V. H. (2013). The Capacity for Self-Governance and Post-Disaster Resiliency. *The Review of Austrian Economics, 27*(3), 301–324. DOI: 10.1007/s11138-013-0210-3.
107. Kelman, I. (2008). Relocalising Disaster Risk Reduction for Urban Resilience. *Proceedings of the Institution of Civil Engineers—Urban Design and Planning, 161*(4), 197–204. DOI: 10.1680/udap.2008.161.4.197.
108. Sewordor, E., Esnard, A.-M., Sapat, A., and Schwartz, L. (2019). Challenges to Mobilising Resources for Disaster Recovery and Reconstruction: Perspectives of the Haitian Diaspora. *Disasters, 43*(2), 336–354. DOI: 10.1111/disa.12318.
109. Lai, C.-H., Tao, C.-C., and Cheng, Y.-C. (2017). Modeling Resource Network Relationships Between Response Organizations and Affected Neighborhoods After a Technological Disaster. *Voluntas, 28*(5), 2145–2175. DOI: 10.1007/s11266-017-9887-4.

110. Oktari, R. S., Shiwaku, K., Munadi, K., Syamsidik, and Shaw, R. (2018). Enhancing Community Resilience Towards Disaster: The Contributing Factors of School-Community Collaborative Network in the Tsunami Affected Area in Aceh. *International Journal of Disaster Risk Reduction*, 29, 3–12. DOI: 10.1016/j.ijdrr.2017.07.009.

111. Sciulli, N., D'Onza, G., and Greco, G. (2015). Building a Resilient Local Council: Evidence from Flood Disasters in Italy. *International Journal of Public Sector Management*, *28*(6), 430–448. DOI: 10.1108/IJPSM-11-2014-0139.

112. Sugino, M., et al. (2014). Issues Raised by Nurses and Midwives in a Post-Disaster Bantul Community. *Disaster Prevention and Management*, *23*(4), 420–436. DOI: 10.1108/DPM-05-2013-0086.

113. Cutter, S. L., et al. (2008). A Place-Based Model for Understanding Community Resilience to Natural Disasters. *Global Environmental Change*, *18*(4), 598–606. DOI: 10.1016/j.gloenvcha.2008.07.013.

114. O'Sullivan, T. L., Corneil, W., Kuziemsky, C. E., and Toal-Sullivan, D. (2015). Use of the Structured Interview Matrix to Enhance Community Resilience Through Collaboration and Inclusive Engagement. *Systems Research and Behavioral Science*, *32*(6), 616–628. DOI: 10.1002/sres.2250.

115. Kruger, L. (2016). The Timing of Agricultural Production in Hazard-Prone Areas to Prevent Losses at Peak-Risk Periods: A Case of Malawi, Madagascar and Mozambique. *Jamba: Journal of Disaster Risk Studies*, *8*(2), 1–9. DOI: 10.4102/jamba.v8i2.179.

116. Lassa, J. A., Boli, Y., Nakmofa, Y., Fanggidae, S., Ofong, A., and Leonis, H. (2018). Twenty Years of Community-Based Disaster Risk Reduction Experience from a Dryland Village in Indonesia. *Jamba: Journal of Disaster Risk Studies*, *10*(1). DOI: 10.4102/jamba.v10i1.502.

117. United Nations International Strategy for Disaster Risk Reduction. (2005). Hyogo Framework for 2005–2015: Building the Resilience of Nations and Communities to Disasters. Retrieved from: http://www.unisdr.org/2005/wcdr/intergover/official-doc/L-docs/Hyogo-framework-for-action-english.pdf.

118. Burnside-Lawry, J. and Carvalho, L. (2015). Building Local Level Engagement in Disaster Risk Reduction: A Portugese Case Study. *Disaster Prevention and Management*, *24*(1), 80–99. DOI: 10.1108/DPM-07-2014-0129.

119. Tadele, F. and Manyena, S. B. (2009). Building Disaster Resilience Through Capacity Building in Ethiopia. *Disaster Prevention and Management: An International Journal*, *18*(3), 317–326. DOI: 10.1108/09653560910965664.

120. Yi, C. J., et al. (2015). Storm Surge Mapping of Typhoon Haiyan and its Impact in Tanauan, Leyte, Philippines. *International Journal of Disaster Risk Reduction*, 13, 207–214. DOI: 10.1016/j.ijdrr.2015.05.007.

121. Grove, K. (2013). Hidden Transcripts of Resilience: Power and Politics in Jamaican Disaster Management. *Resilience, 1*(3), 193–209. DOI: 10.1080/21693293.2013.825463.

122. White, R. K., Edwards, W. C., Farrar, A., and Plodinec, M. J. (2015). A Practical Approach to Building Resilience in America's Communities. *American Behavioral Scientist, 59*(2), 200–219. DOI: 10.1177/0002764214550296.

123. Howe, P. D. (2011). Hurricane Preparedness as Anticipatory Adaptation: A Case Study of Community Businesses. *Global Environmental Change, 21*(2), 711–720. DOI: 10.1016/j.gloenvcha.2011.02.001.

124. Orhan, E. (2016). Building Community Resilience: Business Preparedness Lessons in the Case of Adapazari, Turkey. *Disasters, 40*(1), 45–64. DOI: 10.1111/disa.12132.

125. Rose, A. and Krausmann, E. (2013). An Economic Framework for the Development of a Resilience Index for Business Recovery. *International Journal of Disaster Risk Reduction*, 5, 73–83. DOI: 10.1016/j.ijdrr.2013.08.003.

126. Carpenter, A. (2015). Resilience in the Social and Physical Realms: Lessons from the Gulf Coast. *International Journal of Disaster Risk Reduction*, 14, 290–301. DOI: 10.1016/j.ijdrr.2014.09.003.

127. Weichselgartner, J. and Kelman, I. (2015). Geographies of Resilience Challenges and Opportunities of a Descriptive Concept. *Progress in Human Geography, 39*(3), 249–267. DOI: 10.1177/0309132513518834.

128. Leykin, D., Lahad, M., Cohen, O., Goldberg, A., and Aharonson-Daniel, L. (2013). Conjoint Community Resiliency Assessment Measure-28/10 Items (CCRAM28 and CCRAM10): A Self-report Tool for Assessing Community Resilience. *American Journal of Community Psychology, 52*(3–4), 313–323. DOI: 10.1007/s10464-013-9596-0.

129. Lisnyj, K. T. and Dickson-Anderson, S. E. (2018). Community Resilience in Walkerton, Canada: Sixteen Years Post-Outbreak. *International Journal of Disaster Risk Reduction*, 31, 196–202. DOI: 10.1016/j.ijdrr.2018.05.001.

130. Pujadas Botey, A. and Kulig, J. C. (2014). Family Functioning Following Wildfires: Recovering from the 2011 Slave Lake Fires. *Journal of Child and Family Studies, 23*(8), 1471–1483. DOI: 10.1007/s10826-013-9802-6.

131. Osofsky, J. D. and Osofsky, H. J. (2018). Challenges in Building Child and Family Resilience after Disasters. *Journal of Family Social Work, 21*(2), 115–128. DOI: 10.1080/10522158.2018.1427644.

132. Lam, R. P. K., et al. (2017). Urban Disaster Preparedness of Hong Kong Residents: A Territory-wide Survey. *International Journal of Disaster Risk Reduction*, 23, 62–69. DOI: 10.1016/j.ijdrr.2017.04.008.

133. Howard, A., Blakemore, T., and Bevis, M. (2017). Older People as Assets in Disaster Preparedness, Response and Recovery: Lessons from Regional Australia. *Ageing and Society, 37*(3), 517–536. DOI: 10.1017/S0144686X15001270.

134. Roth, M. (2018). A Resilient Community is One that Includes and Protects Everyone. *Bulletin of the Atomic Scientists, 74*(2), 91–94. DOI: 10.1080/00963402.2018.1436808.

135. Delicado, A., Rowland, J., Fonseca, S., de Almeida, A. N., Schmidt, L., and Ribeiro, A. S. (2017). Children in Disaster Risk Reduction in Portugal: Policies, Education, and (Non) Participation. *International Journal of Disaster Risk Science, 8*(3), 246–257. DOI: 10.1007/s13753-017-0138-5.

136. Nahayo, L., et al. (2018). Extent of Disaster Courses Delivery for the Risk Reduction in Rwanda. *International Journal of Disaster Risk Reduction,* 27, 127–132. DOI: 10.1016/j.ijdrr.2017.09.046.

137. Shelton, T. and Lodato, T. (2019). Actually Existing Smart Citizens. *City, 23*(1), 35–52. DOI: 10.1080/13604813.2019.1575115.

138. Aldrich, D. P. and Meyer, M. A. (2015). Social Capital and Community Resilience. *American Behavioral Scientist, 59*(2), 254–269. DOI: 10.1177/0002764214550299.

139. Arneson, E., Deniz, D., Javernick-Will, A., Liel, A., and Dashti, S. (2017). Information Deficits and Community Disaster Resilience. *Natural Hazards Review, 18*(4). DOI: 10.1061/(ASCE)NH.1527-6996.0000251.

140. Kontou, E., Murray-Tuite, P., and Wernstedt, K. (2017). Commuter Adaptation in Response to Hurricane Sandy's Damage. *Natural Hazards Review, 18*(2). DOI: 10.1061/(ASCE)NH.1527-6996.0000231.

141. Jones, K. B., James, M., and Mastor, R.-A. (2017). Securing Our Energy Future: Three International Perspectives on Microgrids and Distributed Renewables as a Path toward Resilient Communities. *Environmental Hazards, 16*(2), 99–115. DOI: 10.1080/17477891.2016.1257974.

142. Borba, M. L., Warner, J. F., and Porto, M. F. A. (2016). Urban stormwater flood management in the Cordeiro watershed, São Paulo, Brazil: Does the interaction between socio-political and technical aspects create an opportunity to attain community resilience? *Journal of Flood Risk Management, 9*(3), 234–242. DOI: 10.1111/jfr3.12172.

Index

A

adaptation, 63, 141, 284, 288
aeration, 225, 227, 233
agricultural production, 116, 121, 126–127
agriculture, 6, 63, 115–127, 133, 135, 137–138, 140
air accumulator, 163
algae, 181, 186–188, 222, 224–226, 230
allergic, 247
antibiotics, 213–223, 225, 228–234
aquatic, 181, 213–215

B

bacterial, 216–217, 230–231
benthic, 180, 189, 193
bioaccumulation, 218–221
bioadsorption, 218–220, 230
biodegradation, 213–234
biodiversity, 138–140, 149, 179–200
biomaterial, 213–234
biosurfactant, 263–277
breakwaters, 179–180, 184, 186

C

carbon dioxide (CO_2), 13, 16, 19, 24, 26, 37–38, 77, 81–82, 161, 173–175, 217, 221, 225–226, 230
civil, 282, 284–288, 299, 301, 303–304
civil society, 282, 284–288, 299, 301, 303–304
climate change, 4–5, 12–13, 28, 37, 39, 62, 81, 83, 85–87, 116–117, 127, 133–134, 141, 268, 282–283, 287–288, 292–293, 295, 301, 303
coastal, 13, 180, 182, 186, 192, 195
community, 6, 21, 23, 41, 44, 51–53, 71, 139, 181–182, 187–188, 194–195, 197–200, 217, 281–304
compressed air energy storage (CAES), 162, 175
contaminant, 5, 13, 15–16, 20, 41, 45–46, 48, 181, 193, 217, 231
contamination, 41, 50, 193, 221, 231, 269
coral, 180–181, 186–192, 196, 198–199
cosmetic, 247–248

D

development, 12, 34, 37, 39, 41, 55, 67, 84, 117, 124, 126, 133–138, 143, 147–149, 160, 162, 180, 194, 196, 216, 220, 282–283, 287, 298, 304

disaster, 141, 282–284, 287–290, 293–304

diversity, 137, 185–186, 189, 191–197, 199, 303

E

ecological, 135, 137–139, 150, 199, 268, 291

economics, 5, 9, 32–33, 55, 62, 66, 82–83, 87, 118, 122, 136–140, 143, 145–146, 150, 160, 266, 285–288

economic sustainability, 140, 144

ecosystem, 48, 132–133, 136–137, 142, 151, 199, 216

energy, 2, 4, 6–7, 9, 11, 26–29, 39–40, 54, 56, 74–87, 125–126, 131–151, 159–176, 226, 268, 299

energy consumption, 56, 75–77, 81, 125, 133, 144–145

energy storage, 26–27, 39, 159–176

environmental, 4, 136, 138, 140–141, 143–144, 147, 149, 161, 181, 183–185, 214, 219, 224, 231, 245–258, 267–270, 273, 276, 290

environmental sustainability, 140, 144, 149–150, 249

epibiotic, 180–181, 185–189, 195

e-waste, 25

F

fish, 13, 180, 182, 186, 194–198, 200, 268

food production, 6, 26, 35, 115–116, 118, 121–122, 126, 138, 140–141

food security, 2, 55, 60, 115–127, 133–135, 137, 139–140, 147–148

G

global, 1–7, 10, 12, 14, 18–22, 24, 28–29, 31, 34–36, 38, 40–43, 46, 48, 52–54, 56–59, 61–64, 67–70, 73–74, 76–78, 80–83, 86–88, 117, 122, 126–127, 133–134, 138, 142–143, 145, 147, 149, 160, 162, 174, 215, 283, 292, 295

global hunger index (GHI), 56–59, 77

governance, 136, 142, 150–151, 282–284, 286, 289, 294, 296–300, 302–304

greenhouse gas, 13, 15–16, 19–20, 145, 149, 160–163, 175, 268, 287

H

health, 5, 14–15, 20–21, 24, 30–31, 35, 38, 46, 54, 66, 71, 84, 142, 246, 248, 257, 264, 267, 277

heat recovery, 162, 165, 172

hydrodynamic, 186–187, 190, 198

I

infrastructure, 28, 41, 51, 53, 65, 76, 83–84, 123, 292, 296, 299–300, 302

irrigated agriculture, 115–127

irrigation, 46, 48–49, 51, 116–126

irritant, 247

L

land, 6, 24, 26–27, 46, 65, 78, 117, 119–120, 122–123, 126, 137, 141, 182

levelized cost of energy (LCOE), 83–84, 86

M

macrobenthic, 192–194, 199

marina, 179–200

mercury, 247–248

microalgae, 217–223, 225–228, 230

microbial, 217, 247

multi criteria decision making (MCDM), 149, 248, 257, 276

N

natural, 5–6, 12–13, 19, 24, 26–29, 31–32, 63, 118, 132, 134–137, 144, 148, 174, 180, 194, 197–198, 222, 224, 248, 288

nature, 1–88, 115–127, 131–151, 159–176, 179–200, 213–234, 245–258, 263–277, 281–305

neoliberalism, 284–288, 303–304

net-zero, 3–4, 8, 24

nexus, 131–151

nitric oxide (NO_x), 36

nuclear, 27, 34, 74, 76, 80–83, 144, 268–269, 276

O

occupational safety and health administration (OSHA), 13

offshore, 186, 196

oxygen-enriched atmospheres (OEA), 13

ozone, 16, 18–20, 31–35, 38

P

pharmaceutical, 214–217

photobioreactor, 221, 227, 232, 234

photodegradation, 217

pier, 186, 189

pollutant, 5, 13, 15–22, 27–28, 30, 33, 35–40, 181, 188, 193, 198–199, 214, 218, 222, 225

pollution, 14, 20, 28, 32–33, 36, 82, 137, 139, 145, 160, 181, 193, 199, 214, 250

polychaetes, 192–194, 199

pontoon, 181–192, 195–196, 199–200

population, 2, 5–7, 10, 12–13, 21–24, 29, 35, 39–44, 48, 52–54, 62–64, 68–71, 74, 77–78, 80, 87–88,

116–120, 122, 126, 132, 137, 139, 145, 160, 191, 200, 246, 291, 295

psychological, 61, 282

R

reclamation, 180, 182, 188

resilience, 136, 144, 150, 282–284, 287–299, 301–304

resilient, 64, 139, 141, 283, 291–292, 301

resilient community, 281–305

resources, 60–61, 74, 76, 87, 116–126, 131–151, 173, 189, 282–283, 298, 303

rural, 23, 32, 52, 59, 69, 78, 132, 144, 249, 291–292, 300

rural communities, 52, 291

S

seawall, 180–182, 184–192, 194–196, 198–199

self-organizing, 293–295, 301

self-reliance, 282

skincare, 245–258

social future, 281–304

social justice, 140, 285

social network, 293–296, 298, 300–301, 304

social sustainability, 140, 144

society, 69, 71, 116, 136, 138, 144, 246, 282–290, 293, 299–304

step-wise weight appraisal ratio analysis (SWARA), 245–258, 264–270, 274

sulfur dioxide, 16, 31, 35

sustainability, 5, 131–151, 217, 248–249, 265

sustainable, 4, 55, 64, 75

sustainable development, 12, 40, 55–56, 134–137, 147–149, 249

sustainable development goals (SDGs), 3, 12, 86, 149

T
thermal energy, 27
thermal energy storage, 162
toxic, 30, 35, 247, 267, 276

U
uncertainty, 19, 116, 160, 282–283
United Nations (UN), 4, 11, 55, 119
urbanization, 64, 69, 88, 119, 282, 287, 301, 303–304

V
volatile organic compound (VOC), 16, 30–34

W
waste, 28, 38, 85, 141, 199, 213–234, 268
waste heat, 160–163
wastewater, 47–51, 53, 213–234
water, 2–5, 7, 9, 13–14, 27, 40–54, 65, 76, 88, 116–127, 131–151, 164–165, 172–174, 180–186, 189–191, 193–195, 197–198, 291
water, energy and food (WEF), 131–151
water purification, 213–234

Z
zero-emission, 161